MW01438651

Scientific Cosmology and International Orders

Scientific Cosmology and International Orders shows how scientific ideas have transformed international politics since 1550. Allan argues that cosmological concepts arising from Western science made possible the shift from a sixteenth-century order premised upon divine providence to the present order, which is premised on economic growth. As states and other international associations used scientific ideas to solve problems, they slowly reconfigured ideas about how the world works, humanity's place in the universe, and the meaning of progress. The book demonstrates the rise of scientific ideas across three cases: natural philosophy in balance of power politics, 1550–1815; geology and Darwinism in British colonial policy and the liberal-colonial order, 1860–1950; and cybernetic-systems thinking and economics in the World Bank and American liberal order, 1945–2015. Together, the cases trace the emergence of economic growth as a central end of states to its origins in colonial doctrines of development and balance-of-power thinking.

BENTLEY B. ALLAN is Assistant Professor of Political Science at Johns Hopkins University.

CAMBRIDGE STUDIES IN
INTERNATIONAL RELATIONS: 147

Scientific Cosmology and International Orders

EDITORS
Evelyn Goh
Christian Reus-Smit
Nicholas J. Wheeler

EDITORIAL BOARD
Jacqueline Best, Karin Fierke, William Grimes, Yuen Foong Khong, Andrew Kydd, Lily Ling, Andrew Linklater, Nicola Phillips, Elizabeth Shakman Hurd, Jacquie True, Leslie Vinjamuri, Alexander Wendt

Cambridge Studies in International Relations is a joint initiative of Cambridge University Press and the British International Studies Association (BISA). The series aims to publish the best new scholarship in international studies, irrespective of subject matter, methodological approach or theoretical perspective. The series seeks to bring the latest theoretical work in International Relations to bear on the most important problems and issues in global politics.

CAMBRIDGE STUDIES IN INTERNATIONAL RELATIONS

146 Peter J. Katzenstein and Lucia A. Seybert
   Protean power
   Exploring the uncertain and unexpected in world politics
145 Catherine Lu
   Justice and reconciliation in world politics
144 Ayşe Zarakol
   Hierarchies in world politics
143 Lisbeth Zimmermann
   Global norms with a local face
   Rule-of-law promotion and norm-translation
142 Alexandre Debs and Nuno P. Monteiro
   Nuclear politics
   The strategic causes of proliferation
141 Mathias Albert
   A theory of world politics
140 Emma Hutchison
   Affective communities in world politics
   Collective emotions after trauma
139 Patricia Owens
   Economy of force
   Counterinsurgency and the historical rise of the social
138 Ronald R. Krebs
   Narrative and the making of US national security
137 Andrew Phillips and J.C. Sharman
   International order in diversity
   War, trade and rule in the Indian Ocean
136 Ole Jacob Sending, Vincent Pouliot and Iver B. Neumann (eds.)
   Diplomacy and the making of world politics
135 Barry Buzan and George Lawson
   The global transformation
   History, modernity and the making of international relations
134 Heather Elko McKibben
   State strategies in international bargaining
   Play by the rules or change them?
133 Janina Dill
   Legitimate targets?
   Social construction, international law, and US bombing

Series list continues after index

Scientific Cosmology and International Orders

Bentley B. Allan
Johns Hopkins University

CAMBRIDGE
UNIVERSITY PRESS

CAMBRIDGE
UNIVERSITY PRESS

University Printing House, Cambridge CB2 8BS, United Kingdom

One Liberty Plaza, 20th Floor, New York, NY 10006, USA

477 Williamstown Road, Port Melbourne, VIC 3207, Australia

314-321, 3rd Floor, Plot 3, Splendor Forum, Jasola District Centre, New Delhi - 110025, India

79 Anson Road, #06-04/06, Singapore 079906

Cambridge University Press is part of the University of Cambridge.

It furthers the University's mission by disseminating knowledge in the pursuit of education, learning and research at the highest international levels of excellence.

www.cambridge.org
Information on this title: www.cambridge.org/9781108416610
DOI: 10.1017/ 9781108241540

© Bentley B. Allan 2018

This publication is in copyright. Subject to statutory exception and to the provisions of relevant collective licensing agreements, no reproduction of any part may take place without the written permission of Cambridge University Press.

First published 2018

A catalogue record for this publication is available from the British Library

ISBN 978-1-108-41661-0 Hardback
ISBN 978-1-108-40400-6 Paperback

Cambridge University Press has no responsibility for the persistence or accuracy of URLs for external or third-party internet websites referred to in this publication, and does not guarantee that any content on such websites is, or will remain, accurate or appropriate.

To my teachers

Contents

List of Figures   page x
List of Tables   xi
Acknowledgements   xii

1 Introduction: Science and the Transformation of International Politics   1

2 Cosmology and Change in International Orders   29

3 Natural Philosophy in Balance of Power Europe, 1550–1815   75

4 Darwin, Social Knowledge, and Development in the British Colonial Office and the League of Nations, 1850–1945   139

5 Neoclassical Economics and the Rise of Growth in the World Bank and Postwar International Order, 1945–2015   207

6 Conclusion: The Future of Cosmological Change   263

Methodological Appendix   285
References   294
Index   331

Figures

2.1 The institutional and discursive structure underlying
international order   *page* 33
2.2 The constitution of international discourses   45
2.3 The recursive institutionalization of international order   54
3.1 The heliocentric model of the solar system, Nicolaus
Copernicus, 1543   76
3.2 Image of the celestial realm and the human body,
Catholic Church, 1556   85
3.3 Drawing of an orrery, 1749   94
3.4 Excerpt from Metternich's proposal for the reconstruction
of Prussia   130
4.1 Geological section from London to Snowdon, William
Smith, 1819   140
4.2 Diagram illustrating variation within and between
species through the stages of development, Charles
Darwin, 1861   149
4.3 Statistical graph showing class differences in caloric and
protein intake, Seebohm Rowntree, 1901   168
5.1 Scientists observing the world's first self-sustaining
nuclear chain reaction in the Chicago Pile No. 1, 2
December 1942, Gary Sheahan, 1957   208
5.2 A simple feedback arrangement, Ludwig Bertalanffy, 1951   214

Tables

4.1	Colonial Office recruitment, 1913–1952	*page* 175
4.2	Growth of Colonial Office advisory committees, 1900–1961	175

Acknowledgements

This book is dedicated to my teachers who have given me their time, energy, and care. First and foremost, I want to thank my dissertation committee: Alexander Wendt, Sonja Amadae, Ted Hopf, Michael Neblo, and Alexander Thompson. Alex was a wonderful advisor. He was generous with his time, blunt in his assessments, and always focused on making my work more important and interesting. He taught me to pursue the question wherever it leads, even if it seems hard or against the grain. Sonja taught me to write and think historically. My only regret is that I didn't understand many of her lessons until after the dissertation was complete. My hope is that her influence is nonetheless revealed in the pages that follow. Ted Hopf taught me to care deeply about social theory. His enthusiastic support for the dissertation helped me believe in it when I had my doubts. Michael's door was always open – to my benefit and his detriment. Michael's generosity, breadth, and passion are a true inspiration. Alex Thompson's hard work on my dissertation was all the more appreciated because my research was so far from his own interests. His line-by-line feedback taught me how to integrate the abstract and concrete in ways that make sense. I often say that my committee was crazy to let me work on something so ambitious. But the lovely truth in that joke is that they trusted me and gave me the freedom to make the project my own. I am grateful to them for that gift.

My time at Ohio State was exciting and engrossing because I shared the experience with other amazing students. Austin Carson and Jason Keiber provided the kind of engaged intellectual friendship I value so much. Austin has been my first call on everything in this book. I don't know how I would have done it without him. Eric Grynaviski showed me the ropes both intellectually and professionally and I continue to lean on him. Srdjan Vucetic took the time to mentor me and I value our continued friendship. All the members of the pizza group contributed to this project in its early stages: Zoltán Búzás, Burcu Bayram, Tim Luecke, Fernando Nunez-Mietz, John Oates, Dave Traven, and Clement Wyplosz. A number of other grad students also took the time to comment on my work: Eun-Bin Chung, Andrew Dombrowski, Kevin Duska Jr., Erin Graham,

Acknowledgements

Chris Kypriotis, Matt Hitt, Josh Kertzer, Marcus Kurtz, Eric MacGilvray, Sebastien Mainville, Eleonora Mattiaci, and Vittorio Merola. I only realize now how special it was that many faculty members outside my committee, and even my area, gave me their time and advice. I am grateful to Bear Braumoeller, Eric MacGilvray, Marcus Kurtz, William Minozzi, Jennifer Mitzen, Irfan Nooruddin, Philipp Rehm, Randall Schweller, and Daniel Verdier for comments and conversation.

I have continued to learn from my colleagues at Hopkins. Writing this book in a dynamic and creative department has made it richer. I would like to thank Kavi Abraham, Yoni Abramson, Jane Bennett, P.J. Brendese, Sam Chambers, Bill Connolly, Steven David, Dan Deudney, Siba Grovogui, Michael Hanchard, Nicolas Jabko, Margaret Keck, Renée Marlin-Bennett, Jon Masin-Peters, Beth Mendenhall, Daniel Schlozman, Sebastian Schmidt, Lester Spence, Tarek Tutunji, and Emily Zackin for engaging with the project. Special thanks to Sam Chambers, Jane Bennett, Bill Connolly, Adam Sheingate, and Lester Spence for extensive comments on my book workshop. I need to thank my department chairs, Dick Katz and Adam Sheingate, for their unstinting support – and I have needed more of it than most. I could not do what I do without the support of the staff in the Political Science Department. Lisa Williams and Mary Otterbein have helped and indeed saved me so many times I have lost count. Also at Hopkins, Angus Burgin in the History Department generously read the World Bank chapter. Bill Rowe, also in History (but his more important position is as a Dean of the lunchtime basketball game), listened to me think through pieces of the argument many times. My students Kathryn Botto, Eric Buck, Ian Gustafson, Michael Hur, and Jarrett Olivo provided essential questions and research assistance.

I was originally inspired to study the growth imperative by Thomas Homer-Dixon when I was an undergraduate at the University of Toronto. Tad was the first professor to trust and reward my instinct to read and think across disciplines. I am grateful for that. Also at Toronto, David Welch reached out to me as an undergraduate in one of his classes, I'll never forget that and I try to emulate his generosity with my own students. I also need to thank Nisha Shah, my TA in David's class, who introduced me to constructivist IR theory, Hedley Bull, and so many more.

This book was the subject of a workshop held at Hopkins. I would like to thank Bud Duvall, Mlada Bukovansky, and Eric Grynaviski for coming to Baltimore and providing excellent feedback on the manuscript. I would also like to thank Dan Nexon and two anonymous reviewers for generous and constructive comments on the whole manuscript. James Der Derian and Jairus Grove invited me to present part of the book in Sydney where the conversation was not just helpful but inspirational. I also need to thank Will Bain, Tamar Gutner, Matthias Staisch, Ron Krebs, Jeff Legro, Julia MacDonald, James Morrison, Stephen Nelson, Michael Poznansky, Fahad Sajid, Robbie Shilliam, and Martin Weber for comments and conversations that improved the arguments herein. Avery White provided invaluable assistance and substantive comments.

At Cambridge, I would like to thank John Haslam and the series editors Evelyn Goh, Chris Reus-Smit, and Nicholas Wheeler. They supported and believed in the book even when it contained more promise than payoff. I would also like to thank the staff at Cambridge University Press and Out of House Publishing who helped with the production process: Toby Ginsberg, Claire Sissen, Helen Flitton, Nanditha Devi, and Gail Welsh.

On a personal note, I would like to thank my first teachers, my parents Phyllis and Howard Allan. They gave me the confidence and the critical mind that has brought me here. I am so grateful for all the love and hard work they poured into me. Thank you as well to my siblings Gillian, Kori, and Blake for the long talks and lessons. I also want to thank some of my earlier teachers: Marilyn Emery, Ted Bryant, Dorothy Cameron, Dale Dixon, Geoff Walker, and, especially, Geoff and Carolee Mason. Geoff and Carolee created a space in a small Canadian town where kids could be nerdy and intellectual. I needed that space and am so grateful for their decades of service to young people. There are some other people whose work made the physical act of writing of this book possible. Mary Ellen Chambers, Morgan Heston, Avendui Locavara, Amos Moecker, Werner Mueller, and Thom Stroschein were wonderful neighbours. They helped me keep my house in good working order so I could spend more hours at my desk. E.P. Taylor was a godsend for helping me out while peppering me with smart questions about science and politics. Kacy Cook gave me her expert assistance and, more importantly, some peace of mind. Mej and Tim Stokes have spent countless hours cooking, building, and caring for my family. I do not know how we would have gotten through these last years without them.

My wife Caitlin is the perfect partner for me. She is so smart and is always willing to talk through whatever intellectual and professional problems are consuming me. Moreover, she makes me feel as if teaching and thinking are truly important work. However, she never lets me get carried away with myself or the ideas and reminds me that life beyond the academy is rewarding and deserving of my attention. I am so grateful for this and for so much else she has given me. I must also thank my daughters, Augusta and Louisa. Both of them have been on my chest for many hours while I wrote this book and they have been the best co-authors anyone could ask for.

1 Introduction: Science and the Transformation of International Politics

Changes in State Purpose

After the Napoleonic wars, the great powers of Europe met in Vienna to forge a peace settlement. After eight months of wrangling and a crisis that almost returned Europe to war, the great powers signed the 1815 Final Act. The main function of the Final Act was to distribute the territories held by the victorious coalition. In so doing, it delineated a balance of power that had been carefully and precisely calculated on the basis of population statistics. However, in a little-noticed clause by which the Russian Empire took possession of the Duchy of Warsaw, the Emperor Alexander reserved the right to conduct the "interior improvement" of the Polish state.[1] The appearance of improvement at the heart of international order marked a transformation in ideas about the goals or purposes of states. In the eighteenth century, international order was premised upon establishing a balance of power conceptualized as a "gigantic mechanism, a machine or a clockwork, created and kept in motion by the divine watchmaker."[2] This mechanistic representation of the balance drew on the new natural philosophy as articulated by Copernicus, Galileo, Boyle, Descartes, Newton, and others.[3] However, at the time of the Vienna Congress, the image of a balance governed by mechanical, deterministic natural laws was being displaced by the notion that humans could harness the power of knowledge to understand and manipulate the laws of nature.

By 1815, states no longer sought to obey the "rational maxims" imposed by the balance. Rather, they sought to construct and change the balance

---

[1] Final Act of the Vienna Congress 1815, 76.
[2] Morgenthau 2006 [1948], 214.
[3] This has long been noted by scholars of International Relations, but the phenomenon has not been systematically investigated and theorized. In addition to the Morgenthau reference above, see Gulick 1955; Butterfield 1966; Keens-Soper 1978; Anderson 1993; Sheehan 1996.

through the application of knowledge to problems of government.[4] Over the course of the nineteenth century, the idea of improvement emerged as a central concept in international politics. Improvement was incorporated into British imperial ideology and appeared in important international treaties regarding trade and colonial conquest. In the latter half of the nineteenth century, Darwinian ideas were used to reconceptualize improvement as a process of evolutionary development. The concept of growth then entered international order after the Second World War as new economic techniques were used to represent state goals in statistical terms as gross national product and, later, gross domestic product.[5] In short, the Congress of Vienna stood between a series of orders based on balance of power purposes and a succession of orders oriented to notions of progress.

We often take the goals of improvement, development, and growth for granted. But these purposes emerged only recently and they are quite different from the ends that underwrote international order in the sixteenth and seventeenth centuries. In the sixteenth century, the central concepts of European political discourse were drawn from aristocratic and religious discourses. God was a political force, ancient laws defined rights to territory, and blood relations conveyed political authority.[6] The reason of state was equated with the glory of the monarch and the dynastic house. Discourses in the sixteenth century also lacked a wide range of basic assumptions, concepts, and practices that now structure the landscape of international politics. States had no institutionalized procedures for using reason or knowledge to enhance power or standing. There was no imperative to govern domestic social and economic problems. Not only was there no idea of "the economy" as an entity distinct from "society," there was no discourse that divided society into a series of objects that could be understood and manipulated by the government. More fundamentally, sixteenth-century European states had no understanding or vision of progress. Indeed, they were more likely to understand time in cyclical rather than in linear terms.

These stark differences raise a question: how and why were the purposes that underlie international orders transformed between the sixteenth and the twentieth centuries? We can think of state purposes

---

[4] On the emergence of governmentality from balance of power discourse, see Foucault 2007, 67–79; McMillan 2010.
[5] Collins 2000; Maier 1987, 1989; Mitchell 2002, 2014.
[6] On the broader discursive shifts in European thought, see Cassirer 1963; Kuhn 1957; Reiss 1982; Tuck 1993; Skinner 1978. On the change in international politics, see Reus-Smit 1999; Philpott 2001; Bukovansky 2002; Nexon 2009.

broadly as the ends to which state power is expected to be used.[7] State purposes are key elements of international orders because they shape and legitimate the practices and rules that organize politics among states.[8] Purposes link the shared normative backdrop of politics to the goals and actions of states. Thus, we can see purposes when policy-makers draw on that normative backdrop to justify their actions.[9]

Despite the importance of state purposes to international orders, existing International Relations (IR) theories are ill-equipped to explain how and why they change. First, many scholars represent international history as a timeless struggle for wealth and power, so they do not recognize or explain variation in state goals.[10] However, as we shall see, there are many different ways to conceptualize and measure wealth and power. Historically shifting discourses and practices steer the pursuit of wealth and power in different directions over time.[11] Second, existing theories of long-run international change focus on order-building moments such as postwar settlements.[12] While these are important elements of any account of international change, a theoretical focus on order-building moments risks missing the slow, cumulative changes in discourses that happen between great power wars. Some recent works have broken free from order-building moments, but they have not theorized how international discourses change over the long-run.[13] To explain change in purposes, we need a theory that can account for both ongoing alterations in international discourses and order-building moments that consolidate and extend ongoing discursive shifts.

Moreover, even existing discursive and ideational theories cannot explain the transformations in international order between the sixteenth and twentieth centuries because they do not recognize the cosmological character of the underlying shifts. Most ideational theories of change in IR aim to explain the emergence of sovereignty or the rise of liberal

---

[7] On state purpose in international politics, see Finnemore 2003; Reus-Smit 1999; Ruggie 1982. My definition is closer to Finnemore and Ruggie's conceptions than Reus-Smit's because Reus-Smit focuses on the *moral* purpose of states, which I see as a specific type of the more general class of state purposes.
[8] Buzan 2004; Phillips 2011; Reus-Smit 1999.
[9] Finnemore 2003, 15.
[10] E.g., Monteiro 2014, 33–34. See also, Waltz 1979; Gilpin 1981; Mearsheimer 2001; Bueno de Mesquita *et al.* 2003.
[11] This variation has important distributional consequences. For example, the rise of economic growth privileges economic representations and policies that ignore inequality and environmental degradation. See Daly 1996; Homer-Dixon 2006; Kallis 2017; Purdey 2010; Sen 1999.
[12] Gilpin 1981; Osiander 1994; Reus-Smit 1999; Ikenberry 2001.
[13] Nexon 2009; Buzan and Lawson 2015.

international order in the nineteenth and twentieth centuries.[14] In doing so, they argue that changes in identity, legal thought, liberal norms, economic ideas, and humanitarian sympathies shaped new international purposes. However, these factors cannot explain the radical, thoroughgoing character of the changes that took place between the sixteenth and twentieth centuries. Only a change in the foundational concepts and categories of political discourses could introduce and legitimate the notions of rational control and human progress that had been so foreign to sixteenth-century political life. These changes in political discourse drew on what I call cosmological shifts in the image of the universe and the role of humanity in the cosmos. Once the cosmological character of the transformations is revealed it is clear that the rise of scientific ideas played a central role in the discursive changes that constituted modern international politics. In short, I argue that cosmological shifts originating in the European scientific tradition made possible and desirable the transformations of state purpose between the sixteenth and the twentieth centuries.[15] In other words, in the absence of scientific ideas, the international pursuit of power and wealth would look very different than it does today.

Rather than seeking to demonstrate every aspect of this transformation, in this book I present three chronologically ordered cases that explain how and why state purposes were reoriented from God and glory to economic growth. First, the emergence of balance of power orders in Europe between 1550 and 1815 reveals how cosmological ideas from the new sciences entered political discourses, displacing and reconfiguring religious-dynastic ideas. Second, changes in British colonial policy from 1850 through 1945 show how the concept of improvement, first institutionalized at Vienna, was transformed into the goal of economic development. Third, the role of the World Bank in the post-Second World War order shows how neoclassical economists created and naturalized the concept of economic growth. Taken together, these cases outline the macro-level transformation of state purposes. But focusing on three cases also provides an opportunity to closely examine the concrete mechanisms and processes that produce macrohistorical change.

---

[14] For example, Reus-Smit 1999, 124–128; Bukovansky 2002, 73, 88–91; Buzan and Lawson 2015, 6–9. See also, Barkin and Cronin 1994; Barnett 1997; Barnett 2011; Bowden 2009; Crawford 2002; Finnemore 2003; Hall 1999; Legro 2005; Philpott 2001; Price 1997.

[15] On the logic of conditions of possibility arguments, see Finnemore 2003, 14–15. I add the term "desirable" here to denote the fact that scientific cosmology did not just make new ideas possible, but made some ideas about purpose more appealing than others. That does not mean that scientific cosmology is a sufficient condition of change in purpose, but it does contribute causal effect beyond establishing a necessary condition.

Each case traces the effects of a cosmological shift that introduced new ideas about what exists, what counts as true knowledge, the nature of time, and the place of humanity in the universe. As such, these cosmological shifts provided the opportunity for individuals and groups to challenge existing ideas about state goals and articulate new purposes. First, early modern natural philosophy from Copernicus to Newton introduced new ideas about motion and matter in a law-governed universe. Second, geological and biological thinkers culminating with Darwin altered understandings of time, development, and human progress. Third, the success of atomic physics and engineering during the Second World War inspired social scientists to model the world as a series of quantitatively defined objects and cybernetic systems. The accompanying rise of economic knowledge disseminated a narrative of scientific and technological progress that bolstered and naturalized the idea of economic growth. As these new cosmological ideas were institutionalized in states, international organizations, and other associations, they slowly transformed the discourses of state purpose embedded in international order.

Explaining Change in International Order

International orders are stable patterns of behaviour and relations among states and other international associations.[16] Although there are empirical differences in the operationalization of the concept, the core idea is that international orders are historical periods characterized by distinct combinations of political, military, and economic practices. For example, the nineteenth-century Concert of Europe is considered a coherent order in which new communicative practices and multilateral institutions underwrote a long peace amongst the European powers.[17] Why do the rules and practices that underlie international orders change over time?

In this section, I argue that previous efforts to explain change in international orders suffer from three weaknesses. First, leading theorists of international order do not provide an account of the mechanisms and processes of ideational change.[18] Second, while there is now a growing

---

[16] Bull 1977, 7–8; Wendt 1999, 251; Reus-Smit 1999, 13; Finnemore 2003, 85. This definition, although widely employed, is not uncontroversial. Some scholars define order not as patterns of behaviour, but as the basic rules or governing arrangements amongst a group of states (Ikenberry 2001, 23; Phillips 2011, 5). See Schweller (2001) on why this definition is preferred to defining order in terms of governing arrangements. On my view, governing arrangements underlie and shape the patterns of behaviour that constitute order.

[17] Ikenberry 2001; Finnemore 2003; Mitzen 2013; Haldén 2013.

[18] Branch 2013; Buzan 2004; Gilpin 1981; Ikenberry 2001; Phillips 2011; Reus-Smit 1999.

IR literature on the micro- and meso-level mechanisms of change, these have not been integrated into a multilevel theory that can explain macrohistorical transformations.[19] Finally, as I pointed out above, neither macro nor lower-level theories can explain why the purposes underlying international orders change because they focus on the rise of liberalism and do not go deeply enough into discourses to see the cosmological origins of new ideas. While other theorists do theorize cosmological elements, they do not incorporate these insights into a theory of change. In short, theorists who argue that ideas are central to change do not theorize cosmology and theorists who recognize the importance of cosmology do not theorize change.

Mechanisms and Processes of Change

First, existing theories of change do not actually provide an account of how ideas spread throughout and become embedded in international orders. For example, theorists from each of the realist, liberal, constructivist, and English School traditions theorize change in international order as a series of order-building moments in which the great powers alter governance arrangements after great power war. Some of these accounts do not take ideas or change in purposes seriously. For example, in Gilpin's account of international change, the primacy of security goals stays constant throughout history.[20] Ideas matter in Gilpin's theory only as elements of governance arrangements and prestige.[21] Postwar settlements redefine the rules governing the system and reorder hierarchies of prestige, but the ends and goals of states remain constant. His model neither recognizes nor explains change in state purposes. Ikenberry expands the role for ideas by taking seriously the beliefs and perceptions of leaders, but he adopts Gilpin's model of change and ignores where the ideas that structure international institutions come from.[22]

Constructivist and English School approaches are best equipped to explain change in purposes, but theorists in this tradition have not theorized the mechanisms and processes by which ideas come to be institutionalized in international orders.[23] Buzan and Reus-Smit both argue

---

[19] Adler and Pouliot 2011; Barnett and Finnemore 2004; Bueger and Gadinger 2015; Finnemore 2003; Goddard 2009; Guzzini 2013; Hopf 2010; Pouliot 2009. Finnemore and Sikkink (1998) is an important exception, but their macro-level account is quite schematic.
[20] Gilpin 1981, 6–8, 19, 23–24.
[21] Gilpin 1981, 30–35, 203.
[22] Ikenberry 2001.
[23] Mahoney and Thelen (2010, 5–7) have levelled this criticism at meso-level discursive theories as well.

that international orders are constituted and supported by a structure of governing arrangements and constitutional norms.[24] As Wendt and Duvall put it, a "hierarchy" of institutions shapes the patterns of activity that comprise international order.[25] The structure is hierarchical because there are constitutive relations between the various institutions. Sovereignty is "deeper" than other institutions because it makes possible or creates the conditions of existence for higher-level institutions.[26] For example, sovereignty is constitutive of an international trading regime premised upon control of the flow of goods across sovereign borders. In Buzan's schema, primary institutions underlie and constitute secondary institutions.[27] In Reus-Smit's account, fundamental institutions define the rules that shape state behaviour.[28] For both Buzan and Reus-Smit the institutional basis of international order in turn rests on values.[29] For example, Reus-Smit argues that fundamental institutions emerge from constitutional structures that define the "moral purpose" of political organizations and the norms of procedural justice. Reus-Smit argues that as the moral purpose of the state changes, there is upward pressure for change in the fundamental institutions of international order.

However, despite the fact that Reus-Smit and Buzan take ideas and the possibility of change seriously, they do not theorize how new ideas emerge or come to be embedded in international order. Reus-Smit argues that institutions are created via a process of communicative action in which states debate "within the context of preexisting values that define legitimate agency and action."[30] However, Reus-Smit does not tell us where or when these debates happen, nor does he explain where the pre-existing values come from. The result is a static model of international order that can describe how international orders are different from one another but cannot explain change from one international order to another.

Phillips has recently built on Reus-Smit's account, arguing that international orders shift when the social imaginaries that support them break down.[31] For Phillips, imaginaries include "our most basic and

---

[24] In this way, they reject the largely materialist accounts of realists like Gilpin (1981) who concede only a residual role for ideas in shaping the behaviour of states. For a synthesis, from the constructivist side, see Phillips 2011.
[25] Wendt and Duvall 1989, 67. See also, Ruggie 1983.
[26] Wendt and Duvall 1989, 64. This is similar to Buzan's (2004) distinction between primary and secondary institutions, which is discussed below.
[27] Buzan 2004. For a review of other work in this vein, see Wilson 2012.
[28] Reus-Smit 1999, 14.
[29] Buzan 2004, 181; Reus-Smit 1999, 15.
[30] Reus-Smit 1999, 27.
[31] Phillips 2011, 43–44. For an important precursor to this argument, see Legro 2005, 29–35.

mostly unarticulated assumptions about social reality, extending even to those that condition our experience of categories as allegedly basic as time, space, language and embodiment."[32] This theorization of social imaginaries helps explain where values and purposes come from and the conditions under which international orders change. My account of cosmological change builds on these insights. Yet Phillips does not theorize the processes and mechanisms of discursive or ideational change.[33] So he does not demonstrate how the structures underlying international order are actually reconfigured. Without a more fine-grained account of how international discourses are reproduced and transformed, we cannot explain why some ideas become dominant rather than others.

Linking Micro, Meso, and Macro

Second, existing accounts of change do not provide an integrated, multilevel theory that shows how and why ideas come to be embedded in international orders. To explain international change we need to combine three elements: a micro-level account of how ideas are formed in the everyday life of social action; a meso-level explanation for why some ideas rather than others take hold in the organizations or associations that carry international order; and a macro-level theory of how the distribution of ideas in the international system shifts. The problem is that existing approaches leave out one or the other of these elements. On one hand, macro-level theories of change focus on snapshots of order-building that leave out micro- and meso-level drivers. However, change in international politics begins in the everyday life of groups and organizations including states, international organizations (IOs), non-governmental organizations (NGOs), firms, epistemic communities, and so on. To see the mechanisms and processes of change, we have to descend from the macrohistorical level to the meso-level of concrete organizations. After all, it is here that micro- and macro-level phenomena meet and are converted into one another.[34]

On the other hand, existing micro- and meso-level approaches leave out an account of what the macro-level is and how it might change. The recent discursive, practice, and relational turns in IR theory have generated exciting insights into the mechanisms and processes of change within groups. But it is not clear how these might translate into a theory

---

[32] Phillips 2011, 24.
[33] The same critique could be applied to many theories that bracket continuous processes of ideational change and so rely on exogeneous changes in ideas during critical junctures. See, e.g., Gilpin 1981; Ikenberry 2001.
[34] Katznelson 1997, 84, 102; Nexon 2009, 61–63. See also Tilly 1984.

of macrohistorical change.[35] For example, Finnemore offers a taxonomy of micro- and meso-level mechanisms to explain change in the purposes of international order.[36] These mechanisms include persuasion, affect, social influence, coercion, legal rationalization, professional capture, social movement pressure, and so on. Finnemore rightly points out that in order for social purpose to change, "widely shared social structures must change."[37] But an explanation for change in social structure requires a specification of what that social structure looks like and how its constituent elements might be altered. In this sense, Finnemore's account of change at the collective level is incomplete because we need both a set of mechanisms and an account of what exactly is being changed and how.

Nexon's relational theory of international structure is an important exception here. Nexon decomposes international structure into a set of interlinked networks such that change at the meso-level is a macro-level change. This is an essential insight that I build upon in the next chapter. However, Nexon downplays the role of structural elements like anarchy or systemic norms.[38] While he is correct that IR theorists should be careful about reifying structural properties, we must retain some way to talk about relatively stable practices and discourses if we want to build a theory of change.[39] After all, change in international order is meaningful and important precisely because we can distinguish relatively stable patterns in international life. Once ideas and practices become embedded in the core sites of international order such as multilateral treaties, postwar settlements, and powerful IOs they are reliably reproduced in a way that makes them structural forces.

However, the solution is not, as theorists of international order have often done, to focus only on order-building moments and leave the rest of history in stasis. Instead, we need a theory that combines the dynamism of recent micro- and meso-level theory with an account of why some ideas become stable elements of the landscape of international politics. But as Nexon argues, this must be done in a way that carefully links the micro-, meso-, and macro-levels of international life.[40] Following Nexon

---

[35] Adler and Pouliot 2011; Barnett and Finnemore 2004; Bueger and Gadinger 2015; Finnemore 2003; Finnemore and Sikkink 1998; Goddard 2009; Guillaume 2009; Guzzini 2013; Hopf 2010; Jackson 2006a; Kessler and Guillaume 2015; Nexon 2009; Pouliot 2009.
[36] Finnemore 2003, 146–161. Finnemore stops short of saying that purposes underlie international order in the same way that Buzan and Reus-Smit do.
[37] Finnemore 2003, 146.
[38] Nexon 2009, 48–60.
[39] See Giddens (1984, 16–18) for a defence of this line of argument.
[40] Nexon 2009, 61–63.

on this point allows us to produce a more fragmented and dynamic conception of international ideational structures.

The Cosmological Basis of State Purposes

Finally, existing approaches cannot explain the transformation in international order over the last five centuries because they neither see nor specify the deepest, cosmological levels of international discourses. I pointed out above that ideational IR theorists argue that change is driven by changes in norms, beliefs, and emotional dispositions. For example, Reus-Smit argues that the difference between pre-modern and modern international politics can be explained by the emergence of a new moral purpose of the state. Whereas the pre-modern order was premised on the divine right of kings, the modern era rests on the liberal norm of popular sovereignty and the concomitant purpose of advancing individual interests.[41] Empirically, this begs the question of how states came to have the idea that they could intervene in the lives of individuals in the first place. In the sixteenth century, there was no sense that the state was responsible for or could even shape individuals' fortunes. Individual welfare was considered to be a product of human nature, not social or state policy. Indeed, it was unthinkable that the state could affect the social forces that determined welfare because there was no conception of society as a set of elements that could be rationally controlled. Thus, the shift in state purposes had to be driven by a more fundamental shift in political discourse than a shift in moral norms could produce.

This book argues that cosmological shifts made possible and desirable new ways of thinking about state purpose that came to be embedded in successive international orders. This argument draws on insights from a number of theorists who have demonstrated the power of epistemic and ontological ideas. Ruggie and Walker demonstrate how geometrical and aesthetic concepts constituted the norms of sovereignty.[42] Similarly, Bartelson's study of *mathesis* argues that a new scientific episteme constituted the conceptual basis of sovereignty and state interests.[43] Grovogui contends that international law remains constrained by deep epistemic structures constructed in the sixteenth century.[44] Scott demonstrates the importance of grids of legibility in state projects to remake the world.[45]

---

[41] Reus-Smit 1999.
[42] Ruggie 1993; Walker 1993. See also, Branch 2013.
[43] Bartelson 1995.
[44] Grovogui 1996, 43–53.
[45] Scott 1998, 39.

As we saw above, Phillips argues that ideas about space and time form part of the normative underpinnings of international orders.[46] These arguments show that ideas about what counts as knowledge (episteme) and what exists in the world (ontology) are central to understanding the fundamental categories of international politics. But they do not reveal the importance of cosmological ideas specifically and do not theorize how these are incorporated into processes and mechanisms of change.[47]

On my conception, cosmologies weave ideas about what counts as knowledge and what exists into broader narratives about the origins and operations of the universe. Defined broadly, a cosmology is "any composition or cultural construct relating to the structure or process of systems of creation: the origins of physical elements of earthly or astronomical spheres, the genesis of the material world, the order and function of the observable universe."[48] Drawing on this definition, I think of cosmologies as comprised of ideas about:

- the fundamental units of matter, the forces that govern them, and categories of representation (ontology);
- the modes and procedures likely to produce reliable or true knowledge of the universe (episteme);
- the nature and direction of time (temporality);
- the origins and history of the universe (cosmogony);
- the role or place of humanity in the cosmos (destiny).

On this conception, cosmologies are not integrated totalities that are understood and internalized by all members of a group.[49] Rather, they are compositions or configurations of cosmological elements that circulate in discourses. Cosmologists then weave together these elements into powerful narratives in specific texts, rituals, or institutions.[50] So in theorizing the effects of cosmologies on international politics, we should focus on how specific cosmological elements are deployed in context.

---

[46] Phillips 2011.
[47] Recent accounts by Bartelson (2009) and Phillips (2011) recognize the importance of cosmologies, but do not theorize their effects systematically. Bartelson (2009) looks at changes in cosmology to explain shifts in ideas about world community, but not how these ideas shaped international orders. Phillips (2011) makes allusions to cosmological elements in his historical cases, but does not theorize cosmological ideas apart from the spatio-temporal aspects of social imaginaries.
[48] Destro 2010, 227. See also, Howell 2002.
[49] In structuralist theories that emerged in the 1960s, cosmologies provided the overarching framework of meanings that ordered and integrated social life (Lévi-Strauss 1963, 1966). This approach was criticized for suppressing the diversity of cosmological beliefs within and between societies (Barth 1987).
[50] Berger and Luckmann 1966; Douglas 1986.

Cosmological elements are central to understanding and explaining change in state purposes. Political purposes, as opposed to mere interests or intermediary ends, are infused with meanings drawn from the broader cosmological ideas. Cosmological elements define purposes by providing a narrative of humanity's place in the universe that suggests what ends political agents should pursue. Technical beliefs or religious rules shorn of their cosmological backing may struggle to motivate or legitimate actions because they cannot be justified as necessary elements of nature or the universe. So, cosmological elements are powerful because they infuse the beliefs and institutions that order collective life with meaning and motivate actors by placing them in narratives that structure time and space.

My Theoretical Approach

In this book, I offer a historically flexible, multilevel theory of change that explains how and why cosmological shifts transform the discourses of state purpose underlying international order. My approach combines the recent focus on the processes of international change with a traditional constructivist interest in the distribution of ideas.[51] I explicitly link the processes and mechanisms of change to a macrostructure of ideas. In particular, I aim to explain change in international discourse. International discourses are comprised of the distribution of ideas across the states, IOs, NGOs, firms, and transnational networks that influence international politics. I call all these groups "associations." The key theoretical move is to argue that change in associations produces change in international discourses.[52] Associations are constantly changing as they combine and recombine ideas, practices, and institutional elements to solve problems and engage in political contestation. So the key question in explaining international change is why the discourses and practices of a large number of associations would change *in the same way at the same time*. This points to structural causes that can operate on multiple associations simultaneously.

I argue that a new scientific cosmology can operate as a structural factor that constitutes new concepts of state purpose for many associations simultaneously. When transmitted by transnational knowledge networks, cosmological ideas arising from the scientific tradition act as a form of productive power that constitutes the discursive landscape of

---

[51] On process, see Jackson and Nexon 1999; Jackson 2006a; Nexon 2009. On the distribution of ideas, see Wendt 1999. I adopt the processual dynamism of Jackson and Nexon while maintaining an important role for ideational structures that, while processually constituted, nonetheless still exhibit some continuity.
[52] Holsti 2004.

international politics.[53] These ideas are powerful because they render some goals, interests, and purposes as natural or inevitable and others as unthinkable or illegitimate. But we cannot simply posit cosmology as a unitary force that operates automatically to redefine values and goals. Instead, we have to specify how and why cosmologies enter the contested realm of international political discourses.

I theorize two pathways by which cosmological ideas became embedded in the discourses underlying international orders. First, in the hegemonic imposition pathway, a powerful state leads the production and dissemination of new cosmological ideas to support the institutionalization of its purposes.[54] Here, a powerful state coerces or induces other states to take up new cosmologies and purposes. As we shall see, it was in this way that the American hegemon drove the institutionalization of a growth-based order in the twentieth century. Second, in a horizontal pathway, a transnational network of knowledge-based actors disseminates new ideas, reconstituting the discourses of many associations in similar ways. In 1815, the Austrians, the British, and the Prussians all shared a similar conception of the statistical balance even though it had not been imposed on them by a hegemon.

The pure hegemonic and horizontal pathways form two ends of a continuum. Most real-world cases fall between them, combining hegemonic and horizontal elements. After all, even the hegemonic pathway requires a transnational transmission belt to spread and institutionalize its purposes and policies.[55] Conversely, the horizontal route still depends on power drawn from states or social classes with the economic and cultural resources to sustain a coherent transnational knowledge network. Put differently, cosmologies and purposes do not float freely. They must be carried by concrete associations that possess the military, economic, and epistemic power to reproduce and disseminate ways of seeing the world.

In highlighting the role of state power in international ideational change, my account builds on the insights of Ruggie and Ashley.[56] Ruggie and Ashley both conceptualized hegemony broadly as encompassing both direct, coercive military and economic power over others as well as the cultural and epistemic power to define the interests and meanings that constitute political landscapes.[57] To construct and maintain orders, hegemons must coerce and cajole states to sign postwar settlements and

---

[53] Barnett and Duvall 2005. See also, Foucault 1970, 2007.
[54] On hegemonic imposition, see Ashley 1989; Gilpin 1981.
[55] This is a key argument of Cox 1987.
[56] Ruggie 1982, 1983; Ashley 1989.
[57] For a clear exposition of "power over" and "power to," see Barnett and Duvall 2005.

abide by the rules. But they must also produce and reproduce a cosmological vision that naturalizes and legitimates the goals and institutions of the order. Moreover, these forms of power must be exercised both in and between order-building moments.

States, IOs, NGOs, and other international associations draw on and reproduce cosmologies to legitimate and naturalize orders. So cosmologies are constantly being pulled into the processes of political contestation that shape the construction and maintenance of international orders. From this perspective, we can think of the history of European and international politics itself as a process of cosmological contestation. Throughout the modern period, European states deployed both Christian and scientific cosmologies to bolster their authority, legitimate colonial conquest, and naturalize self-serving rules. These moves were resisted by religious authorities, conservatives, colonized peoples, and a wide variety of traditionalists from the Luddites to environmental activists. So, the advance of cosmological ideas can never be taken for granted. Their rise and consolidation must be explained as a fundamentally political phenomenon.

Rethinking Science in International Relations

This book provides a macrohistorical account of how scientific ideas entered into and transformed the discourses underlying international order. In making this argument, I adopt a historical and plural definition of science informed by the recent historiography and sociology of science and technology.[58] I define and operationalize science in historical terms as the Western tradition of disciplined inquiry into the operations of nature since the sixteenth century.[59] This definition includes the modes of "natural philosophy" and "natural history" that organized inquiry from the sixteenth century up to the eighteenth century, as well as the modern "natural sciences" of biology, physics, chemistry and so on that were consolidated in the nineteenth

---

[58] For a review of related approaches in IR, see Mayer *et al.* 2014. For good introductions to the historical and sociological literatures, see Bloor 1991 [1976]; Cunningham and Williams 1993; Latour 1987; Jasanoff 2004; Park and Daston 2006; Shapin 1996; Shapin and Schaffer 1985.

[59] Park and Daston 2006, 2–3. This is a good working definition, but to be made meaningful the definition has to be related to concrete historical eras which each have unique ways of defining what counts as "disciplined inquiry" and the "operations of nature." The goal of a broad, historicist definition is to include traditions and practices that were perceived by contemporaries as scientific. Such a conception avoids a post-hoc, normative definition that restricts science to our present standards of what counts as scientific. See Bloor 1991 [1976] for discussion.

century.⁶⁰ I am also interested in applications of these traditions in the social sciences and their precursors because these applications spread the core concepts and beliefs of science.

My conceptualization of science is plural in the sense that it recognizes a variety of movements and discourses within the scientific tradition.⁶¹ Even within the modern, Western tradition there is no single, hegemonic "scientific method" or "scientific discourse." Rather, there are multiple scientific enterprises, norms, values, and methods promoted by disparate, sometimes competing, communities of practice. Moreover, Western science is not entirely "Western" because it was influenced and shaped by non-Western knowledge and practices. Islamic science played an important role in the development of early modern natural philosophy.⁶² Science should be conceptualized as plural and historical because scientific practices and ideas are always embedded in temporal and cultural contexts.⁶³ So the modes and discourses of the scientific enterprise change over time and across contexts.

This account of science allows me to theorize scientific knowledge as a source of cosmological meanings that were used to redefine state purposes. Scientific ideas are not the only source of cosmological meanings, but from 1550 onwards they were persistently deployed to alter and challenge dominant conceptions of the universe and associated understandings of political order. This view of science as a transformative, cosmological force is distinct from existing understandings of science in the discipline of IR. Scholars in IR have variously referred to science as a historical curiosity, a minor factor in international change, or an instrumental means that allows states to accomplish their goals.

Realist, English School, and constructivist histories contain scattered references to the importance of scientific ideas, but these have never been integrated into an overarching study of how these ideas transformed international orders. For example, Morgenthau, Gulick, and Butterfield each suggest that eighteenth-century balance of power thinking was based on mechanical and systemic ideas inspired by Newton.⁶⁴ Morgenthau blames Enlightenment rationalism for the rise of nineteenth-century liberalism and the apolitical optimism of the League of Nations era.⁶⁵ In the course of an argument focusing on diplomatic

---

⁶⁰ Cunningham and Williams 1993; Shapin 1996.
⁶¹ Cunningham and Williams 1993; Daston 1995.
⁶² See, for example, Saliba 2007.
⁶³ Cunningham and Williams 1993; Latour 1987; Park and Daston 2006; Shapin and Schaffer 1985.
⁶⁴ Morgenthau 2006 [1948], 214; Gulick 1955, 24. There is a small historical confusion here: Newton himself did not expound the mechanical philosophy. In fact, mechanists criticized his notion of gravity as an "occult" force.
⁶⁵ Morgenthau 1946, 13–29; Morgenthau 2006 [1948], 41–49.

and legal institutional developments, Keens-Soper pauses to recognize the importance of the "shift from the theocentric 'metasystem'" to a metasystem of "matter and motion."[66] More broadly, Ernst Haas spent decades trying to make sense of the role of science in international politics.[67] In his late work, Haas argues that in addition to serving instrumental purposes, scientific knowledge has deep, constitutive effects on political discourses:

[T]he very mode of scientific inquiry infects the way political actors think. Science, in short, influences the way politics is done. Science becomes a component of politics because the scientific way of grasping reality is used to define the interests that political actors articulate and defend. The doings of actors can then be described by observers as an exercise of defining and realizing interests informed by changing scientific knowledge.[68]

Over time, Haas suggests, "the intellectual commitments of the seventeenth-century scientists and mathematicians penetrated the way political economists and their disciples in governments began to see the world."[69]

The fundamental insight contained in these claims is that scientific ideas operate as a form of productive power that defines the central metaphors and categories of international politics. This idea, while intuitively plausible, has not been translated into a systematic empirical demonstration of the effects of scientific ideas on international order.[70] Indeed, major studies of macrohistorical change in international order assign science only a minor role. Reus-Smit argues that eighteenth-century scientific and political theorists helped dismantle the divine foundations of dynastic order and produce the atomistic basis of liberal order.[71] However, he does not specify which scientific ideas mattered or how they entered political discourses. Bukovansky suggests that Enlightenment ideals supported the spread of rational administration and the emergence of strategic, *realpolitik* calculations in international politics.[72] But her overall focus is on showing how Enlightenment legal and civil thought changed

---

[66] Keens-Soper 1978, 31.
[67] See Allan 2017b.
[68] Haas 1990, 11.
[69] Haas 1990, 22.
[70] There are also discussions of scientific ideas in intellectual history and international political thought. In what follows, I draw on this literature, but with a few exceptions this literature does not make systematic connections between the history of ideas and the ideas of leaders and policy-makers. So, I only engage this literature where it touches on the narrative I develop here. See Ashworth 2014; Bell 2007; Bell and Sylvest 2006; Devetak 2011; Keene 2005; Skinner 1978; Tuck 1993.
[71] Reus-Smit 1999, 124.
[72] Bukovansky 2002, 86, 90.

the bounds of legitimate statehood, not explaining the associated shift in state purposes. Buzan and Lawson trace the origins of the present international order to the nineteenth-century emergence of industrialization, rational statehood, and ideologies of progress, but their analysis begins too late and does not go deep enough into political discourses to see that the changes they identify began in the sixteenth century with the rise of scientific ideas.[73]

While others have thought more systematically about how science shapes politics, they conclude that its effects have been limited. Literatures in military history,[74] hegemonic transition,[75] and realist institutionalism[76] all suggest that science is subordinate to state interests. In each case, science serves states by increasing military capabilities or advancing economic growth. This tendency can also be seen in studies of science and expertise in global governance. Peter Haas shows how an epistemic community of atmospheric scientists and bureaucratic officials persuaded government leaders that the emerging hole in the ozone layer posed a threat to the health of their citizens.[77] This is an important argument, but it amounts to the claim that scientists can aid states in realizing their interests by supplying information those states would not have otherwise had.[78] For Haas, the expected effect of epistemic communities "remains conditioned and bounded by international and national structural realities."[79]

Skolnikoff's ambitious history of science and technology in international politics also conceptualizes science in instrumental terms. He argues that while the effects of science and technology are interesting, a fundamental transformation has been elusive:

[T]he evolution of the *details* of international politics due to the interaction with advancing science and technology has been as impressive and astonishing as common rhetoric proclaims. But the more general impact on the underlying concepts and assumptions that govern the relationships among nations has in fact been considerably less marked. There are some important exceptions, but they do not counterbalance the realization that fundamental changes in the international system as a whole have been quite limited.[80]

---

[73] Buzan and Lawson 2015, 6–9.
[74] Gilpin 1968; McNeill 1982; Parker 1988; Van Creveld 1989; Friedberg 2000; Lieber 2005; Boot 2006; Horowitz 2010; Hymans 2012.
[75] Thompson 1990; Drezner 2001; Taylor 2012.
[76] Krasner 1991.
[77] Haas 1992b.
[78] Haas 1992b, 193.
[79] Haas 1992a, 7.
[80] Skolnikoff 1993, ix–x. Emphasis original.

In short, Skolnikoff denies that science and technology have altered the core dynamics that govern and shape international politics. Thus, he concludes that "dramatic" examples notwithstanding, science and technology remain closely tied to "national goals."[81]

IR scholars largely deploy a thin, instrumental conception of science as a means to ends. This view shares some affinities with the Weberian image of science as a force for rationalization and disenchantment. Weber argues that the spread of scientific rationality drained meaning out of politics. On one hand, rationalization entailed the belief that "we are not ruled by mysterious, unpredictable forces, but that, on the contrary, we can in principle *control everything by means of calculation*."[82] On the other hand, rationalization produced "the disenchantment of the world" embodied in the fact that "the ultimate and most sublime values have withdrawn from public life."[83] This followed from the fact that science can only help achieve ends and thus cannot answer the questions, "What should we do? How shall we live?"[84] So Weber also presents us with an instrumental and anti-cosmological view of science.

By contrast, I demonstrate that scientific ideas have done more than serve as instrumental means; they have laid the groundwork for the transformation of state purpose. Scientific ideas allowed individuals and groups to reimagine their relationships to the cosmos. In so doing, they inspired new ways of thinking about what political life could and should be about. Scientific ideas were thereby transformed from means to ends. Far from draining the world of meaning, scientific discourses have been used to naturalize a number of ends and purposes since the sixteenth century. Indeed, Weber himself presupposes that science has operated as a *negative* cosmological force that reduced the meaning of life and death to moments in the organic life cycle. Moreover, he doubts that progress itself can have "an intrinsically meaningful end."[85] However, the inability of science to provide meanings that are logically deduced from scientific principles has nothing to do with the fact that scientific ideas have nonetheless been used to define humanity's place in the universe.

Another aspect of the problem is that IR scholars are used to characterizing the political orders of non-Western societies as drawing on cosmological beliefs, but less likely to think of Western political orders in

[81] Skolnikoff 1993, x.
[82] Weber 2004, 12–13. Emphasis original.
[83] Weber 2004, 13, 30.
[84] Weber 2004, 13, 17. The questions are Tolstoy's.
[85] Weber 2004, 13.

those terms.[86] This is part of a broader orientalist tendency to see Western political orders as rational and progressive while viewing non-Western societies as backward.[87] As we shall see in Chapter 4, this tendency was built into the social sciences by colonial anthropology.[88] Weberian ideas about rationalization are also bound up in this discourse. Weber argues that "traditional" societies rest on "the sanctity of orders and powers of rule which have existed since time immemorial."[89] For Weber, the process of rationalization in the West eroded magical thinking and disrupted the ritualistic basis of traditional rule. As a result, Western political orders came to be based on the impartial administration of fixed, rational rules.[90]

We might seek to avoid thinking in these orientalist terms by discarding the cosmological analysis of political orders altogether. However, this would leave us unable to understand and explain shifts in political purpose in the West. Instead, we can push back on the cosmological–rational dichotomy itself by retelling Western history in cosmological terms. Thus, in contrast to the instrumental and Weberian views of science, I conceptualize the Western scientific tradition as carrying and expressing cosmological elements that have been used to infuse the world with meaning.

As John Meyer and his colleagues in the World Polity School argue, "science operates as the secular equivalent of a 'sacred canopy' for the modern order, generating a modern, rational interpretation of world order and offering this logic as a secular interpretive grid for natural and social life."[91] In short, science describes both nature and society as knowable, calculable, law-governed domains. In so doing, it provides ontological and cosmological support to legitimate the modern idea that actors (individuals, states, and organizations) can rationally harness knowledge to their ends. The role of science in world order further bolsters ends of justice (equality) and progress (economic growth).[92]

On one hand, this argument usefully extends and modifies Weber's rationalization thesis. For the World Polity School, rationalization has not drained the world of meaning but has simply replaced older cultural frames with modern scientific ones.[93] On the other hand, the argument reproduces the weakest aspect of Weber's schema: the idea that science and technology exhibit a singular, rational logic that produces the same

---

[86] For a critique of this view in IR, see Inayatullah and Blaney 2004.
[87] Said 1978; Guha 1997.
[88] Asad 1973; Stocking 1987, 1995.
[89] Weber 2004 [1922], 135.
[90] Weber 2004 [1922], 134.
[91] Drori et al. 2003, 23.
[92] Meyer et al. 1997, 174; Drori et al. 2003, 23.
[93] Meyer et al. 1997, 166; Drori et al. 2003, 2, 7. In this regard Drori et al. (2003) follow Berger and Luckmann (1966).

effects everywhere throughout the world. The result is that the varied political effects of scientific ideas are folded into monolithic, abstract processes like modernization, rationalization, and commodification.[94] This universalistic conception of science is at odds with the plural and historical conception of science that has emerged from the last forty years of historical and sociological studies of science. Indeed, the singular conception of science in the World Polity School was designed to explain isomorphism and is poorly suited to explain change in international discourses.[95]

To be adapted to the historical analysis of science in international politics, the World Polity School's narrow, universalistic conception of science must be replaced with the plural and historical view. In light of the fact that the meanings and methods of the scientific enterprise change over time, it does not make sense to refer analytically to "science" as a single enterprise with uniform effects on international politics over the course of 450 years. Instead, I trace the effects of specific scientific movements on the history of the present international order. So, rather than produce a general theory of how science has shaped political institutions, my approach seeks to demonstrate the effects of three specific cosmological shifts on the discourses of state purpose underlying international order. This approach decomposes a macrohistorical process, the emergence of scientific ideas in politics, into a series of contested moments of institutionalization.

The Argument: From Means to Ends

The main argument of the book is that cosmological elements arising from the Western scientific tradition made possible and desirable a transformation in the discourses of state purpose underlying international orders. The argument demonstrates the effects of three cosmological shifts. A cosmological shift occurs when new ideas about what exists, what counts as knowledge, time, the origins of the universe, and the place of humanity in the cosmos are introduced into political discourses. In the three cases that follow, these ideas emerge from major developments in natural philosophy and, later, the natural sciences.

First, I trace the effects of the cosmological shift from Copernicus to Newton on balance of power thinking. Second, I analyse how the historical sciences of geology and biology in the age of Darwin reconfigured ideas about human progress and development. Third, I show that the

---

[94] On this broader tendency in Weberian thought, see Horkheimer and Adorno 2002.
[95] Finnemore 1996a, 338–339.

rise of a cosmology centered on cybernetic-systems thinking reshaped twentieth-century political discourses by representing the world as a series of objects. Together, these cosmological shifts slowly reconfigured the discourse of state purpose, reorienting it from God and glory to economic growth backed by scientific and technological progress.

The theoretical argument explains how these ideas entered into and transformed the discourses of state purpose underlying international orders. My account combines meso-level mechanisms of change into a macro-level model of transformation I call *recursive institutionalization*. The model is recursive in the sense that macro-level change emerges from an iterative process of micro- and meso-level changes.[96] So change begins as new ideas arising from scientific, religious, or other cultural traditions create a *cosmological shift* in broader societal discourses. Second, the states, international organizations, and other associations that constitute international order undergo a process of *associational change* as they take up the new cosmological ideas into international politics.[97] The process of associational change unfolds in two steps: (1) associations *strategically deploy* scientific ideas to solve problems and fulfil their interests; (2) the introduction of these ideas sets off processes of *discursive reconfiguration* that alter the ends and purposes of the associations. As associational changes accumulate in the international system, they change the discourses of state purpose and cosmological ideas that underlie primary and secondary institutions. This creates incentives and opportunities to change those institutions so they are supported and legitimated by the underlying discourses. So, finally, states *institutionalize new purposes* by altering the rules and norms of the international orders. This could proceed in an order-building moment after great power war or through incremental changes in treaties and international organizations that are always ongoing.

In the empirical chapters, I show how and why scientific cosmology came to be embedded in international order by moving back and forth between meso-level studies of associational change and macro-level snapshots of international order. The meso-level studies of associational change show how the strategic deployment of scientific ideas produced new purposes in important associations. The macro-level analyses demonstrate that these purposes were embedded in the core sites of international order. Together, the meso- and macro-level analyses illustrate the

---

[96] See Giddens 1984, xxiii, 2, 25; Wendt 1987.
[97] I conceptualize this as a process of institutional or organizational change as articulated in historical institutionalism (Fioretos *et al.* 2015; Katznelson 1997; Lieberman 2002; Schmidt 2008) and organizational sociology (DiMaggio and Powell 1983; Douglas 1986; Selznick 1984[1957]; Zucker 1977).

processes of recursive institutionalization: international discourses shift as associations import new cosmological elements and work to embed new purposes in the core sites of international order. In short, the strategic deployment of science as a means has unintentionally transformed the ends underlying international order.[98]

The historical narrative begins in 1550 when political discourses were dominated by a cosmological discourse of divine providentialism on which God controlled the universe. The universe, further, was understood in Aristotelian terms as divided into earthly and celestial realms. I start in 1550, seven years after the publication of Copernicus' *On the Revolutions of the Heavenly Spheres*, to clearly demonstrate the cosmological shift that followed. Copernicus' heliocentric universe challenged the dominant Aristotelian cosmology because it challenged the earthly-celestial division by depicting all bodies as matter in motion. Over the course of the seventeenth century, Mersenne, Gassendi, Harvey, Boyle, Hobbes, and others transformed Copernicus' insights into a "mechanical philosophy." Newton, while departing from mechanist orthodoxy in some ways, offered the image of the universe as a harmonious natural order governed by unchanging laws. Inspired by these ideas, the early political economists William Petty and William Temple articulated a materialist view of the polity as reducible to "number, weight, and measure."[99] They used this idea to challenge providentialism and articulate the rationalist cosmological claim that the world could be controlled. These new forms of knowledge interacted with the spread of capitalist markets and the bureaucratization of early modern states to elevate the political significance of territorial and economic issues in European politics.[100]

As the new cosmological ideas entered political discourses they weakened and displaced the old religious and aristocratic purposes. Materialist and mechanical ideas entered the processes of recursive institutionalization through a variety of meso-level channels. First, they were imported into states by aristocratic brokers who were embedded in both scientific and political networks. Second, the new ideas were taken up into political discourses when eighteenth-century states expanded foreign ministries and created statistical offices to harness the power of knowledge. As these ideas set off associational changes in states, they reoriented the discourse of state purpose from God and glory to the balance of power. Further, the balance of power came to be portrayed as a natural law and obeying the balance was considered a "rational maxim."

[98] On means-ends change, see Barnett and Finnemore 1999, 720.
[99] See Petty 1690. For a complete history of materialist thinking, see Deudney 2007.
[100] Katznelson 1996, 20–21. See also, Deudney 2007; Mann 1986; Mukerji 1997; Tilly 1992.

During the eighteenth century, the view of the world as a harmonious, mechanical field of forces was linked to the rationalist idea that the field of forces could be controlled through the discovery and manipulation of its laws. By the time of the Congress of Vienna in 1815, Castlereagh and Metternich believed they could intentionally construct the balance of power through the careful division and allocation of land and peoples. At Vienna, the rationalist idea that politics could be controlled can be seen in the role of the statistical commission. The statistical commission enabled the delegates to precisely calculate the balance in terms of the number of "souls" on each parcel of land, allowing them to reach a difficult political compromise.

In the early nineteenth century, another cosmological shift was initiated by developments in geology, botany, and zoology. These historical sciences imported historical themes from the Enlightenment human sciences to challenge the young earth depicted in Christian cosmology.[101] Cuvier, Buffon, and Hutton depicted a prehistoric earth that changed slowly over millennia. This new image of the earth drew upon Newtonian ideas about time as an absolute, open plane without beginning or end. Absolute time served as the basis for linear conceptions of progress that displaced earlier cyclical notions. The idea of linear progress was then naturalized as thinkers like Robert Chambers and Charles Darwin argued that development was a law of nature. At the same time, shifts in capitalist production and colonial law decreased the value of conquest for European power. The onset of the industrial revolution in the late eighteenth century made science, technology, and energy-use more important elements of power than territorial acquisition.[102] These discursive and political-economic changes created an opportunity to shift notions of state purpose away from increasing land and people in the context of the balance of power.

The British hegemon that emerged from Vienna spent the nineteenth century constructing a laissez-faire liberal order cribbed from the pages of Adam Smith and J. S. Mill. This order was premised upon a cosmological formation I call natural providentialism. Natural providentialism challenged the divine cosmogony of earlier discourses. However, it maintained the determinist element of divine providentialism, transposing it into materialist, scientific terms. According to natural providentialism, improvement was an automatic process of moving through stages of civilization that was fuelled by natural laws. All that was required was free commercial contact, which unleashed a linear process of civilizational

---

[101] Rudwick 2005.
[102] Buzan and Lawson 2015; Deudney 2007.

development. After 1850, this providential vision of progress was taken up by anthropologists. Inspired by Darwin, sociocultural anthropologists argued that societies went through a process of evolution that could be facilitated and guided by Western trustees. The British hegemon led the recursive institutionalization of this cosmological vision at the Berlin Conference of 1885 and again at Versailles in 1919. Under British leadership, these conferences embedded the purposes of improvement and civilizational development in core sites of international order.

At the meso-level, these ideas were taken up and applied in the British Colonial Office. Informed by sociocultural theories of evolution, colonial officers believed that the process of colonial improvement and development would be automatic. However, after 1920 the idea that primitive peoples would naturally proceed through the stages of sociocultural evolution was challenged. By 1935, it was replaced by the idea that evolution must be guided by knowledge and expertise. The emergence of new experts in political economy, labour, nutrition, soil, public health, and so on bolstered the modernist notion that the problems of colonial development could be resolved by knowledge.[103] In short, these new experts used techniques from the natural sciences to represent and measure political reality in ways that made it possible to imagine new forms of state intervention. In this period, ideas about colonial development emerged alongside the ideology and practices of the European and American welfare state. Together, they articulated a modernist, interventionist understanding of politics that displaced evolutionary theories of human nature and society.[104] These ideas formed the basis of the discourse of state purpose institutionalized at Bretton Woods.

After the Second World War, the idea of development was reconfigured under the influence of a cosmological shift led by physics and engineering. Wartime collaborations between engineers, physicists, and economists inspired cybernetics, spread practices of mathematical modelling, and produced the world's first computers. Mathematical and computational modelling depicted social and political reality as a series of cybernetic systems or objects like the "economy," "society," "climate," and so on.[105] In addition, they extended and intensified the modernist idea that knowledge could be used to control an unending future. In this context, neoclassical economists promoted the ideal of unlimited economic growth and naturalized it as scientific and technological progress. Thus,

[103] On this history, see Mitchell 2002.
[104] On the importance of modernism in politics, see Scott 1998. One of the contributions of this study is to provide a history of Scott's high modernism. See Chapter 4 for more details.
[105] Here I am generalizing the framework of Daston (2000) and Mitchell (2002). See also Allan 2017a, 2017b.

an ontology and episteme oriented to the control of objects and systems were bound up into a cosmological narrative in which the history of humanity is the history of harnessing science and technology. Thus, the success of humanity and state purpose were redefined as the endless production of prosperity unleashed by scientific knowledge.

The cybernetic, object-centered view of the world was institutionalized in the post-Second World War order by the American hegemon. Here the rise of the idea of growth combined with changes in the wartime American economy and developments in the electoral dynamics of Western democracies to create a powerful political-economic system premised on expanding production.[106] Once embedded in international order, the goal of economic growth was inscribed in other states, international organizations, and non-governmental organizations through the channels of associational change. I illustrate these meso-level changes in a case study of the World Bank. Over the course of the 1970s, World Bank President Robert McNamara set out to reform the Bank. He argued that the Bank had become too focused on economic growth and that it should design projects to directly improve the lives of the poor. However, efforts to institutionalize direct poverty alleviation achieved only limited successes and by the 1980s growth-oriented policies returned to the top of the agenda. I argue the marginalization of direct poverty alleviation and the privileging of economic growth was bolstered and naturalized by the representational constraints imposed by scientific management techniques and the rise of neoclassical economics in the Bank. Over time, growth came to be valued as scientific and technological progress and pro-poverty policies came to serve the ends of growth, rather than the other way around.

The World Bank case is emblematic of a larger process: the recursive institutionalization of a growth-based order under the auspices of American hegemony. The goal of growth as scientific and technological progress is now embedded in the key sites of international order. The result is that many states are invested in the purpose of growth bolstered by a scientific and technological cosmological backdrop. Time has been redefined as a linear, unending plane upon which economic progress unfolds. Economic progress is achieved by unleashing human potential through the means of scientific and technological progress. The success of this programme is itself calculated in scientific, statistical measures of human well-being.

Although my empirical argument is intended to show the importance of scientific ideas in the transformations of international orders, there

[106] Ikenberry 1992, 316–317. See also Maier 1987.

are other important political-economic factors at the meso- and macrolevels.[107] In the above narrative I highlighted the rise of capitalism, the bureaucratization of the early modern state, the industrial revolution, wartime exigencies, and democratic electoral incentives. The logic of my argument is configurational, so it does not hinge on showing that scientific cosmology would have had the same effects in the absence of these factors. That is, scientific cosmology does not have to be sufficient to explain the transformation of state purposes. Rather, I aim to show that cosmological shifts arising from the Western scientific tradition are a necessary component of any explanation of why one set of ideas came to structure state purposes rather than others.

However, discourses do not select and institutionalize themselves. In each case, the institutionalization of new purposes depends on the backing of powerful states and IOs. The emergence of balance of power discourse depended on the power politics of uniting European states against the French hegemonic threat in 1713 and 1815. The rise of development discourse depended on the power and drive of the British hegemon. In the twentieth century, the rise of economic growth must be accounted for by paying close attention to American military-economic and cultural-epistemic hegemony. The reinstatement of growth at the World Bank in the 1970s and 1980s was also bolstered by the rise of neoliberal preferences of the United States and other Western countries.[108] But the importance of powerful actors and political-economic incentives to the story here does not obviate the need for interpretivist, contextual analysis of the actual mechanisms and processes of ideational change. In the end, social scientific explanation must refer to the actions and practices of concrete historical agents and so depends on reconstructing how they see the world. This book shows that this reconstruction must proceed to the deepest, cosmological levels of discourse to see and make sense of macrohistorical change in international politics.

The Plan of the Book

I begin in the next chapter by building a multilevel theoretical framework for the analysis of international change. First, I argue that international

---

[107] These factors are often described as "materialist." I prefer not to use that term because even purportedly material factors like economic pressures and military power are ontological hybrids in the sense that they combine physical, ideational, and other elements. Instead, I refer to these as political-economic or military-economic factors. I use the term "political-economy" as distinct from "political economy," which I use to designate the intellectual tradition of Petty, Smith, and others.

[108] Babb 2009; Nielson and Tierney 2005.

order is constituted by an institutional and discursive structure. Second, I argue that alterations in this structure begin with meso-level changes in states, IOs, and other associations. Third, I embed this meso-level theory in a dynamic, macro-level theory of change I call the recursive institutionalization of international order. I then use this theory to explain the transformation of state purposes between 1550 and 2015 in three empirical chapters. Each empirical chapter combines original discourse analyses of primary documents (to demonstrate change in state purpose) with process-tracing of the main mechanisms (associational change and hegemonic or horizontal institutionalization).

In the first case, spanning 1550 to 1815, I examine the emergence of balance of power politics in Europe. I show that this was premised upon a change from religious and dynastic ends to rationalist goals rooted in control of the material elements of power. I argue that ideas from natural philosophy and political economy eroded the discursive basis of the divinely ordained dynastic order and constituted the materialist and quantitative basis of balance of power politics. The chapter culminates in an analysis of the 1815 Congress of Vienna that reveals the decisive role of the statistical commission in finely tuning the balance of power.

In the second case, spanning 1850 to 1945, I demonstrate how Darwinian ideas in the British Colonial Office made possible the idea of evolutionary development that entered international order in the League of Nations. I begin by showing how the British hegemon built a liberal-colonial order on commercial and civilizational purposes from the 1860s through the 1880s. I then show how the strategic deployment of sociocultural anthropology and social scientific expertise constituted and reconstituted ideas about colonial development. First, under the influence of Darwinism, colonial development was conceptualized as an automatic process of evolution up through the 1930s. I trace how this discourse entered into the Colonial Office and became embedded in the discourses underlying the League of Nations. Second, under the influence of new forms of colonial science and social thought colonial development was reconceptualized as a modernist enterprise of harnessing knowledge to intervene in native societies. It was this new discourse of state purpose that would enter the US-led order after the Second World War.

In the last case, spanning 1945 to 2015, I investigate how development came to be understood as economic growth and measured in terms of gross domestic product. I illustrate the dynamics of associational change with an in-depth analysis of World Bank policy. The case shows the rise of neoclassical economists and the theory of trade-led growth. As growth theory entered the World Bank and other institutions in the 1970s, it pushed out alternative concepts of well-being and naturalized the idea that the central goal of states should be to pursue unending economic growth, understood as scientific and technological progress. In a discourse analysis of international organizations in 2015, I show that the same ideas that emerged in the Bank are now widely institutionalized in key sites of international order.

The Conclusion aims first to generalize the theory of recursive institutionalization and identify the conditions under which we might expect major macrohistorical changes. I then use this generalized version of the theory to speculate about the prospects for future cosmological changes in the age of the anthropocene. I conclude by outlining some of the implications for understanding the role of the social sciences in political processes.

2 Cosmology and Change in International Orders

Explaining International Change

The longue durée of international politics is often presented as a history of order-building moments in which the great powers sat down after major wars to rewrite the rules of international order.[1] This focus on order-building moments is unsatisfying because it does not account for changes that happen between great power wars and landmark treaties. Such an approach cannot explain which ideas matter in order-building moments or where these ideas come from. Second, while recent accounts have theorized change at the meso-level of groups and institutions, these insights have not been scaled up to the macro or international level.[2] These accounts contain the tools necessary to produce a more dynamic theory of change in international order, but they need to be integrated into a macro-level framework. In short, we need a theory of international change that can capture both ongoing shifts at the meso-level and the institutionalization of these changes in the core sites of international order.

My aim in this chapter is to construct a theory of international change to explain the transformation in discourses of state purpose over the last five centuries. Whereas existing theories privilege either order-building moments or processual changes, my account combines these elements in a recursive process that integrates the micro-, meso-, and macro-levels of international order. The core of the theoretical model is simple. The source of change in international politics is constantly shifting knowledge about how the world works. These ideas have cosmological implications for how we understand the role of humans in the universe. As

---

[1] Realist, liberal, and constructivist accounts of the history of international order centre on these moments: Gilpin 1981; Ikenberry 2001; Organski and Kugler 1980; Osiander 1994; Reus-Smit 1999. See also, Widmaier *et al.* 2007; Drezner and McNamara 2013.
[2] On the importance of the meso-level, see Nexon 2009, 61–63. See also, Adler and Pouliot 2011; Braumoeller 2013; Goddard 2009, 2010; Guzzini 2013; Jackson 2006a; Pouliot 2009; Sending 2015. For precursors to this kind of argument see Koslowski and Kratochwil 1994; Ruggie 1989.

powerful international actors take up these ideas, they set off changes in ideas about state purpose. This slowly alters the composition of international discourses and these discursive changes are later institutionalized in multilateral treaties or international organizations.

I counterpose the model to Robert Gilpin's theory of international change.[3] Gilpin's account is focused on how powerful states reconstruct the governing arrangements of international orders after major power wars. However, Gilpin's model has less to say about *which* ideas and rules will shape international order. By contrast, I argue that cosmological ideas are powerful sources of the raw ideational material that is used to build international orders.[4] In addition, whereas Gilpin focuses on order-building in the wake of great power war, I draw on insights from theories of continuous change to create a dynamic account in which change happens both between and in order-building moments. While I counterpose my theory to Gilpin's account, I retain an important role for state power and hegemony. Following the seminal work of Richard Ashley and John G. Ruggie, I argue that international politics is structured by the fusion of power, knowledge, and purpose.[5] My theory draws on Ashley and Ruggie's insights to build a more historically flexible and contingent model of historical change through a critique of realist thought. The payoff of this approach is twofold. First, it offers a theory that can account for change both in and between order-building moments. Second, it can explain which ideas matter to international politics and trace their emergence from the micro- to the macro-level.

In what follows, I first present a static image of international order as resting on cosmological ideas. Here I extend the English School framework advanced by Reus-Smit and Buzan. The next three sections put this structure in motion by introducing a multilevel model of international change. I begin at the micro-level, where I explain how discourses structure individual action. I then move to the meso-level to explain how international actors like states, IOs, NGOs, firms, and epistemic communities change international discourse by deploying new ideas. At the macro-level, I integrate these elements in the model of recursive institutionalization.

I then apply this general model of change to the case of scientific cosmology in European and international politics since 1550. With a dynamic theory of international change in hand we will be able to

---

[3] Gilpin 1981.
[4] I am grateful to Austin Carson for this formulation.
[5] Ashley 1989; Ruggie 1982.

rethink and challenge the dominant Weberian narrative that science was a rationalizing force that eliminated cosmology from Western politics.[6] In addition, the model of recursive institutionalization provides a way to analyse the influence of scientific ideas on political discourse without relying on abstract processes like rationalization or modernization. Instead, my account decomposes macrohistorical scientific change into micro- and meso-level processes and thereby allows the effects of scientific ideas on political discourses to be studied inductively. The result is a non-instrumental account of science in international politics that can be used to explain change in state purposes over the longue durée.

The Cosmological Elements of International Order

International orders are patterns of state behaviour (foreign policies and transaction flows) shaped by an underlying structure of institutions and discourses.[7] As we saw in the Introduction, Buzan argues that primary and secondary institutions are the central elements that constitute international orders. These institutions contain the formal and informal rules and practices that guide and channel state behaviour. Primary institutions are "durable and recognized patterns of shared practices rooted in values held commonly by the members of interstate societies."[8] Buzan defines these as primary because they define the rules of the game, constitute principal actors, and structure relations between actors.[9] Primary institutions include the norms that structure statehood, sovereignty, diplomatic relations, economic transactions, and so on.[10] Primary institutions make possible and constitute secondary institutions. Secondary institutions are formal agreements and organizations such as treaties (the body of international law generally) and IOs.[11] Secondary institutions

---

[6] See Weber 1946, 1978; Horkheimer and Adorno 2002; Habermas 1970, 1984, 1987. In the international context see Meyer *et al.* 1997; Drori *et al.* 2003.

[7] The label "international *order*" is somewhat misleading because, as Lieberman (2002, 701) puts it, "any political moment or episode is situated within a *variety* of ordered institutional and ideological patterns." Order emerges from "a regular, predictable, and interconnected pattern of institutional and ideological arrangements that structures political life." So, any given order is fragmented and multiple rather than homogeneous and unitary.

[8] Buzan 2004, 181.

[9] Buzan 2004, 179.

[10] Buzan also includes the balance of power as a primary institution, but on my reading, the eighteenth- and nineteenth-century balance of power is codified in international law and has such specific and well understood rules that it functions essentially like a secondary institution. For further discussion of definitional and categorizational issues, see Wilson 2012. My inductive approach to international history is consistent with Wilson's call for a grounded, empirical solution to the problems of definition and categorization.

[11] Buzan 2004, 164–167. On the debate about how to distinguish primary and secondary institutions, see Wilson 2012.

contain the explicit rules that prescribe state behaviour. Primary and secondary institutions shape and influence but do not determine actual patterns of practice.

Buzan states that this institutional structure rests on the values of member states. On this point, Buzan follows Reus-Smit who argues that the fundamental institutions of international order depend on state purposes.[12] For Reus-Smit, the moral purpose of the state is the set of "reasons that historical actors have for organizing their political life into centralized, autonomous political units."[13] In any given era, the dominant moral purpose of the state constrains the fundamental institutions of international politics because those institutions must enable states to fulfil their moral purpose. Understanding the moral purpose of the state, the set of reasons that legitimates the existence of the state, is important to understanding the constitutive rules of fundamental institutions. However, if we want to understand change in international orders over time we have to broaden our conception of state purpose.

With Ruggie and Finnemore, I define purposes broadly as the ends to which state power is expected to be deployed.[14] On this definition, there is never a single purpose of the state, but rather a discourse of state purposes in which multiple, competing goals circulate throughout the international system. Nonetheless, discourses have topographies, slopes, and hierarchies, so some ends will be more natural, legitimate, or prominent than others.[15] So the discourse of state purpose is an international discourse carried and reproduced by states, IOs, and other international associations. Different associations advance different ideas about what states should do and how international order should support those ends, producing a discourse of state purpose.

As I argued in the Introduction, Buzan and Reus-Smit's accounts suffer from two weaknesses. First, they do not explain how and why international orders change. Second, they do not explain where the values and purposes that structure international orders come from. In the remainder of this section, I develop a solution to the second problem by extending

---

[12] Reus-Smit 1999, 13, 30–32.
[13] Reus-Smit 1999, 31.
[14] Ruggie 1982, 386. Ruggie applies the concept primarily to economic purposes, but Finnemore (2003, 1–4) adapts the concept to the security realm, defining it as the ends to which force is expected to be used. The use of passive voice here reflects the fact that the holder of this expectation is ambiguous. There are two key audiences for discourses of state purpose. First, there are domestic audiences who must believe state power is being used properly. Second, there is an international audience of states and other associations that demarcate some goals as legitimate and others as illegitimate. I am more interested in the latter because it is more central to the state purposes that become embedded in international order, but both are important.
[15] Taylor 1985, 73–75.

The Cosmological Elements of International Order 33

```
┌─────────────────────────────────┐
│      Secondary Institutions     │
└─────────────────────────────────┘
┌─────────────────────────────────┐
│       Primary Institutions      │
└─────────────────────────────────┘
┌─────────────────────────────────┐
│     Discourse of State Purpose  │
└─────────────────────────────────┘
┌─────────────────────────────────┐
│      Cosmological Elements      │
└─────────────────────────────────┘
```

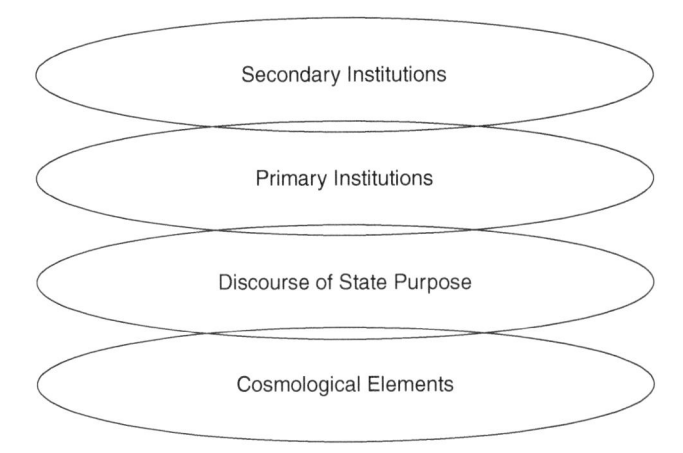

Figure 2.1 The institutional and discursive structure underlying international order

Buzan and Reus-Smit's theoretical framework to include the cosmological level of international discourses. In the following sections, I put this framework in motion by building a multilevel theory that specifies the micro- and meso-level processes that drive macrohistorical change.

My central claim here is that cosmological ideas provide the discursive resources that constitute discourses of state purpose. The claim that cosmological ideas underlie purposes, which in turn underlie the primary and secondary institutions of international order can be arrayed in a "generative structure" (Figure 2.1). A generative structure is a hierarchical structure in which the deep levels constrain possibility at higher levels. As Ruggie puts it, in a generative structure "the deeper structural levels have causal priority, and the structural levels closer to the surface of visible phenomena take effect only within a context that is already 'prestructured' by the deeper levels."[16]

The central premise of my generative structure of international order is that beliefs, norms, goals, and values draw their meaning and appeal from ideas about ontology, episteme, and other cosmological elements. These ideas shape beliefs about what to value and which goals to pursue, which in turn influences which rules and institutions will be seen as the necessary means to those ends. The levels of the generative structure are linked via chains of constitution and justification. First, lower levels form the conditions of possibility for the constitution of higher level ideas.

---

[16] Ruggie 1983, 283.

Second, lower levels provide discursive resources to justify and legitimate the ideas and rules at the higher levels.

From the perspective of the generative structure, most constructivist history explains change in the norms of primary and secondary institutions.[17] But these are merely the surface reflections of deeper, more powerful ideas that shape international politics. If IR theorists want to understand the origins of change, they must pay attention to the cosmological elements that make new purposes and institutional rules possible and desirable. This framework builds on the insights of other theorists who have shown that ontological and epistemic ideas about space, time, and knowledge structure the institutions and purposes that underlie international orders.[18] Ontologies specify what exists in the world.[19] They contain the available sets of categories and classifications offered by a language to represent the world. Ontologies also outline how entities emerge, interact, and change. So ontological elements identify the stuff that populates the universe and explain how it moves or acts. For example, the ontology of Christian discourses in early modern Europe divided reality into spiritual and temporal domains. Whether an entity was spiritual or temporal determined its properties, such as whether or not it was eternal or changing. Ontologies are necessary for politics because any group requires "a common language to talk about social reality" and specify the "objects in the world."[20]

Epistemes specify what counts as true or authoritative knowledge and establish the procedures for producing reliable knowledge.[21] Whereas ontologies structure representations of the world, epistemes provide the

---

[17] Price 1997; Reus-Smit 1999; Hall 1999; Philpott 2001; Bukovansky 2002; Crawford 2002; Barnett and Finnemore 2004.

[18] Adler and Bernstein 2005; Bartelson 1995; Grovogui 1996; Jackson 2006a; Phillips 2011; Ruggie 1993; Walker 1993. This literature draws on a more general tendency in political and social thought to focus on the ontological and epistemic elements of political discourses. Consider Foucault's (1972) analysis of epistemes and Taylor's (1985, 1987) emphasis on the common ontology of politics. See also the more recent ontological investigations in anthropology following Povinelli (1995). I am indebted to Sam Chambers for conversations on this point.

[19] See Taylor (1987) and Searle (1995) for philosophical ontologies that move beyond brute facts or experiential datum to include social facts.

[20] Taylor 1987, 51.

[21] In the literature following Foucault, an episteme specifies the rules and practices of signifying systems. Foucault defined an episteme as "the total set of relations that unite, at a given period, the discursive practices that give rise to epistemological figures, sciences, and possibly formalized systems" (1972, 191). Applications of the concept to IR have been more schematic, but they are consistent with this. Ruggie conceptualizes an episteme as delimiting "the proper construction of social reality" (1975, 570). For Adler and Bernstein, an episteme is "productive of what social reality is" and so "helps constitute the order of global things" (2005, 295). All of these conflate episteme and ontology, which I separate for the sake of clarity.

The Cosmological Elements of International Order 35

rules for building good ontologies and models of the universe. For example, the transition from the medieval to the modern episteme was a transition away from a patterning or resemblance scheme. On the patterning episteme, knowledge was produced via analogies and connections within discursive or symbolic systems. On the modern episteme, knowledge was produced by creating true representations or maps of the reality.[22] This was a change in the rules about how knowledge was produced from methods centred on analogical reasoning to those centred on external observation and verification.[23] Ontologies and epistemes constitute one another because ideas about what exists depend on the methods used to establish what exists and vice versa. Epistemes are politically powerful because they help to establish who or what will be regarded as authoritative and deserving of obedience.[24] As such, they are a foundation of political legitimation and naturalization.

Epistemes and ontologies are important elements of any analysis of political order. However, on their own, epistemes and ontologies tell us little about where purposes come from. Purposes, as more than just interests or goals, depend on ideas about what is worth doing or what must be done to satisfy the gods, or reason, or the public. To understand where purposes come from, we have to look at other cosmological elements. On my conception, cosmological discourses include ontologies and epistemes, but they also include ideas about the structure and direction of time (temporality), the origins and nature of the universe (cosmogony), and the place of humanity in the universe (destiny).[25] Cosmological discourses organize epistemic and ontological elements into images, narratives, and rituals that constitute purposes. For example, ideas about time orient individuals to historical cycles or conceptions of progress. Cosmogonies place humans in a narrative of creation and development that defines the meaning of existence. Ideas about human destiny highlight and legitimate certain forms of agency. These elements do more than present an ontology or a worldview. They advance and even require some actions and goals, depicting them as natural while others are portrayed as impossible or despicable. That is, cosmological narratives constitute purposes by establishing the cosmic order and placing humans within it. So while cosmologies depend on ontologies and epistemes, they are not exhausted by them and are an important element of political analysis.

---

[22] Reiss 1982. These terms are explained in more detail in Chapter 3.
[23] Of course, it was Popper (1959) that consolidated a half-century of challenges to the modern, correspondence episteme, beginning with Frege.
[24] See Bartelson 1995; Foucault 1970; Weber 2004, 133–145.
[25] Destro 2010, 227.

36   Cosmology and Change in International Orders

The claim that cosmologies play a central role in the construction of political orders is supported by a variety of sources in the sociology of religion and the sociology of institutions. For example, Weber argues that all social and political orders rest on a dominant interpretation of reality or an authoritative definition of what exists.[26] After all, Weber maintains, human beings are thrown into a confusing totality and must, in order to go on, attach meaning to some subset of reality.[27] While this process of assigning meaning to parts of the totality is relatively arbitrary and contingent, it must not be experienced as such. Instead, an order is premised upon group members believing in the necessity of adopting the order's sacred beliefs and values.[28] Fundamentally, any group's purposes must be legitimated and justified by a normative backdrop that infuses representations and meanings with significance and obligation. Thus, Weber conceptualizes symbolic power over the sacred as a basic source of power alongside political, military, and economic power.[29]

Similarly, Berger and Luckmann, working from Durkheim's basic insights, argue that institutional orders and shared values must be legitimated within an all-encompassing frame of reference that contains claims about the nature of the cosmos.[30] For them, all complex societies feature an epistemic division of labour that includes, say, a caste of priests. These priests provide for the legitimation of social and political order by drawing support from the "symbolic universe" which depicts the order as a product of "the nature of things."[31] Douglas, also drawing on Durkheim, argues that established institutions must be legitimated by the nature of the universe. On her account, an element of social order is fully institutionalized when the final answer to the question, "Why do you do it like this?" makes reference to "the way the planets are fixed in the sky or the way that plants or humans or animals naturally behave."[32] To be legitimate and stable, the rules and purposes of institutions must be supported by ideas about what exists, what human nature is, and how we should go on in the world. This naturalizes rules and purposes because it renders them part of the cosmos. Cosmological naturalization is integral to the construction of values and purposes because we

---

[26] Weber 2004, 133–145, 182–194. My reading here follows Poggi (2006).
[27] Poggi 2006, 21, 41–43.
[28] Weber 2004, 135. This is why Weber's own contrast between the magical basis of traditional society and the rational basis of the modern state is at odds with his own political anthropology. By moving up to the cosmological level, we can see that magical and rational modes both express cosmological themes.
[29] Poggi 2006, 38–39.
[30] Berger and Luckmann 1966, 89–91.
[31] Berger and Luckmann 1966, 89.
[32] Douglas 1986, 46–47.

value things as ends-in-themselves when we can give no further reason for why we accord importance to something. We form values when we connect concepts to cosmological elements including our beliefs about what humans are, what our place in the universe is, and how we define progress.

These sociological arguments are also supported by historical work on comparative international systems. For example, Liverani shows that the basic cosmological principles of Mesopotamian religion shaped their political purposes and institutions.[33] Mesopotamian cosmology posited a division between order, located in the stable and predictable realm of the city, and disorder, associated with the dark and chaotic forces of nature beyond the gates. This constituted two purposes: the maintenance of harmony within the polity and establishment of universal empire beyond the gates. The dream of universal empire, Liverani argues, emerged from the desire to banish the forces of darkness that threatened the city.[34]

Phillips argues that Luther created a crisis in the international order of Latin Christendom by challenging the epistemic, ontological, and cosmological basis of Catholic discourses.[35] Luther presented a new Augustinian view of humanity alongside an ontological and epistemic image that replaced the intermediary function of the Church with the "invisible priesthood of all believers linked directly to God by their faith."[36] While the particularistic cosmology of Reformation Christianity underpinned a divided order of sovereign states, the holistic and paternalistic cosmology of Confucianism legitimated and naturalized an East Asian order built upon hierarchical and familial relations.[37] In the Sinosphere, "the purpose of collective association was to achieve a temporal state of peace, fairness and harmony (*ping*) in accord with the rhythms of a larger cosmic order."[38] Here the cosmological dimension of Phillips' account is clear: what produced political purpose was the need to render *political* order consistent with *natural* order. Thus, political order was to be governed by an emperor "conceived as the Son of Heaven" presiding over an East Asian order built on super- and subordinate

---

[33] Liverani 2001.
[34] However, power projection capabilities could not support such a purpose, and core–periphery tension was a central dynamic of the Ancient Near Eastern system. This case shows that dreams of empire can be supported by multiple cosmologies, but my historically inductive account would suggest that the purposes and institutions built on differing cosmologies will vary significantly.
[35] Phillips 2011, 86–96.
[36] Phillips 2011, 87.
[37] Phillips 2011, 171.
[38] Phillips 2011, 155.

relations of benevolence and obedience.[39] Suzuki makes a similar argument. As Suzuki puts it, East Asian "constitutional structures ... were primarily the extension of universalist Confucian philosophy."[40] On his reading, the fundamental institution of the tribute system bolstered the purpose of maintaining "appropriate social hierarchies that would reaffirm China's superior moral standing."[41]

As these cases show, the history of non-Western international orders has been theorized in cosmological terms, but the cosmological history of the West is thought to end with the collapse of Latin Christendom. However, on the broad definition of cosmology offered above, the Western scientific tradition is a rich source of cosmological meanings. Anthropologists now recognize that this way of counterposing rational Western modernity to cosmological non-Western societies is a product of colonial anthropology. Perhaps as a result of this, cosmological analysis fell out of favour.[42] However, if we abandon cosmological analysis we ignore the important role that cosmological elements play in constituting purposes and goals. My approach, by contrast, provincializes Europe by revealing the cosmological basis of its own political discourses.[43]

From this vantage point, the rise of scientific ideas in the West did not drain political discourses of cosmic meaning. Rather, it inaugurated a series of shifts in the cosmological discourses that underwrote European and international orders. As the natural sciences developed, they generated and disseminated new ideas about the nature of the universe, time, and destiny that were used to redefine political purposes. At first, these ideas emerged within European states, but were spread and enforced through European colonial domination. We can think of the rise and spread of cosmological elements drawn from Western science as a truly global process of cosmological conflict that helped to establish the dominance of scientific cosmology.[44]

But none of this should leave us with the sense that there is a unitary Western cosmology in global politics today. First, all social discourses are incomplete, fragmented, and multiple. Discursive stability and dominance is always a temporary achievement. But second, and more

[39] Phillips 2011, 155–157.
[40] Suzuki 2009, 36.
[41] Suzuki 2009, 36.
[42] It has, however, recently returned in a dynamic literature on cosmological conflicts between the West and non-Western peoples in late modernity. Povinelli 1995; de Castro 1998; de la Cadena 2010; Blaser 2013.
[43] On provincialization, see Chakrabarty 2000.
[44] Anghie 2004; Chatterjee 1986; Grovogui 1996; Inayatullah and Blaney 2004; Kalpagam 2000.

importantly, while the rise and spread of scientific ideas under the auspices of European colonialism altered cosmological discourses, it did not eliminate or replace existing discourses. Rather, it added new ideas that generated the reconfiguration of discourses. Thus, cosmological conflicts and negotiations remain significant today in global environmental politics, the war on terror, and ethnic conflicts.[45]

Thus, although I refer to "cosmology" as a shorthand, in reality we are dealing less with a cohesive overarching interpretive framework than a set of cosmological elements that are variously used to shape purposes, justify institutional rules, and legitimate political hierarchies. So, when I talk about a scientific or religious cosmology, I am referring to a set of elements that circulates with varying coherence in political and social discourses. After all, in the course of political action cosmological elements must be taken up by particular historical actors who weave cosmological narratives that place humanity in the arc of the universe. Sometimes these cosmological claims are accepted as offering a true account of the nature of things. But these cosmological narratives are often contested or rejected by others. As we shall see, sometimes relations of force decide this process of cosmological contestation, enabling the strong to maintain and impose a relatively coherent set of cosmological elements and their attendant purposes on others. But the outcomes of these contests are uncertain and thus the histories of cosmologies are contingent and nonlinear.

The generative structure of international order tells us where we can expect change in purposes and institutions to originate: new cosmological ideas will disrupt the chains of legitimation that link the institutions of international order to underlying discourses. New cosmologies can denaturalize taken-for-granted purposes and make possible new ways of representing and valuing the world. However, this schema offers a static representation of order that does not explain how change happens. A theory of ideational international change must specify the practices and processes that produce change in international order.

Micro: The Discursive Contexts of International Order

In the Introduction I argued that if we want to understand the actual mechanisms and processes of international change, we need a multilevel theory that places the micro-, meso-, and macro-levels in systematic relationship to one another. In my schema, the micro-level refers to the level of individual social action, the meso-level to the level of groups and

---

[45] Atran 2016; Atran and Axelrod 2008; de la Cadena 2010; de Castro 1998; Hurd 2015.

organizations, and the macro-level to the structure formed by the relations between the meso-level actors.[46] My goal in this section is to ground the international discourse of state purpose in a careful exposition of the micro-level of individual action. Even though this micro-level theory is not central to the empirical analysis that follows, it is necessary in order to understand how the meso- and macro-level mechanisms work.

Put differently, in this section I build an image of international structure on a microfoundational account of how ideas influence everyday political life. Constructivist IR theorists have been less likely than rationalists to use the language of microfoundations, but a concern for microfoundations is simply a concern with "the micro-level mechanisms" that drive social action.[47] Earlier constructivist microfoundations relied on the logic of appropriateness or the force of argument. On these accounts, agents consciously act in accordance with internalized identities or norms.[48] However, these logics have been trenchantly critiqued and replaced by a series of relational, practical, and discursive accounts of agency.[49] While the micro and meso literature that has grown up in place of the earlier logics is exciting, its insights have not been translated into a clear account of how we should conceptualize the distribution of ideas at the international level.[50]

My account of social action begins with an agent embedded in a variety of physical, relational, discursive, and institutional contexts.[51] On this view, social action is the assembling of meaningful actions from the resources at hand.[52] I define discourse as the set of shared symbols,

---

[46] Jepperson and Meyer 2011. This schema maps onto the three levels of analysis (Singer 1963) or the three images (Waltz 1979), except there, the first level is usually conceptualized in psychological terms and the second level is always the state. Here, following Ruggie (2004), Nexon (2009), and Corry (2013), the meso-level of the international system is any group or organization that participates in international politics. Moreover for me, the micro-level is less about psychology, than about specifying a theory of social action, which must draw on at least a folk psychology (as in, e.g., Berger and Luckmann 1966), but which places agents in a discursive-social-phenomenological context.

[47] On constructivist microfoundations, see Wendt 1999, 33–38, 152–153; Fearon and Wendt 2002. See also Kertzer 2017.

[48] Finnemore 1996a; March and Olsen 1998; Wendt 1999; Hurd 1999; Risse 2000.

[49] See Hopf (2002a, 279–282) and Sending (2002) for important critiques. See Schmidt (2008, 314) for a discursive institutionalist critique.

[50] For a review, see Guzzini 2016. Again, as I said in the introduction, Nexon (2009) has scaled his relational insights up, but is not interested in theorizing the distribution of ideas or any other structural property. See also Adler and Pouliot 2011; Barnett and Finnemore 2004; Bueger and Gadinger 2015; Goddard 2009; Guillaume 2009; Guzzini 2013; Hopf 2010; Jackson 2006a; Kessler and Guillaume 2015; Pouliot 2009.

[51] Sewell 2005, 197–199.

[52] This is Swidler's (1986) central insight, which has been deployed in IR by a variety of discursive, relational, and practice theoretic approaches. See, Adler and Pouliot 2011,

meanings, rules, assumptions, and practices that delimit the imaginable, knowable, sayable, and doable in a given social, historically situated group.[53] Within discourses, we can identify specific *discursive configurations*.[54] Discursive configurations or formations are collections of concepts and ideas that hang together and tend to be reproduced over time. A discursive configuration may come to dominate a given discourse, but it will constantly compete with other formations that challenge its representations and valuations of the world. For example, in the empirical chapters the formation of the discursive configuration of economic growth which links together an object-based ontology, a mathematical-representational episteme, a cosmological narrative rooted in scientific and technological progress, and so on. Growth is a dominant goal in international discourses, but it exists alongside and in competition with other ends.

Discourses are relational and configurational: the meaning of concepts, symbols, rules, and practices are constituted by the connections between the elements.[55] We understand what poverty means by linking it to similar concepts like deprivation, or income, and by distinguishing it from wealth, and so on. For heuristic purposes, we can think of discursive configurations as network structures in which concepts are connected to supporting semantic clusters.[56] It follows that discursive change is a change in the configuration of conceptual elements. To transform the meaning of a concept is to alter the conceptual network that a concept is embedded within and the links it has to other concepts. Poverty changes as it is linked and delinked to other concepts in the semantic neighbourhood. In political discourse, connections between concepts are forged in political practice. Meaning is constructed when actors link concepts to another. So the study of discursive change in international politics is the study of how ideas are combined and recombined by individuals in meso-level political contexts.[57]

Discourses both enable and constrain agents. On one hand, discourses provide agents with resources from which to construct meaningful

---

6–7; Barnett and Finnemore 2004, 18; Bueger and Gadinger 2015; Goddard 2009; Guzzini 2013; Hopf 2010; Pouliot 2009; Sending 2015.
[53] Angenot 2004, 200; Hopf 2002a, 21–22; Barnett and Duvall 2005, 55; Campbell and Pedersen 2001, 220.
[54] On configurations in general, see Jackson 2006b; Katznelson 1997; Weber 1949.
[55] This is the convergent insight of recent work in philosophy (Brandom 1994; Brandom 2000), semantic linguistics (Borge-Holthoefer and Arenas 2010; Homer-Dixon *et al.* 2013; Motter *et al.* 2002; Solé *et al.* 2010), cognitive anthropology (D'Andrade 1995), and political psychology (Taber 1998; Morris *et al.* 2003).
[56] See Motter *et al.* 2002; Homer-Dixon *et al.* 2013.
[57] This is the implicit logic of Weldes (1999), which I am indebted to.

action.[58] They make action possible by presenting a world to act upon and by providing resources for the constitution of agents' habits, dispositions, and interests.[59] On the other hand, discourses delimit agency because they present only a small portion of reality and render some actions as thinkable or natural while rendering others as unthinkable or illegitimate.[60] But there are limits to such constraints because agents are constituted by multiple discourses. They can engage in creative action by drawing resources from a variety of discourses and social sub-universes.[61] For example, scientists, clerics, and other specialists work in relatively autonomous sub-discourses with their own languages and modes of thought. When ideas from these sub-discourses are transposed into other domains, they introduce new elements that allow agents to see and act upon the world anew.

This contextual view of agency depicts a world in which social action is constantly reconfiguring and changing discourses. But the churn of ongoing ideational change is also fuelled by broader meso-level patterns of political and social contestation. Individuals and groups challenge dominant concepts and seek to advance alternatives. Political, economic, administrative, and epistemic power play central roles in these contests. Groups backed by the power of the state, expert authority, a broad-based social movement, or large organizations are more likely to shape which discursive configurations dominate public discussion and policy debate.[62] So discourses contain a wide variety of discursive configurations backed by competing social groups that seek to impose a way of seeing and valuing the world. A discursive configuration may come to dominate a given discourse, but it will constantly compete with other formations that challenge its representations and valuations of the world.

From the Micro- to the Meso- and Macro-Levels

A micro-level account of how discursive contexts shape action is necessary in order to carefully specify what the international system looks

---

[58] Swidler 1986; Sewell 2005.
[59] Hopf 2002a; Hopf 2010.
[60] Katzenstein 1996, 21; Katznelson 1997, 83; Barnett and Duvall 2005.
[61] Following Sewell (2005, 205–209), this conception of the agent builds change and difference into social foundations. As Sewell puts it, "if we assume that subjects are formed by structures, a multiple concept of structure is capable of explaining the existence of persons with widely varied interests, capacities, inclinations, and knowledge" (2005, 209). I see my theory, with Sewell, as attempting to incorporate both sensitivity to the instability created by ongoing agency and an account of constraining, structural effects.
[62] As Sewell (2005, 259) puts it, successful ideas receive "authoritative sanction" from institutional nodes in which power is concentrated and which command sufficient geographical and social scope.

like and how it is reproduced. To do this, we need to scale up from the micro-level to the meso- and macro-levels. My approach breaks with existing systems theories by abandoning state-centric assumptions.[63] Instead, I follow Nexon who characterizes the international system as a set of meso-level groups and sites where individual level actors and system level dynamics meet.[64]

My image of the international system does not begin with the state or any other statically conceived unit. Instead, my starting point is to specify *associations* as the central actors in the international system. I conceptualize an association as any meso-level group that constitutes, carries, and reproduces international discourses and other elements of the international system. This includes states, empires, IOs, NGOs, firms, transnational classes and networks, as well as churches and trade unions.[65]

We need a broad term such as associations to build a theory of change that covers five centuries of international history. First, the central units of the international system change over time, so we need a concept that is historically flexible.[66] Second, we do not want to pretheorize where change might arise in the international system. It might arise in states, but it also might arise in activist networks or epistemic communities. On my account, change could emerge from any association that works to introduce a new purpose into the system. Finally, I like the term association because it foregrounds the relational and configurational insight that all groups, including states, are constituted by transactions and processes.[67]

Moving to the macro-level, my claim is that associations carry and reproduce international discourses. That is, the organizational cultures

---

[63] On the need to revise state-centric assumptions, see Czempiel and Rosenau 1989; Milner 1991; Ruggie 1993, 2004. On the need to abandon a strict anarchic ordering principle, see Buzan (2004), Corry (2013), Donnelly (2009, 2012), and Nexon (2009).

[64] Nexon 2009, 61–63.

[65] Generalizing Corry (2013), we might say that a system is a collection of actors oriented to the international and focus on "international associations." Indeed, associations that explicitly aim to influence international politics like a foreign policy think tank in a major state, are more likely to shape international discourses than bowling clubs. However, the discourses and practices of non-international actors matter for international politics because they constrain and shape the discourses that states and international associations operate within. On the importance of masses and everyday life in world politics, see Hopf 2002a, 2013.

[66] From this perspective, even the term "international" is anachronistic because the model of sovereign states in anarchy really only emerged in the nineteenth century (see Buzan and Lawson 2015). It was at this time that actors themselves began to refer to the "international system" (Sending 2015). So, properly speaking, we should use the term "interpolity" to talk about the international, but I follow convention and refer to all interpolity systems as international in this chapter. See Branch 2013, 18–19.

[67] Katznelson 1997; Nexon 2009.

and representational frames of associations are central components of international discourses. Associations provide the infrastructure – bureaucratic reports, letters, speeches – that carry and reproduce discourses. However, international discourses are not reducible to an aggregation of associational cultures and frames.[68] First, interactions and relations between associations are important elements of structure that are not captured by a reductionist aggregation. Moreover, important elements of international discourses lie outside associations because they have been embedded in treaties, shared informal practices, and other sites.[69]

In applying this schema to international history, we have to take a historical, contextual view of associations and the broader international system. In the sixteenth and seventeenth centuries, the European international system was dominated by states, which were ruled by a transnational aristocratic class.[70] New ideas about state purpose had to be channelled through that transnational class. Over the course of the nineteenth century the aristocratic monopoly on European political life was challenged by revolutions, the rising political power of the middle class, and legitimation crises. States bureaucratized, taking on the features and procedures of formal organizations.[71] A variety of institutional and associational forms sprung up in civil society, constituting and widening the public sphere. The emergence of bureaucratized states and formal civil society organizations changed the associational composition of the international system. Then, during the twentieth century, international order came to be centered on the IOs at the heart of the US-led hegemonic order, including the United Nations (UN), the International Monetary Fund (IMF), and so on.[72] So the associational basis of the international system and the channels by which knowledge enters the system both change over time.

Nonetheless, we can outline the ways in which associations link discursive contexts to international structure. Figure 2.2 summarizes the

---

[68] That is, emphasizing the importance of microfoundations does not entail methodological individualism, but macrohistorical explanations should rest on social actions at more concrete levels. For a variant of this argument in an IR context, see Guzzini 2013, 252. For statements in philosophy, sociology, and anthropology see, Mayntz 2004; Jepperson and Meyer 2011; Barth 1987, 8.

[69] Moreover, discourses fail the supervenience test because some associations are more important to the content of international discourses than others. So, the distribution of ideas would shift significantly if certain states or IOs ceased to exist. See Wendt 1999, 155–156. But I am also hesitant to use the language of supervenience because it is linked to intersubjectivity and I am trying to theorize the structure of ideas without relying on that concept as a first principle. Rather, I think intersubjectivity, or the degree to which ideas are shared, is an empirical not an ontological issue. See Grynaviski 2014.

[70] Bukovansky 2002.

[71] Silberman 1983; Rueschemeyer and Skocpol 1996.

[72] Mazower 2012.

Micro: Discursive Context 45

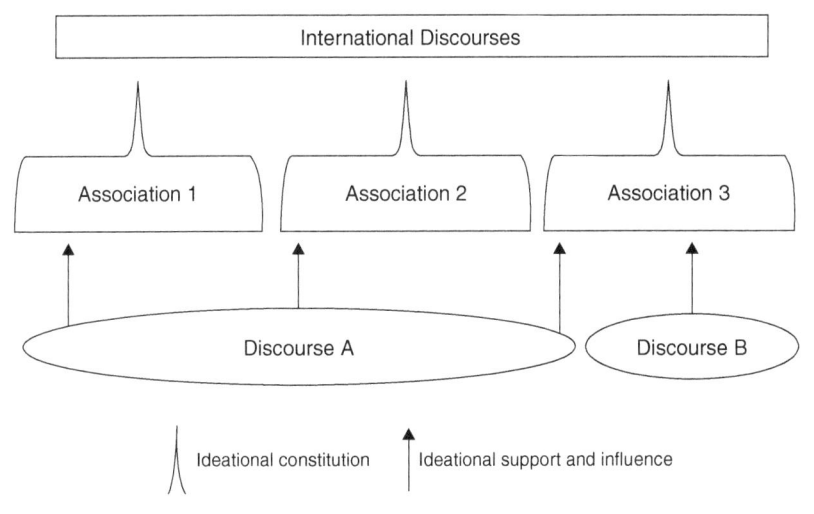

Figure 2.2 The constitution of international discourses

central points. Associations are embedded in a variety of discursive contexts. These discourses might be national discourses centred on governments (Discourse B) or they might be carried by transnational networks that span associations (Discourse A). Discursive elements are taken up by and combined in associations. The distribution of ideas across these associations then constitutes international discourses. So international discourses reflect the distribution of ideas across international associations. We can think of Figure 2.2 as underlying Figure 2.1. It explains where the discourses underlying primary and secondary institutions are located and how they are constituted.

In the empirical chapters that follow, I seek to explain change in the international discourse of state purpose, which I define as the internationally distributed set of ideas about the ends to which state power should be deployed. With Figure 2.2 in hand, we can get a sense of how cosmological ideas emerging from scientific networks can transform the discourse of state purpose. As associations take up new ideas from their discursive contexts and reconstitute their understandings of state purpose, the content of international discourses change. That is, since international discourses are constituted by the distribution of ideas across associations, as associations change so too do international discourses.

It is in this sense that Giddens describes social systems as recursive: "[b]y its recursive nature I mean that the structured properties of social activity ... are constantly recreated out of the very resources which constitute them."[73] In this case, the discourses and institutions of international order are constantly recreated out of the ideational resources and meaningful practices that constitute them. More technically, in recursive mathematical formulas, a given function $f(x_n)$ is defined in part by the output of the function $f(x_{n-1})$. So international discourses are constituted and reconstituted out of the associations that carry international discourse, *as those associations undergo constant change*. This begs the question to be explored in the next section: how and why do associations change their ideas about purposes?

Meso: Associational Change and the Emergence of New Purposes

In the last section, I provided a multilevel architecture that outlines what international discourses are. The central point is that associations carry and reproduce international discourses such that change in associations constitutes change in international discourse. As organizations and groups import new ideas, they alter the international distribution of ideas. In this section, I introduce some dynamism into that architecture by theorizing associational change. Why do associational ideas about state purposes change under the influence of cosmological shifts?

Associational change is a change in the discourses and rules of the states, IOs, NGOs, epistemic communities, and firms that participate in international politics. I offer a simple two-stage model. First, associations strategically deploy a new idea in order to solve a problem, address a crisis, or engage in political contestation. Second, the deployment of new ideas leads to discursive reconfigurations that, under some conditions, generate new purposes forged through links to cosmological elements. While it is easiest to think of these as chronological phases or stages, this division is for heuristic purposes only. In empirical cases, deployment and reconfiguration overlap and recur.

My account here draws on insights from historical institutionalism, organizational sociology, and diffusion theory. Historical institutionalism argues that changes in states, IOs, and other organizations emerge from complex configurations of ideational and institutional elements in path-dependent sequences.[74] The strength of the sociological and discursive variants of historical institutionalism is in explaining individual cases in

---

[73] Giddens 1984, xxiii.
[74] Blyth 2002; Fioretos *et al.* 2015; Hall and Taylor 1996; Katznelson 1997; Lieberman 2002; Pierson 2004; Thelen 2004; Schmidt 2008. My approach is closer to the sociological and discursive variants of historical institutionalism than the rationalist variant introduced to IR in Fioretos 2011.

a way that combines strong theory with contextual, heterogenous causal factors.[75] However, historical institutionalism is not well equipped to explain why many associations would take up similar ideas and change in similar ways.

One way to get at this would be to think of the rise of a new cosmology and accompanying purposes as a diffusion process.[76] A diffusion process is one in which the probability of a given actor adopting a practice or rule is conditioned by prior adoption by other actors.[77] The literature posits a number of mechanisms to explain why states, IOs, NGOs, or firms would take up the same ideas or practices: coercion, competition, learning, and social emulation.[78] First, coercion exploits military, economic, and information power asymmetries to compel policy adoption.[79] Second, competition drives convergence when associations believe that policy changes will increase their resources. For example, states may adopt liberal economic policies to increase foreign investment and bolster export market share.[80] Third, learning is a change in beliefs in response to new information or experience.[81] Learning generates convergence when a number of associations draw similar conclusions from experience. Finally, emulation is the spread of intersubjective models via social pressures that drive actors to adopt "appropriate" or fashionable goals and policies.[82] These models are constructed as socially acceptable and legitimate. Thus, they are adopted voluntarily even in the absence of evidence that the policies and goals are effective or superior.

In some variants, mechanisms of diffusion are driven by rationalist or functionalist logics on which adopters take up a policy or practice because they need to in order to avoid selection pressures. On this view, we do not need to investigate the meso-level processes of decision-making because we can explain or predict associational behaviour with reference to the environment and the structure of incentives alone. While this approach is able to explain why many associations would change in similar ways at

---

[75] Katznelson 1997.
[76] E.g., Finnemore and Sikkink 1998. See also, Simmons *et al.* 2008; Solingen 2012.
[77] Strang 1991.
[78] Simmons *et al.* 2008, 10–35. Campbell and Pedersen (2014) offer the same list. Solingen (2012) helpfully expands the list to include persuasion, emotional motivations, and more. On these, see Finnemore 2003, 152–158.
[79] Simmons *et al.* 2008, 10.
[80] Simmons *et al.* 2008, 17.
[81] Simmons *et al.* 2008, 25. The weak version of this explanation is that the association only perceives that the newly adopted ideas will cause failure or success. The strong version asserts that learning requires that the new beliefs actually improve the success or effectiveness of the association. My argument is compatible with weak learning, but I do not assume that prior adoption matters. See Levy 1994.
[82] Simmons *et al.* 2008, 32–33.

the same time, it relies on the unrealistic assumption that actors have full information about the success and failure of all other actors.[83] Moreover, if we introduce even the weak assumption that these actors exhibit bounded rationality, then the internal cultures of organizations as well as the distributions of ideas in their environment are an important part of any explanation for diffusion.[84]

That is, on a non-functionalist reading of diffusionary pressures, the mechanisms of coercion, competition, emulation, and learning are compatible with ideational theories of change in associations. However, it is important not to limit our empirical analysis to a search for these four mechanisms of diffusion. First, associations might adopt new discourses or practices for a wide variety of other motivations. The list excludes communicative mechanisms like persuasion, emotional motivations such as shame, empathy, or affection, and the mechanisms of productive or constitutive power.[85] Epistemic constitutive mechanisms are particularly important for our purposes: an organization or individual might, in virtue of relying on a given body of scientific or expert knowledge, be constituted so as to adopt or recreate certain policies of their own accord.

Moreover, it is difficult in real-world cases to separate the mechanisms from one another because they overlap. For example, if an association imports the latest scientific findings or hires a cadre of expert counsellors, it might be hard to tell if this was to bolster its epistemic authority in a competitive field or an attempt to learn from experience.[86] Or, the association might be doing something that looks like learning, adopting new practices in response to a crisis, but those efforts set off unintended consequences in discourses and practices.[87] It would be misleading to characterize this as learning, but it nonetheless follows from some sort of strategic intent to solve a problem. In addition, in a historical theory of change it is important not to pretheorize the motivations of actors. Instead, we should allow motivations to emerge inductively from the empirical analysis. Some of these motivations may not be captured by diffusion mechanisms in the technical sense that they do not depend on prior adoption by other states. Or, it might be hard to tell if the policy was diffused or simply constituted in multiple associations around the same time. My study is not designed to resolve these issues, but they influence my theoretical choices.

Rather than rely on a restricted menu of mechanisms, I conceptualize associational change as driven by *strategic deployments* of new ideas.

[83] For a good discussion, see Kahler 2002.
[84] See Meseguer and Gilardi 2009, 530–531.
[85] See Finnemore 2003, 152–158; Solingen 2012, 634.
[86] On such difficulties, see Meseguer 2005; Meseguer and Gilardi 2009.
[87] Haas (1990) calls this "adaptation" to distinguish it from learning proper.

A strategic deployment is when an association imports a set of ideas or a class of professionals in order to help solve a problem. Rather than assuming that associations will take up scientific ideas because they are coerced, undertaking a cost–benefit analysis, emulating another actor, or learning from experience, I simply assume that associations seek to strategically deploy new ideas to solve a problem or engage in political contestation. Strategic deployment is distinct from diffusion because it is not necessarily shaped by prior adoption in other associations. Deployments might be a response to diffusionary pressures, but they also might arise from patterns of constitution, internal organizational dynamics, or the idiosyncracies of individual decision-makers. In any case, deployments must be mediated by the internal, path-dependent discourses and practices of the association.

Strategic deployment is distinct from rationalist or functionalist mechanisms because it does not assume that decision-makers engage in a careful search for information or that they optimize. Rather, it assumes only that humans are active, creative problem-solvers. As such, they seek to "close the gap between what they perceive as reality and what they want to be or achieve."[88] In doing so, they combine and recombine the resources at hand into new ideas and practices.[89] Nonetheless, this act of recombination is still strategic in the sense that it is intended to resolve a crisis or improve an association in some way. The problem might be a technical problem of governance such as how to build more agile ships or increase agricultural yields. Or, the problem might be a political need that requires engaging in symbolic contestation such as a legitimacy crisis or a fight over the proper way to define the identity of a group.

Strategic deployments might take a number of forms. In general, the idea is that agents embedded in associations serve as brokers that transpose ideas from other scientific, religious, moral, and practical domains into international associations.[90] In the epistemic communities literature, networks of experts, scientists, and policy-makers carry causal and normative beliefs into states and IOs where they shape interests and policies.[91] In the literature on transnational social movements, activists push ideas up into states and IOs, altering agendas and priorities.[92] Conversely, new ideas might emerge via creative combinations of concepts within a state or IO, which then transposes it into transnational networks that

---

[88] Adler 1987, 4.
[89] Galvan 2004, 28; Berk and Galvan 2009.
[90] On brokers and change, see Goddard 2009.
[91] Adler 1992; Haas 1990; Haas 1989, 1992a.
[92] Wapner 1996; Keck and Sikkink 1998; Carpenter 2007.

spread them throughout the system.[93] Or, a new group may seize the reins of government in a powerful state, installing and promoting a new discourse.[94]

Strategic deployments import new ideas into associations. This sets off further changes because as the new ideas are contested and legitimated, the broader discourse they are situated within changes, generating new meanings and purposes. This transforms meanings because it reconfigures semantic networks. But the effects of these changes cannot be easily predicted or controlled. So it is important not to overplay the extent to which strategic deployment satisfies or enacts associational interests. Although strategic actors intentionally introduce new ideas, they cannot possibly compute the ripple effects of their actions throughout semantic networks. So, sometimes agents do not get their way. Moreover, in some cases agents are engaging in truly creative acts that even they cannot see the implications of.

Strategic deployments are creative acts because they are rooted in and lead to new combinations of ideas. At the micro-level, this creativity arises from the fact that agents are situated in multiple, overlapping discourses.[95] Since each agent has access to a distinctive set of beliefs and ideas, they combine ideational and institutional elements in unique ways.[96] As we shall see, discursive brokers played a key role in transposing scientific ideas into political institutions. Such brokers operated in two communities at once, transmitting ideas from one domain into another.

This process of creative recombination is important to understanding how and why strategic deployments produce new purposes. By linking two concepts, discourses, or practices together, an actor unintentionally links the supporting semantic networks of those two concepts. In turn, the new configuration can produce new meanings, values, and goals. For example, Weldes argues that national interests are socially constructed by forming connections in foreign policy discourse:

> [M]eaning is created and temporarily fixed by establishing chains of connotations among different linguistic elements. In this way, different terms and ideas come to connote or to "summon" one another ... With their successful repeated articulation, these linguistic elements come to seem as though they are inherently or necessarily connected and the meanings they produce come to seem natural, come to seem an accurate description of reality.[97]

[93] St. Clair 2006a, 2006b.
[94] Hopf 2012.
[95] Sewell 2005, 190–204.
[96] Goddard 2009.
[97] Weldes 1999, 98–99.

Each of the central purposes we will encounter in the coming chapters are composites or configurations of cosmological elements forged through articulations. The idea of economic growth, for example, first emerged in the 1940s when the new concept "the economy" was articulated or connected to the idea of colonial development. Economic growth was later naturalized through articulations to cosmological narratives of scientific and technological progress.

The Conditions for Stable Associational Change

The introduction of new ideas into an association always creates some amount of associational change. However, these changes may leave the dominant policies and purposes in place. Strategic deployments are likely to meet resistance from individuals and groups that disagree with the new ideas or stand to lose from new policies. This initiates a process of symbolic competition over the terms of discourse. Even in the absence of competition, associations often assimilate new concepts, but leave them peripheral to the core ideas driving policy. So when are strategic deployments likely to produce significant and lasting alterations to associational discourses? Ideas are likely to be prominent and stable when they are backed by organizational power, are perceived to solve an important problem, and can be linked to cosmological discourses that naturalize and legitimate the new ideas.[98]

First, historical institutionalist scholarship teaches us that the success of reform initiatives depends on the ability of a leader to build a coalition within the organization and the availability of power or resources to effect change.[99] New discourses need to be embedded in associational life and this requires organizational power. Leadership changes, administrative reforms, and professional changes are key mechanisms in the cases that follow because they each induce ideational shifts. However, each of these requires taking control of or using organizational resources. As we shall see, ideational change was initiated when: European states initiated administrative reforms and created new statistical offices in the eighteenth and nineteenth centuries; the British Colonial Office hired experts and formed a constellation of scientific committees in the 1920s and 1930s; and the World Bank hired large numbers of neoclassical

---

[98] Legro (2005, 35–38) argues that the consolidation of new ideas is a function of how many competitors it has and the perceived results of adopting the idea. These factors help explain the success of strategic deployments, but they don't tell us when to expect stable changes in purpose.

[99] Chwieroth 2010; Hopf 2012; Mahoney and Thelen 2010; Nielson et al. 2006; Thelen 2004; Weaver 2008.

economists in the 1960s and 1970s. In each case, associations changed their professional or intellectual composition, which initiated further discursive changes as the newly embedded experts continued to import concepts.[100]

Second, the capacity of a leader to build a reform coalition or produce lasting change depends on her ability to align the solution and problem, such that the alternative discourse seems like a natural and inevitable successor to the discredited dominant discourse. This is a key lesson of Blyth's study of the rise of neoliberalism in Western states.[101] For Blyth, there are no objectively given crises. Rather, political-economic situations must be interpreted such that they are felt as crises in need of resolution.[102] The construction of a crisis discredits the existing institutional order and its associated discourses creating an opportunity for challengers.[103] Challenger discourses are more likely to be compelling candidates for forming the new orthodoxy when they can diagnose the causes of the purported crisis, reduce uncertainty, and provide a blueprint for positive action. Moreover, when new ideas are perceived to have solved the associated problem, they are likely to be stable.[104]

Finally, lasting change depends on the naturalization of new ideas through links to cosmological elements. This helps explain when new configurations of ideas generate new *purposes*. At first, new goals arise because new combinations of concepts, policy instruments, and administrative practices generate representational constraints on goals. When an organization deploys a new set of rules or policy instruments, it adopts a set of frames, categories, and classifications that restrict its view of the world and makes some goals easier to incorporate into its problem-solving system than others.[105] For example, Scott argues that early modern European states developed scientific practices of cartography, metrical standardization, and statistical classifications that allowed them to render society "legible" or readable.[106] This in turn oriented states to certain interventions and projects rather than others. States and organizations and groups will adopt goals and purposes that appear prominently in their representational ontologies.

But any of these improvised or imposed ends will only be translated into lasting purposes if they are legitimated and naturalized by a wider

---

[100] Suddaby and Viale 2011.
[101] Blyth 2002.
[102] Blyth 2002, 10, 165–166.
[103] See also Legro 2005, 29–35; Phillips 2011, 43–44.
[104] Legro 2005, 36.
[105] Scott 1998; Barnett and Finnemore 2004.
[106] Scott 1998. See also Mitchell 2002, 83–93.

order of meaning. As we saw above, a stable institutionalized practice must be legitimated to the myths and narratives in the "symbolic universe" that defines humanity's place in the cosmos.[107] I conceptualize this as a process of naturalization. I define naturalization, following Hopf, as "a social process by which politically contestable outcomes, policies or practices come to be regarded as given."[108] As we learned from Douglas earlier, naturalization proceeds by articulating concepts to beliefs about what is natural.[109] A naturalized idea can still be challenged or contested. But challengers will argue "uphill" against the discursive slope structured by ideas rooted deep within the symbolic universe of the social group or organization. Defenders of the natural order will be able to call upon powerful rhetorical arguments rooted in the operations of the universe. The process of naturalization creates new values and purposes by linking concepts to cosmological elements that circulate in discourse.

Macro: The Recursive Institutionalization of International Order

In this section, I integrate the micro- and meso-levels of discursive change into a model I call the recursive institutionalization of international order. While I focus here on outlining how new purposes become embedded in international orders, in principle the model could be applied to other forms of international change. The key question is how meso-level changes are converted into macro-level transformations in the international discourse of state purpose. The answer, in short, is that the macro-level changes *both* as new purposes spread via associational changes and as new purposes are embedded in treaties and IOs. But the particular form or pathway of recursive institutionalization differs in historical cases. Thus, in what follows, I first present the core model of recursive institutionalization and then outline some variations to better capture the historical cases that follow.

My model integrates the main points of the structure of international order (Figure 2.1) and the constitution of international discourses (Figure 2.2) introduced above. Recall that on my account, international order is structured by a constellation of primary and secondary institutions that draw on underlying discourses that contain state purposes and cosmological elements. These international discourses change as the associations that carry and reproduce those discourses combine and

---

[107] Berger and Luckmann 1966, 95–97.
[108] Hopf 2002b, 409.
[109] Douglas 1986, 46–47.

```
┌─────────────────────┐   Cosmological Shift    ┌─────────────────────┐
│ International Order │ ──────────────────────▶ │   Associational     │
│        $t_n$        │                         │      Change         │
└─────────────────────┘                         └─────────────────────┘
           ▲                                               │
           │                                               │
           │                                               ▼
┌─────────────────────┐                         ┌─────────────────────┐
│ Institutionalization│ ◀────────────────────── │   Circulation of    │
│   of New Purposes   │                         │    New Purposes     │
└─────────────────────┘                         └─────────────────────┘
```

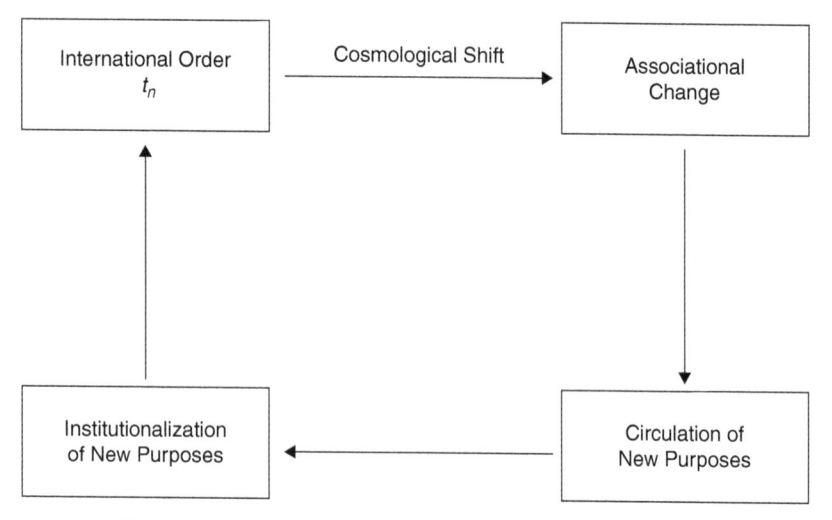

Figure 2.3 The recursive institutionalization of international order

recombine discursive elements. Accordingly, the model (Figure 2.3) begins at time t_n with an international order supported by a given arrangement of institutions and discourses. There are no functional or deterministic relations between the practices, institutions, and discourses that constitute international orders. However, the institutional order does depend on the underlying discourses in two ways. First, the institutional order incorporates discursive categories and ideas into its rules and beliefs. Second, the institutional order draws on purposes and cosmologies to legitimate and justify the rules and beliefs embedded in international order. So there are chains of meaning and legitimation that connect the institutional to the discursive levels.

First, moving across the top of Figure 2.3, a cosmological shift introduces new ideas about the universe. Second, as associations strategically deploy the new cosmological ideas, they adopt new representations and epistemic practices that set off discursive reconfigurations that in some cases produce new purposes. Thus, new purposes emerge in one or more associations, supported and shaped by cosmological ideas. Third, the new purposes then circulate throughout the international system, altering the distribution of ideas. This circulation can take any number of forms. A transnational, knowledge-based network may spread new cosmological ideas slowly throughout the system. Alternatively, a new purpose may be worked out within a single powerful state that seeks to impose it on other states. Finally, new purposes are institutionalized in the core sites of international order.

This act of institutionalization alters the rules and norms of primary institutions, which in turn reshapes the practices and flows that constitute international order. The model is recursive in two senses. First, it recurs in the colloquial sense of unfolding through repeating cycles of associational change and institutionalization. Second, the construction of order in the model unfolds recursively in the technical sense that it is constantly reproduced and changed by the associational discourses and practices that constitute it.

In this model, the term institutionalization means more than the formalization or codification of rules and norms in treaties or IOs. In sociological theory, institutionalization depends on a constellation of background ideas that legitimate the institutional order. For Berger and Luckmann, a fully institutionalized order is backed by cosmological beliefs.[110] Thus, in sociological terms to institutionalize is to both create a new rule and infuse that rule with "value beyond the technical requirements of the task at hand."[111] Similarly, in this instance institutionalization is the creation of new rules and norms backed by cosmological elements.

The model can be usefully counterposed to Gilpin's theory of international change.[112] In Gilpin's model, the system begins in equilibrium because powerful states are satisfied with existing territorial allotments, economic rules, and political institutions.[113] The law of uneven growth disrupts this equilibrium as differential rates of political, economic, and technological change redistribute power among states.[114] This introduces a disjuncture between governance arrangements and the distribution of power.[115] Gilpin argues this disequilibrium will generate change because it creates incentives for rising states to challenge the rules and raises the hegemon's costs of maintaining the status quo.[116] Rising powers emerge as agents of change. In the face of persistent challenges, the dominant power or powers will eventually fall into fiscal crises to be resolved by hegemonic wars that allow the victors to rewrite the rules of the system. The logic of the argument depends on equilibrium dynamics: laws of supply and demand for order predictably and reliably create pressure to restore equilibrium between the distribution of power and the governance of the system.[117]

---

[110] Berger and Luckmann 1966, 95–97. See also Zucker 1977.
[111] Selznick 1984 [1957], 17. See also, Finnemore and Sikkink 1998, 905; Katzenstein 1996, 21.
[112] Gilpin 1981.
[113] Gilpin 1981, 11.
[114] Gilpin 1981, 94.
[115] Gilpin 1981, 14.
[116] Gilpin 1981, 157.
[117] Gilpin 1981, 105.

For Gilpin, the importance of ideas is reflected only in the lingering prestige of the victors and the form of the governance arrangements.[118] But prestige and the form of governance arrangements are reducible to military and economic power, so ideas are not an autonomous source of change.[119] As a result, the argument has two weaknesses. First, it cannot explain why some ideas rather than others are institutionalized and thus why international orders have the form and content they do. As Ruggie and Ashley have pointed out, in order to understand the actual rules and purposes that shape and constrain behaviour, we have to study the worldviews and values of powerful states.[120] In contrast to Gilpin's equilibrium model where ideas follow power, recursive institutionalization relies on contingent constitutive mechanisms: the rise of cosmological ideas in associations produces new purposes that make possible and desirable new ways of arranging and legitimating international order. These mechanisms cannot be reduced to Gilpin's claim that ascendant states make the rules, because that fails to explain why those states promote one set of rules rather than another.

Second, Gilpin's theory does not account for changes in institutions or discourses that take place between major power wars.[121] The model of recursive institutionalization aims to account for order-building moments and continuous processes of change. It can accommodate the idea of order-building moments because an act of institutionalization might take the form of a postwar settlement or major multilateral treaty. However, the model can also accommodate ongoing shifts in international discourses and institutions. In these cases, the institutionalization of a new purpose might take a number of forms. I said above that new purposes are institutionalized in the core sites of international order. These core sites include treaties and other secondary institutions (including IOs), and informal shared practices or conventions. So changes in purpose might not come in the form of a postwar settlement. They might be the result of a series of bilateral treaties, the creation of a number of IOs over time, or alterations to shared practices. Each of these modifies the associations or treaties that carry and reproduce primary and secondary institutions. As such, they influence the patterns of behaviour that constitute international orders in the aggregate.

[118] Gilpin 1981, 30–35, 203.
[119] Gilpin 1981, 30.
[120] Ruggie 1982; Ashley 1989.
[121] Gilpin 1981, 45.

Despite these differences, there are a number of similiarities between my approach and Gilpin's. Like Gilpin's model, recursive institutionalization is a reiterative, recurring process.[122] In addition, Gilpin's fundamental intuition that order represents the fusion of power and ideas is a central insight shared by Ashley, Cox, and Ruggie.[123] I seek to build on Gilpin's insight that powerful states are agents of change. But in doing so, I show that powerful states themselves are embedded in recursive processes of ideational change. Moreover, I theorize the mechanisms and processes of ideational change in more detail, so that that we can account for specific changes in the form and content of international order over time.[124] So my account runs parallel to Gilpin's, aiming to explain how and why powerful states change governance arrangements in the way they do. To do this, we have to give more autonomy to discourses and institutions than Gilpin did.

While power and ideas explanations are often counterposed, the importance of power in the model of recursive institutionalization is consistent with discursive, contextualist principles. Economic wealth and military power are important factors in the production of international discursive structures. Ideas must be produced and spread by practices such as paying authors, printing pamphlets, funding radio stations, and providing side-payments in multilateral fora. Thus, in the absence of state and societal power to support transnational discourses, we would expect the diversity of cultures and the constantly shifting ground of discursive context to fragment international structure and undermine attempts to build international order. Power is a first principle for constructivism.[125]

Pathways of Recursive Institutionalization

The model of recursive institutionalization as presented in Figure 2.3 describes the processes by which international change happens. However, it does not tell us what drives these processes or what determines why some ideas rather than others become institutionalized. The temptation is to answer these questions with a functional, structural level mechanism.[126] A realist version might be that the ideas have to conform to the

---

[122] Indeed, the form of Figure 2.3 is borrowed from Gilpin's (1981, 12) model.
[123] Ashley 1989; Cox 1987; Ruggie 1982. More recently, see Finnemore 2009.
[124] For this critique of realism and materialist approaches generally, see Ruggie 1982, 1983; Kratochwil and Ruggie 1986.
[125] Power here should be read in a straightforward but expansive sense as possessing the political, economic, military, and cultural-epistemic resources to produce effects in other actors. See Barnett and Duvall 2005.
[126] On this problem, see Nexon 2009, 62.

instrumental needs of the most powerful states.[127] A constructivist variant might posit the existence of "tension" between underlying discourses and the primary and secondary institutions of international order.[128] But such explanations beg the question of how instrumental pressures or ideational tensions impinge on the specific individuals and organizations that actually embed new purposes in the core sites of international order.

Instead, my argument combines power and ideas into a contextual, contingent account of how associations change international orders. While power and purpose both shape which ideas matter, they are combined in different ways in different historical cases. To account for this, in this section I introduce two pathways of recursive institutionalization: hegemonic imposition and horizontal change. For heuristic purposes, we can think of these as forming a continuum. Real-world cases will fall somewhere in between these two schematic pathways.

In the *hegemonic imposition* pathway, a new purpose emerges in a dominant state. The dominant state then spreads the purpose amongst other associations and leads the process of institutionalizing it in international order. In this pathway, the hegemon uses a mixture of military-economic coercion, institutional position, and cultural-epistemic power to spread new purposes. This might be accomplished in a postwar settlement or through bilateral negotiations between wars. These are often conceptualized as two distinct moments, but it is important to note their dynamic interaction. For example, the rise of growth after 1940 followed this pattern. The idea of economic growth emerged during the war as economists in the US Department of Commerce refined national accounting techniques in order to measure and legitimate the American war effort.[129] After the Second World War, the United States used military and economic power to institutionalize growth in the Bretton Woods institutions. These institutions then helped spread growth-based policies and discourses to other countries. These efforts constitute a recursive dynamic because as states and IOs adopted the new purpose, they altered the distribution of ideas, creating further pressure for the spread of growth.

While military and economic power is essential to hegemonic imposition, cosmology plays an important role. Any hegemonic order depends on cosmological elements because hegemons must infuse international order with value beyond the requirements at hand. My conceptualization here builds on Ashley's theory of how hegemons produce order.[130] For

[127] Gilpin 1981.
[128] Cortell and Davis 1996; Checkel 1999; Bially Mattern 2005; Risse 2010.
[129] Lepenies 2016. See also Coyle 2014.
[130] Ashley 1989.

Ashley, the central problem of international order is generating common purpose from the diversity of cultures and motivations that constitute the international domain.[131] The production of unity must be sustained and reproduced by a combination of power and knowledge. In particular, Ashley argues, international purpose must rest on hegemonic projects that ground purpose in the appearance of extrahistorical, universal, and necessary truth.[132] Only this can naturalize and legitimate international life under hegemonic rule.

Relatedly, the spread of hegemonic ideology is not a purely coercive process. It depends on cultural-epistemic power rooted in knowledge production networks centred in the dominant state. Cultural-epistemic power to define reality through knowledge is transmitted transnationally by academic disciplines, epistemic communities, and IOs. Under hegemony, transnational knowledge tends to support the hegemonic state because the production of transnational knowledge is steered by the dominant state and its wealthy civil society. So new purposes are carried into the system by both the agencies of the hegemonic state and associated scientific, cultural, social, and political networks. While these networks serve as transmission belts that tend to reproduce the cosmology and purposes of the hegemon, their influence is not directly controlled by or functionally dependent on state power. Rather, state power and the production of knowledge co-produce one another in an ongoing dynamic that alters both the path of scientific development and the representations and goals of the state.[133]

The interests of the hegemon help explain why some purposes spread throughout the international system and others do not. Undoubtedly, the desire of the hegemon to see its own purposes reflected in the practices of other associations, and its willingness to use coercion to that end, are central factors in the spread of new cosmologies and purposes. But we cannot reduce cosmology to hegemonic interests. First, as outlined above, not all the effects of hegemonic power are directed and controlled by the state itself. The effects of epistemic, ontological, and cosmological constitution proceed around and beyond the agencies of the state. Second, not all of the hegemon's influence is reducible to conscious policies and strategic action. In eras of cosmological change, the hegemon is also being actively reconstituted by discursive reconfigurations that reorient its behaviour in subtle ways. So we have to recognize that efforts by the hegemon to impose new purposes on other

---

[131] Ashley 1989, 253–254, 257–258. See also Inayatullah and Blaney 2004.
[132] Ashley 1989, 255, 265.
[133] Allan 2017a.

associations and the effects of cosmological change on the hegemon itself reflect the influence of indirect forms of power.

Moreover, hegemony is not a necessary condition for international change. Historically speaking, the construction of hegemonic international orders under the auspices of Anglo-American power was a unique event.[134] These are the only two powers to have constructed what we might call an international or global hegemony. Hegemons in other times and places dominated only regional systems and did not have nearly as broad or as deep effects on the discourses of other societies and groups.

In addition, an emphasis on hegemonic power hides the important role of a receptive audience for hegemony. A hegemon cannot lead states that resist or challenge its leadership. Thus, other great powers and associations that would follow or obey the leading state must already be in a position to believe its rule is legitimate.[135] That is, they must share elements of the cosmological backdrop that naturalizes and universalizes the goals and practices offered in the hegemonic order. Imposition depends on at least some prior changes in discourses to make the broad institutionalization of purposes possible and desirable. So despite the power of hegemony in processes of change, we need to theorize other pathways of change.

Horizontal change is when a number of states and other associations take up the same purpose from transnational knowledge networks or epistemic communities. The key here is that discursive change does not originate in one state that then imposes it on the others. Rather, a number of states initiate similar associational changes by deploying similar ideas to solve problems. Horizontal change is not necessarily a process of diffusion because the associational changes that drive it are not always dependent on prior adoption. It is not that one idea moves from one association to another, but that many associations incorporate the same idea by drawing from the same transnational networks. As laid out in the theory of associational change above, in doing so associations might be variously motivated by competitive pressures, a desire to emulate success, internal learning processes, or discursive shifts that reconstitute means and ends.

---

[134] On the restrictive conditions for the construction of hegemony, see Saull 2012. Indeed, it seems clear American hegemony was able to free-ride on British leadership of Western colonial order to establish a growth-based hegemonic order. Moreover, without a persuadable group of newly decolonized states it is not clear that the United States would have been able to produce global convergence in the early Cold War era.

[135] This aspect of order-building is ignored or downplayed in the realist and liberal literature on hegemony: Gilpin 1981; Ikenberry 2001; Keohane 1984; Organski and Kugler 1980.

Whatever the motivation, associational changes are only likely to converge on new purposes if they are driven by transnational networks of experts and policy advocates.[136] To have this kind of influence, networks need to have captured or be institutionalized in the bureaucracies of important states. For example, transnational epistemic communities in the environmental domain have had success where they capture or get access to multiple governmental agencies.[137] Anglo-American economists shape policy in many countries because they are embedded in every major state and IO.[138] The extent of embeddedness or institutional capture helps explain the depth and breadth of transnational networks' influence.

In the horizontal pathway, the new purpose needs to emerge more or less fully formed within a transnational network to produce convergence at the macro-level. In the hegemonic imposition pathway, the new purpose need only emerge fully formed in the dominant state before being disseminated through the system. But here, since different states and associations need to agree on the new purpose without overwhelming power to generate unity, convergence needs to be underwritten by a tightly knit transnational network reproducing a common policy discourse. This is an inversion of Ashley's argument that the hegemon needs to create unity. In effect, a transnational network can have a similar effect as hegemony does in changing the distribution of ideas.[139]

As many associations deploy the same ideas, the distribution of ideas in the international system shifts, creating favourable conditions for the institutionalization of new purposes. Backed by cosmological force, the new purposes now circulating in the distribution of ideas will seem desirable, even necessary. States and other associations are likely to feel that the new purposes deserve institutional expression and support. Older purposes are now less prominent in discourses and the rules they supported are less likely to be defended. Existing institutions will not automatically decay or collapse. The creation of incentives and opportunities does not guarantee change. However, all things equal, states are likely to inscribe the new purposes in international order and let older ones decay.

As I said above, hegemonic imposition and horizontal change form a continuum that real-world cases fall between. Take for example the

---

[136] For seminal statements, see Haas 1990, 1992b. For more recent accounts, see Simmons and Elkins 2003, 281–282; True and Mintrom 2001, 37–38; Seabrooke 2014.

[137] Haas 1989; Evangelista 1999; MacDonald 2015.

[138] Fourcade 2006, 2009; Chwieroth 2010; Nelson 2017.

[139] Ashley 1989. My argument here is analogous to Keohane (1984) and Snidal's (1985) claim that other actors or entities can perform the functional equivalent of hegemony.

construction of the nineteenth-century liberal-colonial order premised upon the standard of civilization. British hegemony played an instrumental role in constituting this order, but it did not need to impose it on other states who were simultaneously taking up Enlightenment ideas about liberalism and human progress from political economy and the natural sciences. British leadership was important in creating a constellation of bilateral free trade treaties and isolating colonial competition from European power politics, but there was no single great power war or multilateral conference that built the order. So the processes of change combined hegemonic leadership and horizontal change to create new purposes and rules. In such cases, the new purpose need not emerge fully formed in either the hegemon or the transnational network. Rather, the form and content of the new purpose is worked out in interactions between the leading state, following states, and transnational networks.

In sum, the variations on the model of recursive institutionalization make it clear that state power plays a key role in explaining why some ideas are institutionalized rather than others. In some cases, this state power is wielded by a hegemon and in other cases it is exercised in concert. In the hegemonic pathway, it is the cosmology and purpose of the hegemon that are institutionalized. But in order to explain why the hegemon understood its goals in one way rather than another, we have to trace state interests back to the discourse of state purpose and the underlying cosmological elements that constitute it. In my conception, these arise from long-run processes of discursive change that are beyond the control of the hegemon. So understanding and explaining hegemonic imposition requires that we embed hegemons in macrohistorical ideational change.

In the horizontal pathway, it is the reach and power of transnational knowledge networks that explains which ideas become embedded in international orders. Moreover, the coherence and legibility of the discourses that transnational networks offer are important elements of the explanation. Both of these elements in turn depend on underlying discourses and cosmologies. Between hegemonic imposition and horizontal change is a range of hybrid pathways. Most historical cases are likely to look more like a hybrid model in which states converge on a purpose that emerged in a specific form within a particular state. In any event, the model of recursive institutionalization is flexible enough to be applied to many historical cases while nonetheless guiding our empirical analysis and the search for causal mechanisms and key factors. That said, I use the language of recursivity in part to denote the fact that change is always overflowing the boxes and mechanisms of the model.

Scientific Cosmology and Change in International Order

In this section I show how the general model outlined above applies to the case of scientific ideas in European and international politics since the sixteenth century. First, I provide a more thorough conceptualization of cosmological shifts. Second, I theorize how and why scientific ideas specifically can transform the discourses underlying international order. I identify three factors. First, scientific ideas are disseminated transnationally and therefore are available to many associations. Second, scientific ideas have authority and help solve problems, so they are likely to be taken up by many associations. Finally, scientific discourses have cosmological implications that can be used to articulate universalist claims that naturalize goals. The transnational reach and cosmological substance of scientific ideas enables them to operate as a structural force, fuelling the processes of recursive institutionalization across the international system.

Cosmological Shifts in Modern European Science

A cosmological shift happens when a community of practice or discursive tradition produces radically new ways of looking at and explaining the natural and social universe. It begins with the creation of new discursive configurations that redefine and recombine ideas about what exists, what counts as knowledge, time, the origins of the universe, and the place of humanity in the cosmos. A cosmological shift is developed within a community or sub-discourse that may be either tightly linked to or far removed from political institutions and policy-makers. That is, a cosmological shift might originate in a caste of priests closely allied with the state or in a marginalized prophetic cult. In either case, for the cosmological shift to become politically important, its central elements need to be transported into core political institutions and discourses. So in international politics, cosmological shifts are only likely to change international orders when they are legible and desirable to many associations at the same time.

In this book, I focus on cosmological shifts arising from scientific traditions. As I pointed out in the Introduction, IR theory has largely ignored the cosmological power of science. Instead, it tends to treat scientific knowledge as a set of causal or technical beliefs that are deployed as instrumental means.[140] On this point, IR reflects the Weberian narrative

---

[140] This tendency is shared by realist and constructivist work on scientific knowledge. See Gilpin 1968, 1981; Thompson 1990; Drezner 2001; Taylor 2012; Haas 1989, 1992a, 1992b; Hymans 2012.

that scientific ideas helped rationalize modern societies.[141] The process of rationalization disenchanted the world by draining it of the sacred, mystical elements that infused the world with meaning. However, this view rests on an idealized picture of science that has been challenged by the last forty years of historical and sociological studies of science.[142]

The disenchantment narrative is misleading because a closer examination of key thinkers and texts in the European scientific tradition reveals that the new sciences that emerged in the sixteenth and seventeenth century were rich with cosmological meanings.[143] Kepler's astronomical work was mathematical and empirical, but he was a committed neoplatonist working within a medieval tradition and conceptual order.[144] So his written work exhibits "an almost mystical fascination with celestial harmonies."[145] Newton famously uncovered the fundamental laws of gravitation and optics, but he probably spent more time poring over apocryphal religious texts and indulging his interests in alchemy than he did on what we would call physics.[146]

In fact, natural philosophers creatively combined ideas from the new science with Christian doctrine. The dominant view in seventeenth- and eighteenth-century Europe was that natural philosophy "reveals the handiwork and purposes of God."[147] As such, natural philosophy was an ally of scripture and divine revelation in the search for religious truth.[148] But even after geology and Darwinian thought broke from this tradition, European science did not cease to possess cosmological significance.[149] Demonstrating how and why that happened is the focus of the empirical chapters that follow. But rather than focus on science in the abstract or treat science as a single, unified movement, I focus on specific cosmological shifts.

In particular, the chapters investigate three shifts that had significant implications for the discourses of state purpose underlying international orders:

---

[141] See Weber 1958, 1978; Horkheimer and Adorno 2002; Habermas 1970 [1968], 1984 [1981], 1987 [1981]. In the international context see Meyer *et al.* 1997; Drori *et al.* 2003.
[142] For good reviews of the history, see Cunningham and Williams 1993; Daston 1995; Shapin 1996; Daston and Galison 2007. For good introductions to the sociology, see Bloor 1991 [1976]; Latour 1987; Jasanoff 2004.
[143] On these connections, see Bennett 2001; Kragh 2007; Connolly 2011, 2013.
[144] Kuhn 1957, 212; Reiss 1982, 141.
[145] Briggs 1999, 173.
[146] Bowler and Morus 2005, 48–49.
[147] Gaukroger 2010, 151.
[148] Gaukroger 2010, 493.
[149] On this story see Gillispie 1951, 222–226; Rudwick 2005.

1. Copernican astronomy and Newtonian mechanics, 1543–1687
2. The historical sciences and Darwinian evolution, 1751–1930
3. Cybernetic-systems modelling and the constitution of objects, 1930–.

Each of these shifts reconfigured the central categories underlying international order. Why did they have these effects? They all had broad scientific and societal implications because they challenged established cosmological beliefs. They contained new ideas about time, space, and the role of humanity in the cosmos and so drove discussions about the purposes of life and politics. In addition, they emerged from particularly productive research programmes. Productive research programmes attract resources, generate discussion, and are likely to inspire many individuals and associations to draw on their ideas. These are not the only shifts that matter in the modern history of European science. Nor do they represent all dynamic or successful research programmes. In the conclusion, I discuss why quantum thought has not revolutionized international politics. However, it was these shifts that produced the core features of international discourses today.

Cosmological shifts do not necessarily replace or eliminate earlier cosmological elements in political discourses. But cosmological shifts do not cumulate in predictable ways either. Instead, cosmological shifts introduce nonlinear ruptures and reconfigurations that disorder and reorder discourses. The cosmological shifts arising from Western scientific traditions each exhibited both continuity and change. On one hand, each shift introduced new ideas. On the other hand, each reworked or developed ideas that had been introduced earlier. In many cases, the cosmologists saw themselves as working within a single intellectual tradition, "science." But this does not mean that it makes sense to theorize the effects of science in a monolithic way. Instead, we should examine how multiple, distinct scientific traditions were used to continually rework a set of foundational questions about what exists, how to produce knowledge, the nature of time, and human destiny. As we shall see, the pattern of change is uneven, but there are some common themes throughout the period.

Scientific Ideas and the Conditions for International Change

When and why do scientific traditions produce cosmological shifts that transform the discourses of state purpose underlying international orders? I outline three factors that explain how scientific ideas drive the process of recursive institutionalization of new purposes. First, since the seventeenth century, scientific knowledge has been carried out and reproduced transnationally, so it is accessible to associations in various parts of the world. Second, scientific ideas are seen to be useful and necessary for solving problems, and so they are likely to be taken up by many associations. Third, once embedded in associations, scientific ideas are likely to change purposes because they possess particularly powerful resources for legitimation and meaning creation. In short, new cosmological ideas emerging from scientific discourses are likely to drive structural changes in international discourses because they are diffused widely and can naturalize ideas.

First, throughout the modern era, scientists and experts have been embedded within the transnational social classes and bureaucratic offices that shape policy-making in great power states and powerful associations.[150] In Sewell's terms, scientific ideas are "densely articulated" to the associational basis of international order. So, as scientific ideas change, the new ideas can be channelled into international discourses, albeit at a lag. Since many individuals and organizations are involved, it takes time for cosmological shifts to produce widespread associational changes. Nonetheless, a change in the dominant norms and purposes of a discipline or profession can generate structural pressure on many associations.[151] As we shall see, changes in postwar economics created pressures for change in multiple international organizations and states around the same time. Transnational networks of economists, trained largely in American economics departments, transmitted new ideas all over the world.[152] Just as a change in the professional norms of economists generates slow, uneven, but nevertheless persistent change in states and corporations, a shift in the cosmological elements of scientific discourses generates slow, uneven, but persistent changes in the associations that underlie international order.

In the case of science, cosmological shifts that set off changes in multiple international associations are likely to begin in the most dynamic and visible scientific research programs of an age. Such movements force widespread changes in academic thought and so their central elements are transported along transnational knowledge networks. As a result of this transnational distribution, these ideas are then available to be taken up into many political discourses. As such, cosmological shifts originating from the natural sciences are likely to impress many associations and so many associations are likely to take up scientific ideas and deploy them in similar ways. As we saw above, this might

[150] Wuthnow 1979.
[151] Suddaby and Viale 2011, 428.
[152] Woods 2006; Fourcade 2006.

follow the pathways of hegemonic imposition or horizontal change, but the existence of a transnational network to disseminate the ideas is a precondition of both pathways. The consistency and reproducibility of scientific ideas within transnational networks underwrites the structural power of science to drive change in international discourses. That is, scientific cosmological shifts are particularly effective sources of international change because they can alter the conceptual structure of many agents and organizations.

Second, scientific ideas are likely to be taken up into associational discourses because many actors believe they will help them solve problems and manage complex realities. My argument here builds on the World Polity School insight that many organizations use scientific and expert knowledge because science is the most authoritative source for useful, rational information and it confers legitimacy on actors who are seen as rational if they deploy scientific means.[153]

In some cases the strategic deployment of scientific ideas is driven by instrumental motives: states and other political actors want to fulfil their interests or achieve their goals and science promises to serve as a means to ends. But there are other reasons for taking up scientific ideas. Dewey argues that humans constantly engage in a "quest for certainty."[154] For cognitive and social reasons, humans must reduce uncertainty to go on in a complex world: "[t]he natural man is impatient with doubt and suspense: he impatiently hurries to be shut of it."[155] Dewey and others argue that centuries of war and disease in early modern Europe created a desire for control that fuelled scientific development.[156] So although motivations vary across time periods and between associations, scientific ideas are often appropriated because they promise to enhance control and impose certainty on a complex world.[157]

Scientific ideas are also used to engage in symbolic contests for power and authority. Associational change might be driven by battles of ideas within organizations in which one side enlists a new array of cosmological

---

[153] Meyer and his colleagues argue that the power of science in the modern world polity is derived from a rise in strategic, rational actors that demand knowledge to enhance their capabilities and a scientific cosmology to legitimate rational actorhood itself, which is premised upon knowing and controlling nature. Drori *et al.* 2003, 10, 31–37; Meyer 2009.

[154] Dewey 1929.

[155] Dewey 1929, 229; see also Dewey 1922, 172–181. Dewey's claim has some support in experimental psychology. Experiments find that stress and anxiety induce a search for causal relations. Further, this search is likely to lead to an increased tendency to form causal attributions and an illusory sense of control. See Friedland *et al.* 1992; Keinan 1994; Keinan and Sivan 2001.

[156] Dewey 1929; Rabb 1975; Toulmin 1990.

[157] Scott 1998.

concepts. Associational change might also be driven by an association's attempt to improve its position in a field of competition. In such fields, states and other political associations often seek to establish control over definitions of reality and the terms of discourse.[158] In either of these internal or external contests, efforts to introduce scientific cosmology are likely to be resisted by traditional authorities. In the cases that follow, scientific ideas were resisted by the Church, early modern vitalists, conservatives (in some cases), indigenous peoples, and environmentalists. So the triumph of scientific ideas is not a foregone conclusion. Nonetheless, scientific ideas often came to be stable elements of political associations when backed by powerful actors with the resources to mobilize coalitions for associational change. Moreover, scientific cosmologies benefitted from the enormous success of the natural sciences and associated technologies in the world. The realities of scientific and technological progress were so impressive that they led many to conclude that scientific concepts and methods could be applied to many domains. But despite these many motivations for drawing on scientific ideas, associations are only likely to take up the same concepts in the same way when they face a common set of problems in the presence of a common solution presented by a transnational network. As in the model of associational change above, it is the alignment between a shared problem and an available scientific solution that enables broad change in international discourses.

Third, once embedded in associations, new scientific ideas are particularly likely to generate the kinds of discursive reconfigurations that produce new purposes. As states and associations bring in scientific ideas or new experts to solve problems, they import a whole set of connected cosmological ideas. This alters the basic discursive and conceptual resources that are used to constitute actions, and naturalize orders. In particular, scientific cosmology drives shifts in purpose by redefining the structure of time, the nature and character of the universe, and humanity's role in the cosmos. Concepts of time are intimately connected to purposes because they place individual and group biographies in narratives that define what has been, what is, and what will come. In so doing, they suggest purposes by offering individuals and groups the possibility of situating their actions within meaningful narratives. As we shall see, the European sciences came to be premised upon a linear notion of time and invested in a linear story of scientific and technological development that together constituted ideas about human progress.

[158] Sending 2015.

Scientific ideas about the nature and character of the universe also shape purposes by constraining representations of what exists. Individuals and groups can only value that which is thinkable and possible. In the early modern period, purposes were defined by the belief that nature was determined by providential natural law. Under the influence of materialist and Newtonian thinking, this was reconfigured into a view that scientific laws determined the operations of matter and motion. In a deterministic universe, meaning was to be found in following God's law or discovering and obeying rational maxims of action. But the dismantling of determinism in the nineteenth century permitted a new vision of the universe as a series of entities and objects that were governed by knowable and controllable mechanisms. In a determinist universe, there is little role for human action except to submit to the law. In an open, progressive universe filled with manipulable entities, human purpose is transformed. This shift allowed new rationalist and modernist visions of human agency to emerge. Thus, scientific and social scientific ideas redefined humanity's role in the cosmos and this had major implications on discourses of political purpose.

The new purposes produced by cosmological shifts are likely to be used to legitimate international orders because they can support universalist and naturalizing claims. As Ashley argues, the production of international order depends on a wide variety of actors coming to believe that the rules governing the system are natural and inevitable.[159] Scientific ideas are in an excellent position to provide this support because scientific claims are universal generalizations that apply everywhere and scientific discourses have a near monopoly on claims about nature. So scientific discourses can be used to convince all states and groups that a given representation of the world is true and that attempts to govern it in one way rather than another are natural and necessary.

My claim is not that cosmological shifts reconfigured international politics in an automatic or monolithic process. Rather, my claim is that concrete associations worked out the implications of cosmological shifts for political purposes in the course of solving problems. So, it was not a foregone conclusion that the Copernican revolution in astronomy would lead states to orient themselves to the balance of power understood in terms of territory and population. Rather, the Copernican revolution made that mechanical, materialist understanding possible and desirable in a number of states at the same time. And then, in a series of negotiations and conferences culminating in the Vienna Congress, policy-makers deployed maps and population statistics in such a way as

[159] Ashley 1989.

to make that understanding of state goals seem natural and inevitable. So cosmological shifts create the structural conditions that allow and push associations to redefine their practices and goals.

Demonstrating the Argument: Research Design, Methods, and Alternative Explanations

My central goal in the chapters that follow is to demonstrate and explain change in the discourses of state purpose underlying international orders. First, I aim to track the history of growth to its precursors in ideas about colonial development, improvement, the balance of power, and reason of state. Second, I want to explain how and why these purposes were institutionalized in successive international orders between the sixteenth and the twentieth centuries. In providing this history, I aim to establish that these changes are best explained by the meso-level mechanisms (strategic deployment, discursive reconfiguration) and macro-level processes (hegemonic imposition, horizontal change) I have presented. To these ends, I offer a multilevel research design that traces change back and forth between the meso- and macro-levels. At the macro-level, I provide snapshots of the discourse of state purpose at regular intervals (1550, 1713, 1815, 1885, 1919, 2015). At the meso-level, I analyse the dynamics of continuous change by tracing the emergence of new purposes in key associations (early modern European states, the British Empire, and the World Bank).

At the macro-level, my argument is that change in international order is a process comprised of many meso-level events. But an empirical analysis cannot capture all the associational changes and order-building moments that drive the recursive institutionalization of international order.[160] So the question is how to select meso-level cases that illuminate the larger macrohistorical process.[161] Case selection here was guided by the goal of explaining where the specific ideas that reconfigured discourses of state purpose came from. Thus, I selected meso-level and macro-level events that demonstrated the history of economic growth and its conceptual precursors.

The result is three case studies that move back and forth between the macro- and meso-levels of international politics to demonstrate the transformation in state purposes from God and glory to economic growth. First, the emergence of balance of power order between 1550 and 1815 is examined through changes in early modern European states.

---

[160] On these issues, see Pierson 2003.
[161] Katznelson 1997, 101.

Second, the construction and reconstruction of liberal-colonial order is investigated through British colonial policy. Third, the establishment and maintenance of the US-led post-Second World War order is demonstrated by a case study of the World Bank.

To demonstrate the argument empirically, each case combines discourse analyses of primary documents and process-tracing of the key mechanisms. I studied changes in purpose and the links between purposes and cosmologies at both the meso- and macro-level by adapting an inductive mode of discourse analysis developed by Ted Hopf.[162] Hopf's method has two core elements: he constitutes a large body of texts and codes these texts inductively, without pretheorizing the categories or meanings he expects to find. Likewise, I proceeded to inductively recover categories of purposes, epistemes, ontologies, and cosmologies revealed in the texts. These categories were recovered from large samples of texts. The discourses analyses establish change in discourses of state purpose at both the associational and international level. But to explain that change, we need a historical analysis that demonstrates the mechanisms and processes of change in action. For these purposes, I use process-tracing to identify the key causal mechanisms that changed associational discourses and shaped order-building moments.[163] More information on the discourse analysis, text selection, and process-tracing can be found in the Methodological Appendix.

How does my account compare to rival explanations? The central argument is that a cosmological shift in scientific ideas made possible and desirable changes in the discourse of state purpose over the course of 450 years. The logic of the argument is configurational: scientific ideas were a necessary component of the discursive, institutional, political, and economic configurations that constituted the purposes of the balance of power, development, and growth.[164] I do not argue that scientific ideas alone are sufficient to explain the changes. In large-scale historical processes, it is clear that a combination of causal factors are necessary to explain change. From this vantage point, the cases do not attempt to refute other important causal factors such as the importance of military and economic power or changes in the political-economy of societies. Instead, on a configurational analysis, considering alternative explanations is less an exercise in self-falsification than an attempt to show how

---

[162] See Hopf 2002a; Hopf and Allan 2016. See also Angenot 2004; Howarth 2000; Milliken 1999.
[163] Mahoney 2003, 360. On process-tracing, see George and Bennett 2005.
[164] On necessity and sufficiency, see Mahoney 2008. Scientific ideas are necessary in the sense that if not for those specific scientific ideas, then the configurations of purpose would have turned out differently and history would have taken a different form.

72 Cosmology and Change in International Orders

ostensibly "competing explanations" actually combine and interact to produce the outcome.[165] Nonetheless, I believe my account is superior to rival explanations because it provides a clear account of where the specific ideas that shaped international order came from. On their own, accounts highlighting military or political-economic factors cannot explain changes in the *discourse* of state purpose that shapes the practices and policies in international orders.

Moreover, even though I concede that many so-called rival explanations are also important or necessary, it makes sense to be as clear as possible about what would make my account wrong. However, that is a more complex task than specifying alternative explanations and testing my account against them. On this point, I consider there to be strong evidence for my claim that cosmological shifts made possible and desirable changes in state purpose when a discourse analysis of primary documents and process-tracing of the relevant mechanisms reveals three things. First, cosmological shifts preceded important changes in associations and international orders. Second, that cosmological ideas were in fact channelled into important associations by strategic deployments. Third, that the resulting discourses of state purpose were constituted and legitimated by cosmological elements. I demonstrate this using discourse analyses that reveal how purposes depend semantically on networks of cosmological meanings. My chapters aim to establish each of these claims. Conversely, I would consider my empirical claims that cosmological shifts led to purposive changes to be wrong if changes in purposes or goals preceded cosmological changes, there was no clear channel by which cosmological ideas could have entered the relevant associations, or if the discourse of state purpose did not draw on cosmological meanings in defending and justifying its ends.

It is worth pointing out that these social scientific techniques are a product of the history of science this book covers. That is, this work is embedded in scientific discourses whose emergence it describes. In particular, my explanatory strategy relies on the metaphors and logic of the mechanistic worldview and the historical sciences.[166] But my explanatory and methodological choices have also been informed by the linguistic or interpretive turn and work at the intersection of social theory and history.[167] These sources of inspiration may seem to be in tension. After all, interpretivists have counterposed their enterprise to positivist approaches that seek to imitate the natural sciences.[168] However, on a

[165] Katznelson 1997, 101.
[166] On the history of the social sciences, see Backhouse and Fontaine 2010.
[167] Rabinow and Sullivan 1987; Foucault 1970, 1972; Sewell 2005; Weber 1949.
[168] The issues here are considerably deeper than this. My own view is that the historical and plural view of science emerging from the revisionist historiography of science

practical level, if mechanisms are simply pathways that connect cause and outcome, then mechanismic explanations are consistent with interpretive and historical approaches that help to reveal the ideational mechanisms driving social action.[169] On a broader level, we cannot help that we are partisans in a macrohistorical process of contesting the terms of scientific and political discourse. But we can approach our role in this process with care, humility, and contingency.

Conclusion

Existing ideational accounts of international change tend to focus either on order-building moments or meso-level processes. In this chapter, I have built a model of recursive institutionalization that includes both ongoing changes and order-building moments in a multilevel theory of international change. In my model, the rise of one set of ideas rather than another is driven by the contingent fusion of power, knowledge, and purpose. In the conclusion of the book, I generalize the theory to illustrate how it might be applied to other cases of international change. By way of conclusion here, I want to outline some of the empirical benefits of the theory.

First, the basic conceptual building blocks of the theory – discursive context, association, strategic deployment, discursive reconfiguration, recursivity, hegemonic imposition, and horizontal change – are general enough that they can be used to illuminate the processes of change in a variety of different international systems. A central problem in IR theory has been that our categories often hide differences across historical time. So my concepts are designed to be historically flexible in the sense that they guide our analysis of the past without pretheorizing the modes of action or forms of discourse that are likely to matter. They allow us to proceed inductively in the search for mechanisms and processes, letting the evidence work on us.

But, second, the basic conceptual building blocks are drawn from discursive contextual foundations. By situating action within discursive and institutional contexts, the theory urges us to follow knowledge: trace the ideas that circulate in the international system back to the transnational networks and knowledge-based groups that produce them. Further, it encourages us to pay attention to discursive configurations and the practices of articulation that alter and modify structures of

---

allows us to feel closer to the natural sciences without adopting positivism in its more ambitious, nomothetic forms.

[169] On understandings of mechanism, see Mahoney 2010.

meaning. It tells us to find the fusions of power and knowledge that drive the institutionalization of new purposes. But most importantly, it tells us that political life unfolds against a cosmological backdrop and if we want to know what political agents cared about and why, we have to explore their changing cosmological beliefs.

Recovering agents' beliefs about the physical world, the nature of time, and the place of humanity in the universe helps us understand how they could believe strange things and how their world was different from ours. But embedding our own forms of knowledge within the longue durée of history helps us to understand why their world is familiar and legible to us, even if our would not be accessible to them. But we can only resist the temptation to erase historical and cultural differences, hiding them within universal models, if we denaturalize our own cosmologies.

3 Natural Philosophy in Balance of Power Europe, 1550–1815

> An age of mechanistic philosophy in which Newton was King, Locke, Voltaire, and Montesquieu the royal advisers, and in which the religious vogue of Deism ... relegated God to the role of retired watchmaker of the universe was anything but hostile to the logic of balancing power.
> – Edward V. Gulick[1]

Introduction

When Copernicus developed his model of the heliocentric universe, he did not intend to reconfigure European cosmological discourses. Rather, he aimed to solve persistent mathematical problems in late medieval astronomy. However, his analysis inspired generations of scholars who challenged the Aristotelian basis of European natural philosophy and theology. Over the course of the seventeenth century Galileo, Descartes, Hobbes, Boyle, Newton and others transformed Copernicus' insights into a mechanist, materialist, and law-governed image of the universe. It was this image that made possible the emergence of balance of power discourses that underwrote a series of international orders from 1700 until 1815. The emergence of balance of power politics in early modern Europe is often taken for granted.[2] However, history shows that the modern European balance of power system is the exception rather than the rule.[3] Moreover, even if the competitive pressures of anarchy drove the rise of balance of power politics, the balance could have taken any number of forms.[4] In order to explain why one balance of power order emerged rather than another, we have to reconstruct the configurations of power and purpose that emerged in the early modern European order.

---

[1] Gulick 1955, 24.
[2] See Waltz 1979; Schweller 2004.
[3] For a comparative analysis of international systems, see Kaufman *et al.* 2007.
[4] Finnemore 2003, 96–101.

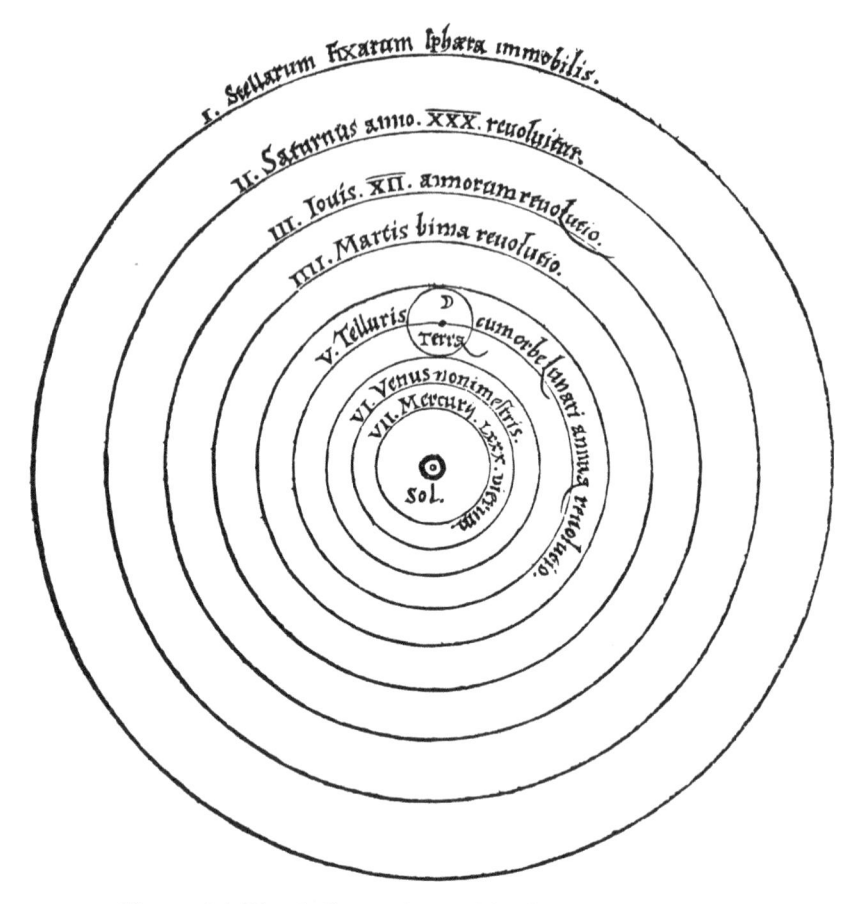

Figure 3.1 The heliocentric model of the solar system, Nicolaus Copernicus (reproduced from Copernicus 1543, 10)

The balance of power discourse underlying eighteenth- and nineteenth-century international orders was constituted by a complex configuration of discursive elements that were not present in European discourses two hundred years before. In 1550, political discourses were dominated by a cosmological discourse I call divine providentialism. Divine providentialism drew upon a Christian worldview in which divine law set the bounds of legitimate action and divine intervention determined outcomes. These cosmological themes were interwoven with aristocratic political ideologies to constitute a European political order premised on divinely sanctioned monarchical rule. In this order, state goals were oriented to the

Introduction

purposes of enacting divine providence, dynastic ends, and the pursuit of glory.[5] But by 1815 governments aimed to increase land and population to obtain a strong position in the balance of power. Implicit in the 1815 Final Act of the Congress of Vienna was a new purpose that pointed beyond the ends of the balance for security's sake: the idea that the world could be improved by human control. This idea was intimately connected to the balance, because in order to occupy a strong, secure position in the balance a state needed to improve its lands and peoples. What explains the shift in purposes between the sixteenth and nineteenth centuries?

In this chapter, I argue that the cosmological shift introduced by natural philosophy from Copernicus to Newton made a transformation in the discourse of state purpose possible and desirable. In short, natural philosophy introduced mechanist, materialist, and measurable cosmological ideas that both eroded support for old dynastic, religious discourses and bolstered the political significance of new rationalist and materialist ideas. Political economists in the tradition of Temple and Petty then transposed these ideas into political discourses, explicitly reorienting state goals from God and glory to land and people. Ideas from natural philosophy and political economy spread throughout the transnational aristocratic class that ruled European states, driving horizontal associational changes.

The recursive institutionalization of balance of power purposes and their underlying scientific cosmology did not have to be imposed by one European state on the others. At the meso-level, European states had independent motivations for appropriating scientific discourse. Rulers financed and strategically deployed scientific ideas for numerous reasons: to legitimate their rule, improve their armies, and demonstrate their power.[6] As a result, new ways of representing and valuing the world emerged in a number of European states in the seventeenth and eighteenth centuries. The new rationalist and materialist ideas were then institutionalized alongside balance of power discourse at Utrecht and Vienna.

The empirical argument here builds on suggestive comments by Morgenthau, Gulick, Butterfield, and others indicating that it could not be a coincidence that the balance emerged in the age of Newton.[7] However, these hints have never been gathered up into a systematic examination of the effects of scientific discourses on balance of power

---

[5] Elliot 1968; Bonney 1991; Blanning 2007.
[6] Wuthnow 1979.
[7] Morgenthau 2006 [1948], 44–45; Gulick 1955, 24; Butterfield 1966; Keens-Soper 1978; Sheehan 1996, 44–48.

thinking in the early modern period. My theoretical argument draws on constructivist and English School theories that conceptualize the balance of power as a shared idea or institution.[8] As Finnemore argues, "[m]aintaining a balance-of-power arrangement required a lot of shared social and cultural baggage. The balance of power existed only because Europeans shared a number of beliefs about what was necessary and good in politics."[9] That is, the balance of power order depended on a discourse of state purpose. My contribution here is to argue that this discourse was made possible and desirable by cosmological ideas emerging from the new sciences.

There are a number of alternative explanations for the emergence of the balance of power in this period. For realists, the balance of power is a universal feature of anarchic systems and so the emergence of this or that set of balance of power ideas is mere historical detail. As Waltz argues, states adopt balancing behaviours because they value survival and, stimulated by fear, calculate that balancing behaviours will preserve them.[10] However, the comparative analysis of international systems demonstrates that balance of power systems are rare and balancing behaviour is not necessarily more or less likely than bandwagoning, hiding, and other strategies.[11] One does not need to be a realist to recognise that the threat of French hegemony throughout the period 1660–1815 played an important role in the rise of balance of power discourse. That said, the balance of power discourse used to consolidate the balance in 1713 was quite different than the one used in 1815. French power and the threat it posed cannot explain the form and content of balance of power order. To explain how balance of power purposes emerged and were embedded in international order, we have to trace the ideational history of those

---

[8] Andersen 2016; Bull 1977; Buzan 2004; Finnemore 2003; Little 2009; Wilson 2012. In the English School, the balance of power is usually conceptualized as a primary institution. However, historically it had a central feature of secondary institutions: clear rules (e.g. compensations) that, while not explicitly outlined in a single treaty, were well understood. Moreover, it is not clear that the balance of power was constitutive in the sense that sovereignty and diplomacy are.

[9] Finnemore 2003, 101.

[10] Waltz 1979, 88, 106, 118–119. Waltz argues that selection and adaptation mechanisms push states towards balancing of power. However, it is not clear that the pressures Waltz supposes were strong enough in this period to compel such fear. Gulick, for example, argues that war had little impact on the composition of the system post-Westphalia: "[I]n the period from 1648 to 1792, there were, generally speaking, no great territorial changes in continental Europe, except for the first partition of Poland ... Wars, an all-too-familiar disfigurement of the seventeenth and eighteenth centuries, repeatedly ended in restoration of either the *status quo* or a close approximation of it" (1955, 39). If survival and territorial losses were not at stake, it seems unlikely that material pressures were strong enough to drive a process of selection.

[11] Kaufman *et al.* 2007; Schroeder 1992; Schweller 2004.

ideas back to the cosmological shifts introduced by natural philosophy and political economy.[12]

In what follows, I begin in 1550 to show what European political discourses looked like before the rise of natural philosophy. Then I provide a short history of the cosmological shift from Copernicus to Newton and show how these ideas were strategically deployed by states. I then demonstrate the reconfiguration of European discourses as states adopted rationalist and materialist ideas in the seventeenth and eighteenth centuries. This reoriented the discourse of state purpose to control of the material elements of the balance of power. The analysis culminates in an analysis of the balance of power as it operated at the Vienna Congress where the final settlement depended on collective efforts to construct a statistically calculated balance of power system. In short, the chapter traces the emergence of the rationalist and materialist ideas that made the Vienna Congress possible.

Cosmology and European Political Discourses, 1550

To see the full effects of the cosmological shift initiated by the new thinking, it is important to begin at the macro-level with an analysis of European political discourses circa 1550, before they were transformed by natural philosophy.[13] The years around 1550 mark the beginning of a shift, which culminated in the widespread dissemination of Newtonian ideas throughout eighteenth-century Europe. In 1550, the new cosmological ideas produced in astronomy, mathematics, optics, and other branches of mechanical natural philosophy had not yet entered broader societal discourses. Yet, European social and political discourses were already entering a transitional phase. The medieval episteme that framed European cosmology in the thirteenth and fourteenth centuries was beginning to unravel.[14] The dynastic-religious social and political order built on the cosmology was also weakening.[15] On the other hand, a coherent successor discourse drawing on ideas from natural philosophy and natural history had not yet been configured. So, core elements of

---

[12] The confluence of scientific cosmology and balance of power thinking may help explain why balancing behaviour was more likely in early modern Europe than it was in other systems and time periods, but that is not my objective here.

[13] I chose 1550 because it is only a few years after the publication of Copernicus' *Revolutions* in 1543 and so political discourses had not yet been changed by the cosmological shift it inaugurated. To constitute a large enough text sample from my sources, I selected texts from the period 1535–1575. For more information on text selection, please see the Methodological Appendix.

[14] Reiss 1982, 106. See also, Foucault 1970 [1966]; Bartelson 1995, 108–109.

[15] Koenigsberger 1987, 61–68.

earlier discourses still circulated in a hybrid discourse featuring configurations of medieval, Renaissance, and early modern ideas. In this period, international order rested on a legal, aristocratic institutional discourse justified and legitimated by a religious, determinist cosmology.

In this section, I provide a discourse analysis of texts from sixteenth-century England and France.[16] At this time, the reading public was quite small, so sixteenth-century French and English texts reflect the beliefs and purposes of the aristocratic classes arrayed around the royal courts. So these texts are likely to capture the associational discourses that underwrote European international order circa 1550.

English and French Discourses, circa 1550

My discourse analysis of French and English texts around 1550 reveals two central discursive configurations: divine providentialism and aristocratic dynasticism. Divine providentialism drew on the cosmological tenet that God created and controls the universe. Texts from 1550 depict God as a present force in everyday life, distributing punishments, and shaping the weather. Although human life is predestined by God, people nonetheless have the ability and duty to seek and obey God's law. The primary end or goal is to honour and further the glory of God, especially by obeying his commandments. At the individual level, people are encouraged to pursue salvation, live a righteous life, and to live in faith.

Divine providence has two sides. First, the world reflects God's will and God often intervenes in everyday life. One should pray to God, know God, and so on, in order to avoid the "miseries of man."[17] Second, God commands that people "keep the law," be "bound unto the law," that all things be "performed according to the law of the Lord."[18] The law of God forms the basis of everyday morality, civic virtue, and the standards of government. These themes come together in the belief that failing to follow God's law exposes one to God's intervention. For example, a news report explains an earthquake as "the horrible punishment and advertisement of God" but it also contends that the survivors escaped "through the help of God."[19] All of this nonetheless leaves room for human agency. For one author, Paul's epistle to the Ephesians shows that God "hath

---

[16] England and France were chosen because there were a large number of early modern texts from these countries available in English. I would have liked to have included German texts, but not enough of those texts are available in English translation.
[17] Viret 1548, 7.
[18] Anon. 1538, 156, 158.
[19] Anon. 1542.

created and made all men and things necessary for them" but also that "all men be of themselves of their own nature, mights, and powers" to accept and fulfil God's law.[20]

This formation manifests itself even in places where we would not expect it. In an account of military campaigns in 1548 and 1549 the author declares that the Scots lost to the English because "God in his Infinite Wisdom permits Calamities of this Nature to attend a People, to Rouse their drooping Faith, and to enforce an acknowledgement of their Sins."[21] The legal component is also evident here as the author declares that this was God's "Punishment" akin to when God permits "Turks and Barbarians ... to be the Executioners of his Justice."[22]

This discursive configuration is evident in both Catholic and Protestant texts. A central feature of Catholic–Protestant contestation in 1550 is an epistemic debate about where divine law comes from and how one can know its truths. The Catholic epistemology privileges authority of Church officials, Aquinas and other philosophers, and the canon itself. The Council of Trent, for example, defends the authority of Holy Council, a body of religious elites, in all matters of scriptural interpretation.[23] The Protestant epistemology privileges individual judgement, stresses the importance of reading and studying the scripture, and dismisses the need for elaborate rituals.[24] The Protestant theologian Beze maintains "all points of difference may be judged and decided according to the simple words of God as contained in the Old and New Testament, since our faith can be founded on this alone."[25]

Aristocratic dynasticism is the other dominant discursive formation. It does not compete with divine providentialism but rather complements it via the divine right of kings.[26] The aristocratic formation rests on the idea that the nobility is entitled to its privileged position in society because it possesses noble blood. The aristocratic classes live according to dynastic laws and aim to preserve or restore these laws.

In French texts, members of the nobility are referred to as "Princes of the Blood" who trace their bloodlines to ancient races.[27] The importance of blood in discourses of 1550 is evident from the variety of contexts in which it appears. In Christian doctrine, the blood of Christ is used

---

[20] Ridley 1540. For similar findings in a different discursive context, see Keene 2005, 111.
[21] de Beaugue 1556, 7.
[22] de Beaugue 1556, 7.
[23] Council of Trent 1965 [1545], 73.
[24] Anon. 1557, 11.
[25] Beze 1969 [1572] 23.
[26] Indeed, some texts suggest that the laws of men are instituted by God's providence. See Ponet 1556, 2–4.
[27] E.g. Estienne 1576, 12, 14. See Bannister 2000.

to wash away the sins of believers.[28] In medical texts it plays a key role in health: "Blood hath preeminence over all other humours in sustaining of all living creatures, for it hath more conformity with the original cause of living ... being the very treasure of life."[29] Blood was sacred and conveyed special life-giving powers. This idea supported the aristocratic idea that the nobility were entitled to their privilege because of their blood. But blood also legitimated the aristocracy because gentlemen were expected to ride into battle and spill their noble blood in service of the king. This obligation was enforced by a strong conception of honour pursued and maintained "merely for honour's sake."[30] The values of reputation and honour are central both to the gentlemen of the aristocratic classes and to princes and kings themselves.

Marriage and other acts of dynastic law are important means in the aristocratic system. Matrimony can seal alliances and reduce conflicts over succession.[31] But the aristocratic classes also live and die in a world of wealth and arms. The importance of resources (troops, money, arms, and land) in political and aristocratic life is taken for granted. War and the need for war is so naturalized that it is rarely justified or thematized in the discourse at all. Instead, the discourse reveals a concern with raising and maintaining troops, protecting and transporting arms, and so on, almost as a way of life. These are clearly strategic acts, but they are taken as a matter of course and not raised to the conscious level of maxims.

In French texts, there is a clear challenger to divine providentialism: natural providentialism. In this discourse, nature is the creator, provider, and source of all things. In one variant, the stars and celestial objects are connected to the earthly realm as a machine, through unseen forces. Astrology can predict the future not through mystical representation of future events, but because astrology maps the stars, which have a direct impact on earthly motions.[32] Nostradamus lists three sources of pestilence: the air brought by the seasons, God's punishment for sin, and the "stars and constellations."[33] The stars, like God, are posited as a real causal force in everyday life. A biographer of Catherine de Medici notes that her parents gathered astrologers who agreed "in case she lived, she should be occasion of great calamities, and of the final and utter subversion of her family and household."[34] In another

---

[28] Anon. 1557, 26.
[29] Elyot 1539, 16. See also Nostradamus 1559.
[30] de Beaugue 1556, 34.
[31] E.g. England and Wales 1554, 5.
[32] Nostradamus 1559; Ferrier 1593 [1549].
[33] Nostradamus 1559, 8.
[34] Estienne 1576, 8.

variant, all creatures – trees, beasts, and man – are produced by nature and the laws of nature guided the actions of all beings.[35] In one popular account "nature hath provided" better for the beasts than for humankind for whom everything requires "great labour."[36] Animals "hath some natural virtue in their affections, in wisdom, strength, cowardice, clemency, rigour, discipline, and erudition."[37] Natural providence is also evident in vitalist, Epicurean theories of matter, in which all entities possess liveliness.

This counter-discourse was strong enough that Calvin takes the time to specifically mention and refute Virgil and "the filthy dog Lucretius" for claiming that "the world which was created for a spectacle of the glory of God, should be the creator of itself."[38] But Calvin was no mere reactionary to humanist and other forms of knowledge. Indeed, although Calvin opposes the cosmological implications of a discourse that removes God from the seat of divine control, he embraces natural philosophy.[39] Calvin argues that natural philosophy is useful and beneficial because it reveals the "cunning workmanship" of God:

[T]o the searching out of the movings of the stars, appointing of their places, measuring of their distances, and noting of their properties, there needeth art and an exact diligence: by which being thoroughly perceived, as the providence of god is the more manifestly disclosed, so it is convenient, that the mind rise somewhat thereby to behold his glory.[40]

Calvin's position is articulated in other texts as well. One author argues that the value of history is that it shows the truth of God's prophecies and reveals that "acts and punishments be the works of God."[41] But the underlying consensus among both the dominant divine providential formation and the naturalist challenger is of an earthly realm governed by otherworldly forces.

There are a number of prominent cosmological themes and discursive configurations that either disappear or are radically altered by 1815. First of all, the dominant epistemic categories in 1550 were informal modes of knowledge. References to "wisdom," "cogitation," "reason,"

---

[35] Boaistuau 1581 [1558].
[36] Boaistuau 1581 [1558], 19–20.
[37] Boaistuau 1581 [1558], 25–26.
[38] Calvin 1561 [1536], 16.
[39] Calvin 1561 [1536], 16–17.
[40] Calvin 1561 [1536], 15.
[41] Lanquet 1548, 29–30. This position would become widely shared in the seventeenth and eighteenth centuries: the value and importance of natural philosophy was in revealing the workings and mind of God, implanted in nature by him at creation. The conflict here, as in the nineteenth century, is between competing cosmological interpretations of science, not between science and religion. See Gaukroger 2007.

and "experience" are more common than references to "natural philosophy," "physick" (medicine), logic, and astronomy. Ancient authorities are invoked as often as new modes of knowledge.[42] Moreover, two medieval modes of obtaining or establishing knowledge are still common. The first is casuistry, the use of biblical anecdote to establish true principles or moral facts.[43]

Second, there are remnants of the medieval patterning episteme in which knowledge of one thing (e.g. the body) is obtained by mapping that thing to another phenomena (e.g. the zodiac).[44] For example, Figure 3.2 reproduces an image from a primer published by the Catholic Church in 1556. Here, the properties of the body are linked to the properties of the zodiac such that knowledge of one illuminates the other. The implication is that the health of people is connected to the movements of the stars. This mode of analogical or resemblance reasoning is common and taken for granted in the discourse. The figure also demonstrates the ontological principle that the position and movement of the stars directly affects the health and well-being of humans.

While spiritual time is eschatological, the dominant representation of earthly time is cyclical, as in the recurrence of the seasons, the rise and fall of empires, and so on. In an "almanac and prognostication" planting instructions are given based on the revolutions of the planets creating recurring natural conditions each season.[45] One history explains that all "mighty empires" and "great kingdoms" have "decayed." This cyclical nature of history allows us to know "the truth of all affairs that have been done in every age."[46] Moreover, changes in societies, people, and infrastructure are not progressive developments, but mutations, as in alchemy. It is not surprising then that there is no conception of progress or improvement in this discourse. Another text foreshadows the stage theory of human development that would emerged in the Scottish Enlightenment by suggesting that societies shift from migratory to sedentary to commercial to legal and religious communities.[47] But this is not described with the progressive zeal that it will be during the Enlightenment period.

Ontologically, the world is not divided into a dualism of mind and body but into a dualism of spiritual and temporal domains. Catholics hold that sacraments crossed the divide and brought the spiritual into

[42] E.g. Boaistuau 1581 [1558], 13, 18.
[43] E.g. Mornay 1969 [1572].
[44] Catholic Church 1556, 2.
[45] Hubrigh 1553.
[46] Lanquet 1548, 2.
[47] Joannes 1554, 10–13.

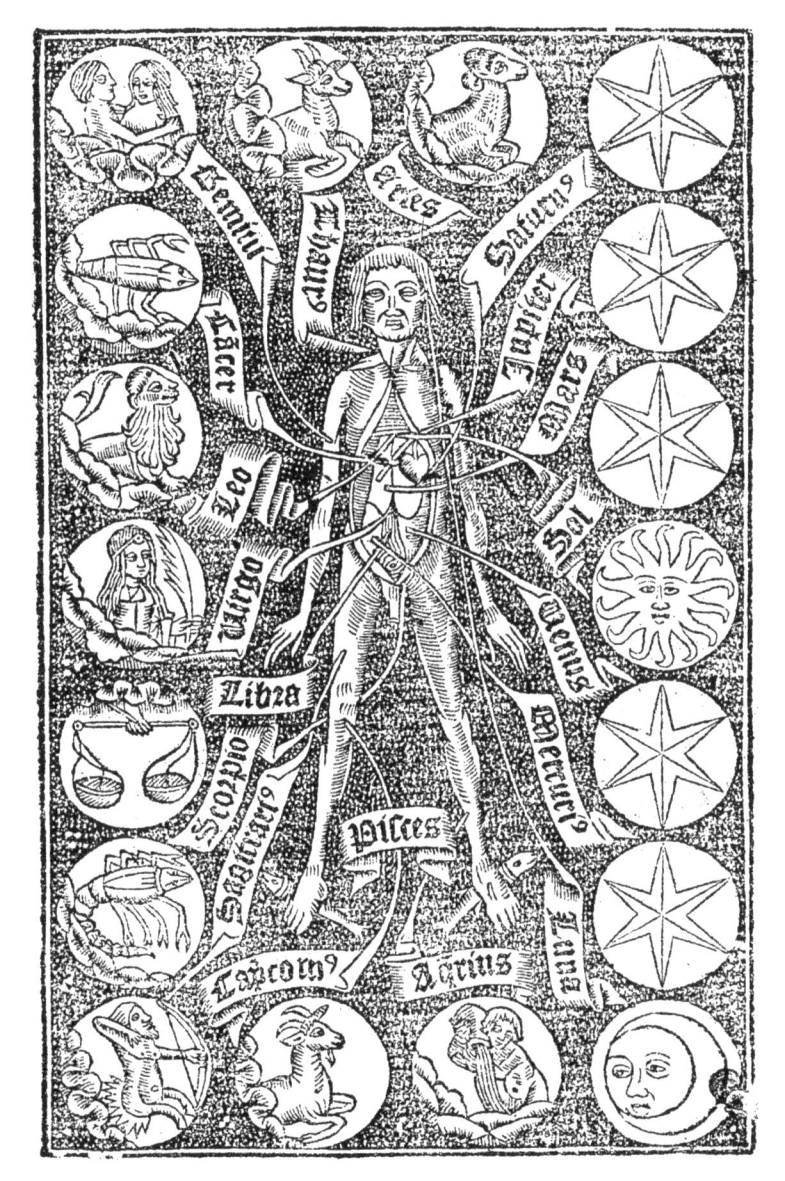

Cosmology and European Political Discourses, 1550 85

Figure 3.2 Image of the celestial realm and the human body (reproduced from Catholic Church 1556, 2)

temporal via transubstantiation. Protestants attack this notion, maintaining the separation of the spheres, but argue that each individual nonetheless had access to the spiritual.[48] The spiritual/temporal distinction in turn supports a disparagement of the temporal and a call to act "without any affection for the world."[49] This is connected to a discourse that criticizes money-oriented action.[50]

Another central ontological theme is elemental naturalism: the world is comprised of earth, water, air, and fire. This Aristotelian doctrine is only expressed explicitly in medical contexts. Nostradamus notes that planetary movements bring bad air, which in turn causes sickness.[51] A medical text explains that elements combine in the body to create "complexions."[52] For good health, the elements must be maintained in "proportion" according to their "natural assignment." Spiritual/temporal dualism and elemental naturalism are interesting because they reveal that there is no division between material and ideal familiar to the modern age. The emergence of materialism in the seventeenth century will allow for a material concept of interest to be formulated but that is not possible in 1550.

In this ontology, there is no clear mechanical concept of cause or causal effect. Mechanism would come to dominate seventeenth-century ontology, and so this is an important baseline finding. "Cause" typically refers to a reason for ("for this cause"[53]) or a stance ("my right and my cause"[54]). Without a clear concept of causation there is no clear idea of mechanical effect, as in the clockwork metaphor that emerged in later centuries.

Finally, divine providentialism and aristocratic dynasticism are both notable in that they are not strategic. They do not prize or feature calculation and, though "design," "management," and other intentionalist categories appear, they are not common.[55] Although "learned men" appear in the discourse, there is little discourse of education, teaching,

---

[48] See Chedsey 1551, 5–9. I see this as creating the bridge to modern mind/body dualism. In the Catholic ontology, individuals can only access the holy/sacred/temporal through sacraments and rituals that make the holy temporal. But in the Protestant view, each individual has the capacity to experience the spiritual realm on their own, via the soul. This capacity of the soul deepens the importance of the spiritual realm in everyday life, but also constitutes a new concept of humanity in conflict with the emergent materialist, mechanist version of humanity as expressed in Hobbes.
[49] Chedsey 1551, 3.
[50] Anon. 1557, 3.
[51] Nostradamus 1559, 4–5.
[52] Elyot 1539, 11–12.
[53] Lanquet 1548, 28.
[54] Coverdale 1550, 11.
[55] Some management is evident in de Beaugue 1556, 28.

or improvement. People have individual "natures," or a divine nature, or a soul, but are not portrayed as growing, improving beings. Labour and hard work are not posited as important means to the ends of life. Only one text advocates "industry" and "labour" and the context is telling: the author argues that only if an individual is born under a certain celestial house will industry and labour lead to profit, otherwise these lead "sooner from thence to loss and damage."[56] Instead, work is portrayed negatively, as God's punishment for the fall.[57] In a world governed by providence, a discourse to govern intentional or strategic action is unnecessary and therefore rare. This of course does not mean people did not act strategically, but it does help to explain why politics looked so different in the sixteenth century.

Religious and Aristocratic Discourses in European Order, circa 1550

After sixty years of war between the Habsburg and Valois dynasties, the 1559 Treaty of Cateau-Cambrésis finally held out the promise of a lasting peace in Europe.[58] Instead, in the following century European politics was overtaken by internal wars of religion that would fuel unrest and conflict on the continent until the Peace of Westphalia. Although the promise of Cateau-Cambrésis was never realized, the treaty illustrates how aristocratic dynasticism shaped state purposes and international order in sixteenth-century Europe.

Cateau-Cambrésis contained two main treaties. The first, between Henry II of France and Elizabeth I of England, recognized the French possession of Calais, which was the last English outpost on the continent. The second, between Henry II of France and Philip II of Spain, solidified Spanish dominance in Italy. The French Valois conceded territory held in Savoy and Piedmont and renounced claims to Milan and Naples in exchange for the rights to three towns, Metz, Toulouse, and Verdun.[59] In France and elsewhere, this must have seemed like a bad deal, but it was to be balanced by two favourable marriages for the Valois dynasty. Henry II's sister would marry the restored Duke of Savoy and his daughter would marry Philip II, creating the possibility of a Valois succession to the Spanish throne.[60]

The Habsburg–Valois rivalry itself and its resolution by two marriages demonstrates the centrality of dynastic discourses and purposes

[56] Ferrier 1593 [1549], 20–21.
[57] Corrozet and Holbein 1549; Boaistuau 1581 [1558], 20–29.
[58] Elliott 1968, 20.
[59] Elliott 1968, 11–17; Bonney 1991, 128–129.
[60] Bonney 1991, 129.

in the sixteenth century. The primary associations in sixteenth-century European political order are not states, but dynastic households or families with non-contiguous territorial claims strewn across Europe. These existed alongside city-states, electorates, duchies, and principalities in a system with multiple, overlapping authorities.[61] So while it would be inappropriate to use a state-centric theory here, an associational perspective allows us to think of all these houses and polities as actors producing and reproducing a transnational political order. These houses competed with one another to expand and aggrandize their families through wars and diplomatic manoeuvres:

> Just as, within states, the great magnates and even less nobles and squires were concerned to win advancement for their own houses by the acquisition of title to great estates, whether by marriage, patronage or seizure, so among states the rulers sought, by skillful matrimony, by ingenious claims, or if necessary by conquest, to add the territories and titles of the royal house.[62]

At this time, the purpose of the dynastic house was to pursue glory in war and dynastic competition with others.

Dynastic means and ends dominated European political order and so it is not surprising that dynastic claims were central to the Valois–Habsburg struggle. The rivalry bankrupted both houses for what amounted to only meagre territorial changes.[63] The Valois had clearly lost the war in 1530, but pressed their dynastic claims in Italy with apparent credibility until 1559.[64] While law could not trump military-economic power, "the pursuit of a ruler's inherited rights ... which may have had no practical application" was a powerful force in sixteenth-century Europe.[65] Moreover, in resolving the conflict, the French Valois were willing to concede territory in exchange for what amounted to a dynastic gambit: favourable marriages would produce family lines that would make the territorial concessions of 1559 a small price to pay in retrospect. After all, dynastic means had bestowed Europe's largest territorial empire since Rome upon Charles V in the first half of the sixteenth century.[66] All the wars of Louis XIV never accumulated as much territory as was bequeathed to Charles by the timely death of well-placed family members. In a dynastic system, marriage was a strategic asset.[67] The hope of a Valois succession in Spain surely motivated Henry to accept the terms of the treaty.

---

[61] Tilly 1992.
[62] Luard 1986, 25.
[63] Luard 1986, 27–28.
[64] Bonney 1991, 80.
[65] Bonney 1991, 80.
[66] Anderson 1998, 89; Yates 1975, 20–27.
[67] Sharma 2005, 15.

Henry II's motivation to reach an agreement, regardless of its poor terms, speaks also to the importance of fiscal and religious constraints on sixteenth-century monarchs. Although the French Valois had increased their revenues and improved their administrative capacity over the course of the first half of the sixteenth century, the interminable pursuit of dynastic claims in Italy had left Henry's war chest depleted and his creditors exhausted. The rise of Calvinism in France also created a domestic threat that demanded Henry's attention. The urgent need to fight heresy and restore legitimacy at home meant the end of a long Valois foreign adventure.

This makes the length and extent of the Habsburg–Valois rivalry all the more puzzling. It is difficult to conclude that the war emerged from any rational, strategic consideration. But the pursuit of dynastic ends to the point of bankruptcy is consistent with a discourse dominated by aristocratic dynasticism embedded in a divine providential cosmological discourse in which neither rational nor strategic thought was encouraged. The discourses of 1550 supported the dynastic quest for familial power and honour without rational restraint or the development of collective norms to restrict violence. Only after the dynastic and providentialist bases of European discourses were eroded by new cosmological ideas would an interest-based discourse oriented to the balance of power emerge.

The same tendencies can be identified in the foreign policy of the sixteenth-century hegemon, the Spanish Empire.[68] In 1556, King Philip II of Spain inherited the world's largest collection of imperial holdings from his father Charles V, who was King of Spain and the Holy Roman Emperor. Charles V left a political testament that offered Philip II strategic advice. Philip II was to: maintain an alliance with his brother, Ferdinand I, the new Holy Roman Emperor; give nothing to the Pope; keep peace and order in Italy; marry into the French royal house; repair the treasury; and defend the Low Countries.[69]

However, Philip II was not a secular strategist in the modern sense. He "believed profoundly that God would provide whatever lay beyond human powers of prediction or execution, and he repeatedly counted on miracles to bridge the gap between intention and achievement."[70] For example, even after he was warned not to engage the English in the famous naval battle of 1588 because his ships were too big and slow, he did so anyway, leading to a calamitous defeat. Moreover, his messianic

---

[68] Parker 1998.
[69] Parker 1998, 80–88.
[70] Parker 1998, 75.

impulses shaped his aims, causing him to adopt over-ambitious goals. Philip II felt he "possessed a direct mandate to uphold the Catholic faith at almost all times and in almost all places."[71] As a result he was unwilling to make peace with Protestants in England or the Netherlands. These costly conflicts contributed to imperial overstretch and the ineffective deployment of resources.[72] Philip's reliance of the "hand of Providence" cost him and Spain dearly.[73] Of course, Philip backed providence with power and planning, but his dependence on God's will "accorded no place to failure" and meant that he did not develop contingency plans and made unnecessary risks.[74]

The Cosmological Shift in Natural Philosophy, 1550–1700

After 1550, the dominant concepts of divine providentialism were reconfigured as cosmological elements from natural philosophy reconfigured religious and Renaissance ideas about matter, nature, and time. Nicolaus Copernicus' (1473–1543) *On the Revolutions of the Heavenly Spheres* was published posthumously in 1543.[75] Copernicus himself was no revolutionary. He worked within the ancient tradition of Ptolemaic astronomy and sought to preserve an ancient, largely Aristotelian image of the universe.[76] With Aristotle, Copernicus argued that the heavens and the earth were distinct realms, but that the motions of the heavens affected motions on earth.[77] But Copernicus was also committed to the Ptolemaic practice of describing the universe in precise mathematical terms. It was this exercise that led him to posit a heliocentric universe that transformed the network of epistemic and ontological beliefs that underwrote early modern European cosmology.

To understand the tremendous effect that *On the Revolutions* had on early modern European thought, we have to take a closer look at pre-modern discourses. As we saw above, the dominant epistemic configuration from the thirteenth through the fifteenth centuries was a patterning or resemblance account of knowledge. The patterning formation

---

[71] Parker 1998, 93.
[72] Parker 1998, 114.
[73] Parker 1998, 107.
[74] Parker 1998, 108.
[75] On the revisionist historiography of the Scientific Revolution, see Cunningham and Williams 1993; Shapin 1996. There is a good case for starting the period here in 1375, when Ibn al-Shāṭir's lunar model, which is nearly identical to Copernicus', was published. Saliba contends that Copernicus must have encountered and built on al-Shāṭir's work (2007, 196–232). See also Ragep 2007.
[76] Kuhn 1957, 135–144; Saliba 2007, 211–215.
[77] Kuhn 1957, 27–28.

ordered the world according to the similarities or resemblances between its elements. Knowledge of the world was not achieved by constructing external representations of reality, but by revealing the non-mechanical connections within the world. Signs and symbols were not separate from the world, but rather they were part of it, given by it.[78] Within the patterning episteme, it did not make sense to provide chronological narration or conduct experiments that assumed isolable, individual entities that move autonomously through time and space. This was because reality was folded in on itself through a complex series of symbolic and organic links between elements.

The patterning episteme was supported by a complex, enfolded view of nature bound up with divine providentialism. Before Copernicus, the world of nature was "still conceived as a living organism, whose immanent energies and forces are vital and psychical in character."[79] If there were mystical and organic connections between all forms of matter and life, then the best way to understand the world was to outline the connections and exchanges among all things. In an enfolded world of patterns, the dominant mode of investigation was akin to hermeneutics or interpretation.[80] The Bible, for example, was a store of historical anecdotes and evidence that revealed God's will. Since God actively intervened in the world, history was "the theater of God's judgment." Casuistry, or reasoning from the case, interpreted texts to find hidden lessons and meanings.[81] Even for some natural philosophers, "nature was something that one learned about from the writings of past masters."[82] Past masters were considered authorities and their texts were deemed to be full of truths.

Copernicus' ideas were transformative not because he aimed to overturn these beliefs, but because his work had interconnected astronomical, scientific, philosophical, and political meanings. His astronomical conjectures were "strands in a far larger fabric of thought."[83] Copernicus' heliocentric universe challenged not just religious ideas about the centrality of humanity, but the whole Aristotelian cosmology upon which Renaissance science and society rested. The Aristotelian universe was built on the temporal-spiritual division of reality into two spheres: a terrestrial realm of change, variety, and decay; and a changeless, eternal celestial realm. Although the two realms had distinct properties, they

---

[78] Reiss 1982, 30.
[79] Collingwood 1945, 95.
[80] Gaukroger 2007, 132–148.
[81] Walzer 1971, 75.
[82] Gaukroger 2007, 136.
[83] Kuhn 1957, 77.

were connected to one another by space-filling spheres. It was a "full universe," without vacuums, in which planets and humans alike were connected by the movements of invisible space-filling spheres. The stability and centrality of the earth played a key role in this account. The earth provided the stable foundation necessary to sustain celestial motion.[84] By the same principle, motions on the earth affected the heavens and vice versa. In the full, two-sphere universe, the motions of the heavens could cause changes in the terrestrial sphere. Therefore, predicting the motions of the heavens could predict the future. As a result, astronomy was intrinsically linked to astrology.[85] The Aristotelian universe was compatible with the idea that God controlled the eternal and changeless heavens, manipulating outcomes in the terrestrial region of decay and variety.

In the sixteenth and seventeenth centuries, Copernicus' conjecture was used to destabilize the ontological and epistemic presuppositions of the Aristotelian cosmology. His idea of a moving earth inspired a new view of the universe. The Italian philosopher Giordano Bruno (1548–1600) soon pointed out that Copernicanism implied that there was no difference between the celestial and terrestrial spheres.[86] While Bruno himself was persecuted, in the seventeenth century the two-sphere distinction was slowly eroded and replaced with an infinite universe created by an infinite God in which every star was a sun.[87] Galileo (1610) extended Copernicus' findings by representing the moon as a body in motion just like other earthly bodies. Descartes' atomistic theory of vortices (1644) was an attempt to develop a new theory of motion consistent with Copernicus' theory of the universe. Hooke's (1666) cosmology radicalized Copernicus, reducing the movements of all bodies to uniform motion.[88]

Furthermore, Copernican debates challenged the patterning episteme and its forms of interpretation. First, Copernicus and Bruno portrayed nature as a "machine" governed by formal and efficient causes.[89] Copernicus depicted a universe in mechanical motion, which Bruno placed in the ether: an "all-embracing and unchanging substance, the matrix of all change."[90] This contributed to the erosion of the organicist, enfolded view of nature that supported the patterning episteme. Once all

---

[84] Kuhn 1957, 78–84.
[85] Kuhn 1957, 92–93.
[86] Collingwood 1945, 99.
[87] Kuhn 1957, 233. See also Koyré 1957.
[88] Kuhn 1957, 218–250.
[89] Collingwood 1945, 97.
[90] Collingwood 1945, 99.

of nature was portrayed as a machine, scientific discourse was able "to grasp an exterior, coming to view itself as a simple translation of objects into a conceptual order."[91] Astronomy and natural philosophy more generally aimed to map reality with a system of quantitative signs. These developments had two epistemic consequences. First, nature itself, and especially the mathematical investigation of nature, came to be considered the source of legitimate knowledge.[92] Second, there arose a host of ideas about how to ensure a true or objective relation between sign and referent. That is, Copernicus and others advanced a representational episteme in which the role of knowledge is to map reality from an objective, external standpoint.

Above all, it was Isaac Newton (1643–1727) who provided the image of a universe in uniform motion, guided by natural laws, divinely created and controlled.[93] Newton's *Principia Mathematica* set out to describe the motion of bodies in geometrical terms. Newton aimed to build a "rational mechanics" that would expound the "science of motions resulting from any forces whatsoever."[94] To this end, Newton presented universal laws of motion to explain the behaviour of earthly and celestial objects alike. Newton's cosmos is a universe of "bodies possessing extension, figure, number, motion, and rest."[95] In this world, time is uniform, absolute, mathematical time.[96] This concept of time is linked to the idea of progress as the unfolding of absolute time, regardless of our understanding of it: "the true, or equable progress, of absolute time is liable to no change."[97]

Newtonian cosmology shared important features with Copernican and Cartesian views. However, Newton explicitly rejected the core tenet of mechanism. Speaking of the complex paths of comets and planets through space, Newton declared "it is not to be conceived that mere mechanical causes could give birth to so many regular motions."[98] The "beautiful system" could only "proceed from the dominion of an intelligent and powerful being."[99] So Newton articulated a variant of divine providentialism. For his critics, this amounted to an explanation dependent on "occult" forces, and was thus unfitting of the label natural philosophy.[100]

---

[91] Reiss 1982, 141.
[92] Gaukroger 2007, 132–145.
[93] Crombie and Hoskin 1970.
[94] Newton 1934 [1729], xvii.
[95] Collingwood 1945, 107.
[96] Newton 1934 [1729], 7–8.
[97] Newton 1934 [1729], 8.
[98] Newton 1934 [1729], 544.
[99] Newton 1934 [1729], 544.
[100] Crombie and Hoskin 1970, 49–51.

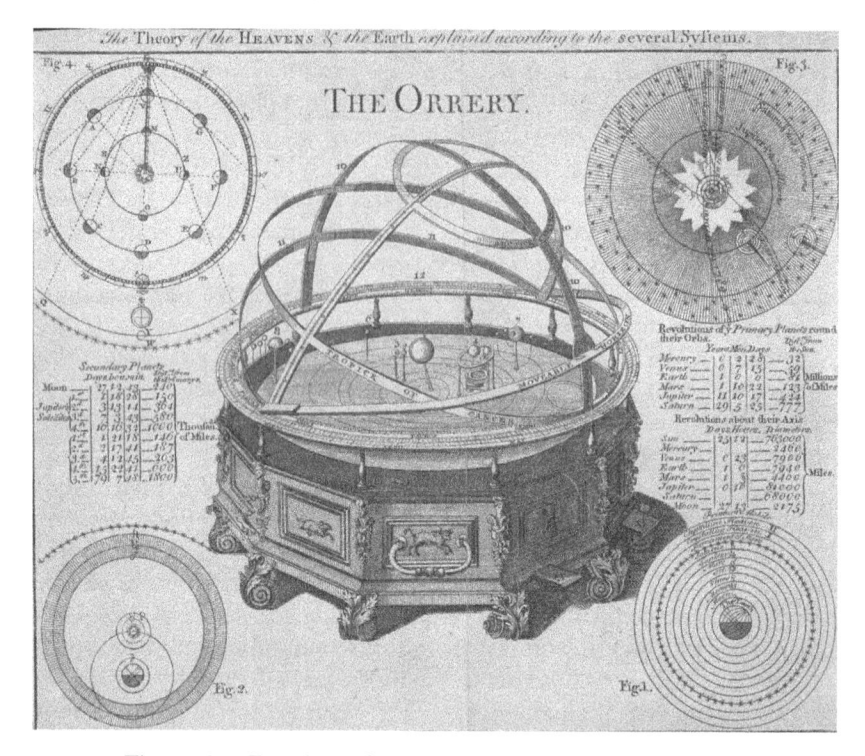

Figure 3.3 Drawing of an orrery (reproduced from *The Universal Magazine*, 1749, 48–49)

Yet, the subtleties of Newton's cosmological disputes with Descartes and others were often ignored. Newtown's system was taken to demonstrate a determinist mathematical model of the universe and he has been portrayed as the father of the clockwork image. Throughout the eighteenth century, the clockwork image of the universe was vividly reproduced in the form of the orrery (Figure 3.3). An orrery is a mechanical model of the planets constructed using the same techniques as clocks. They were mostly used to teach astronomy and demonstrate wealth. The orrery depicted in Figure 3.3 provides an interesting counterpoint to Figure 3.2. Whereas Figure 3.2 linked the zodiac and the planets to the human body, here they are incorporated into a machine, demonstrating the rise of mechanical thinking.

In this context, the "Newtonian method" became synonymous with the search for natural laws or "dispositional regularities" in the world, mechanistic or otherwise.[101] Newton's name was invoked in countless

---
[101] Rudwick 2005, 103.

attempts to advance knowledge in areas as diverse as history, political economy, and geology.[102] These attempts sought to specify the predictable, determinist principles that were believed to lie beneath all reality.[103] Newtonians believed that uncovering these laws or regularities was the first step to revealing the operation of a harmonious, orderly system that would explain phenomena.

While Newton himself posited a central role for God in his system, Newton's followers articulated a form of natural providentialism. On this view, God created the laws of the universe, which determine all events via mechanical causation. However, God had receded to heaven and was no longer actively intervening in human affairs. As a result, Newton was interpreted as revealing the will of God by uncovering the principles that controlled His creation.[104] The Newtonian world was often portrayed as the product of otherworldly forces, accessible to natural philosophy, but not yet subject to human control. However, the determinist implications of Newtonian laws meant that there was little room for a providential God century that intervenes in daily life. God was relegated to the role of a perfect, divine designer of a clockwork universe.[105]

From Copernicus to Newton, natural philosophers constituted a cosmological discourse with three main elements: a materialist and mechanist ontology; a representational episteme; and a new concept of time as an absolute, open plane. Natural philosophy also redefined humans, not as God's subjects, but as reasoning, knowing beings capable of building cumulative knowledge of the world. While this new cosmological configuration was rendered compatible with divine providence, it was used to challenge and reconfigure political discourses. This cosmological shift in natural philosophy soon rippled through the fabric of thought, altering ideas in economic thought, political philosophy, and practical administration.

Strategic Deployment: Aristocratic Brokers and Natural Philosophy, 1600–1700

Moving to the meso-level, we can see that the cosmological shift inaugurated by Copernicus was slowly imported into European political discourse by a series of strategic deployments. In this case, recursive institutionalization followed a horizontal pathway in which materialism and mechanism were spread amongst European states by the transnational

---

[102] Buchwald and Feingold 2013; Redman 1997; Rudwick 2005.
[103] Rudwick 2005, 136.
[104] Gaukroger 2007, 455–457, 508.
[105] Shapin 1996.

aristocratic class that financed and consumed natural philosophy.[106] The European aristocracy, though spread across divided territories and principalities, shared a common language, lifestyle, and outlook. They were bound together by dynastic familial ties and communication networks along which new ideas and political ideologies could spread. In this context, scientific ideas were channelled into European political discourses via informal contacts and discussions between natural philosophers and their friends and patrons in the salons and dining rooms of the European aristocracy.

Rulers and state officials financed and appropriated new ideas from natural philosophy for a variety of reasons. First, deployments often took the form of explicit efforts to improve the military or economic abilities of states in order to compete with others. For example, European militaries imported scientific epistemes and ontologies when they, like Maurice of Nassau, adopted "close order marching" that arranged soldiers into ordered, clean, geometric formations.[107] In addition, European states imported mechanistic thinking when they, like Jean-Baptiste Colbert, undertook engineering schemes to improve fortifications, build canals, and advance shipbuilding techniques.[108]

However, we cannot explain the associational changes that drove the rise of materialism and mechanism in early modern Europe with instrumental conceptions of science alone. First, through the seventeenth century and into the eighteenth century, science, especially of the cosmological sort, was simply not that useful in consequentialist terms.[109] However, science was important as a source of legitimation for European monarchs. Louis XIV, for example, used his relationship with astronomers and natural philosophers to justify the absolutist power of a mercantilist state in epistemic terms.[110] In this way, rulers sought to convince their courts, citizens, and rivals that the state "was capable of protecting its people and territory and of defending its corporate identity and honor, that it was organized rationally and therefore was likely to behave predictably."[111] Thus, states drew on the authority of scientific cosmology to legitimate their rule and bolster their prestige.

Finally, in some cases, deployments took the form of transpositions from transnational scientific networks that were not necessarily motivated

---

[106] Bukovansky 2002, 77–82; Bonney 1991; Koenigsberger 1987; Wuthnow 1979.
[107] McNeill 1982, 125–127. See also Parker 1988.
[108] Brown 2016; Guerlac 1986.
[109] Gaukroger 2007, 472; Wuthnow 1979, 222.
[110] Burke 1992, 53–54; Lebow 2008, 301–302; Wuthnow 1979, 224. See also Mukerji 1997.
[111] Wuthnow 1979, 225.

by competition, coercion, or learning. Here, state officials acted as brokers who were embedded in both scientific and political associations. For example, Francis Bacon and William Temple were polymaths that made contributions to natural philosophy, medicine, and political thought.[112] Further, Bacon served as Chancellor of the Exchequer in the early seventeenth century and Temple served the Queen Elizabeth's secretary. In such roles, Bacon, Temple, and others imported ontological, epistemic, and other cosmological ideas into political discourses both consciously and unconsciously. The importation of natural philosophy into political discourses could also be more indirect. For example, Hobbes explicitly deployed mechanist ideas as the foundation of his political theory. While he was not directly involved in politics, Hobbes was widely read amongst the aristocratic ruling classes and likely played a key role in disseminating natural philosophic ideas.[113]

In this vein, England and France both established national academies of sciences comprised of "gentlemen scientists" who circulated in networks of power and privilege. In England, the Royal Society of London for Improving Natural Knowledge was established in 1660 by King Charles II. In France, the Académie des Sciences was founded in 1666 by Louis XIV. These academies maintained close contact with state officials who attended experiments and read their official journals.[114] One estimate suggests that one-third of the founding members of the Royal Society were crown officials.[115] Such close ties between states and scientific academies surely meant that state imperatives shaped the development of natural philosophy. However, these societies maintained their autonomy because their members had independent sources of legitimacy and wealth.[116]

By various means and to various ends, the new ideas in natural philosophy circulated widely throughout early modern Europe. Dynastic states imported these ideas through a series of strategic deployments designed to solve military and political problems. These strategic deployments unintentionally reconfigured early modern European discourses by introducing new epistemes, ontologies, and other cosmological elements that made possible new ways of thinking about state purpose.

---

[112] In the language of Science and Technology Studies, these actors served as "boundary objects" that bridge communities of practice. Here, Bacon and Temple bridge natural philosophy and political thought, moving concepts and ideas back and forth. See Star and Griesemer 1989.
[113] Rogers 2007; Shapin and Schaffer 1985.
[114] Burke 1992, 53–54; Shapin and Schaffer 1985, 131–139.
[115] Wuthnow 1979, 220.
[116] Wuthnow 1979, 217–219.

Discursive Reconfigurations: Cosmology and the Balance, 1600–1800

In the early seventeenth century, the strategic deployment of cosmological ideas from natural philosophy reoriented state purposes in a series of recursively institutionalized shifts. First, state purpose was reconceptualized in the language of interests. Second, state interests were defined in material terms as taking up a position within the European balance of power. At the time of Utrecht, to act rationally as a statesman was to follow the rational maxims of the balance. Throughout the eighteenth century, the balance was legitimated and naturalized as a natural law. However, the discourse of the balance would shift again by the early nineteenth century.

The Rise of Interests, 1630–1713

Over the course of the seventeenth century, the dominance of dynastic state purposes gave way to the rise of "interests."[117] For Bartelson, the rise of interests was made possible by a new representational episteme that displaced the patterning or resemblance episteme.[118] The analysis of state interest "proceeds by orderly enumeration and the representation of differences between individual objects, whose empirical reality and identity it thereby establishes; it renders them comparable within a taxonomy; it renders their interests calculable by measuring their differences on an ordinal scale."[119] Once states are defined as isolable, self-contained objects they can be placed in a table of comparison with other states, revealing their similarities and differences.[120] A table of similarities and differences, Bartelson argues, permits the science of states to proceed as an "analysis of interests" in terms of geography, resources, religious composition and so on.

The analysis of interests would have been impossible in the patterning episteme. First, the patterning episteme had no concept of representation as mapping a sign onto reality. Moreover, it did not order objects into comparative tables, but into relations of analogy and resemblance. The analysis of interests drew on the new representational episteme on which true knowledge was founded on accurate depictions of external reality. Interests were objectively given properties that could be mapped by knowledge of geography, resources, and so on. As such, the discourse

---

[117] Bartelson 1995; Skinner 1996.
[118] Bartelson 1995.
[119] Bartelson 1995, 171.
[120] On this story, see Branch 2013; Ruggie 1993; Walker 1993.

of interests also drew on the new materialist ontology, which was used to represent the state and its constituent elements in material terms.[121] The analysis of interests disrupted the providential and dynastic basis of early modern politics both by constituting states as distinct, separable sovereign entities and by defining interests as legible and predictable. If the interests of a state are externally given and knowable, then "[f]oreign policy is less prey to the whims of Fortuna" because "the motives of the other become more transparent."[122]

The discourse of state interests emerged in sixteenth-century France from a reappropriation of the Renaissance idea of reason of state.[123] Renaissance political philosophy in the fifteenth and sixteenth centuries challenged medieval political discourses that were dominated by religious ideas.[124] Works of political philosophy in this religious tradition advised princes and kings to act as good Christians, allowing God's providence to bestow glory on the kingdom.[125] This advice was in contrast to the Renaissance reason of state tradition. In this tradition, Machiavelli, Botero, Guicciardini, and others offered a new set of maxims that were derived without regard for God, morality, or justice.[126]

French political thought introduced the concept of state interests by combining the Renaissance tradition with a new rationalist episteme drawn from natural philosophy. French political philosophy in the 1620s and 1630s was centred on the royal house and its first Minister, Cardinal Richelieu.[127] Richelieu's political testament unites the power of reason and the idea of state interests. Richelieu argues that the greater and more important a man is, the more he should avail himself of "the masculine virtue of making decisions rationally."[128] Thus, sovereigns should "make reason sovereign."[129] For Richelieu, reason is God's endowment to man: "man, having been endowed with reason, should do nothing except that which is reasonable, since otherwise, he would be acting contrary to his nature and by consequence, contrary to Him."[130] A prince is rational when he

---

[121] Mukerji 1997.
[122] Bartelson 1995, 181.
[123] Keene 2005, 109–114. See Devetak 2011; Skinner 1978; Skinner 1996; Tuck 1993. Hobbes is often given credit here, but he was influenced by French thought when he incorporated self-interest into his political philosophy. Hobbes' earlier work argues that people are primarily motivated by reason and it is only after his period of exile in France that Hobbes becomes concerned that self-interest may override reason. See Skinner 1996, 426–428.
[124] Keene 2005, 110. See Skinner 1978 on the transition.
[125] Keene 2005, 111.
[126] Keene 2005, 113–116.
[127] Skinner 1996, 426.
[128] Richelieu 1961 [1635], 95.
[129] Richelieu 1961 [1635], 71–72.
[130] Richelieu 1961 [1635], 71.

knows how to "find the right instant to attain his ends," which should be aligned with the "interests of the state."[131] Interests are not defined in material, strategic terms, but as the prudential pursuit of "glory."[132]

Louis XIV, like Richelieu, equated interests with the glory of the state. In his advice to his heir, Louis XIV argues that glory is the highest goal and that the primary means to glory is reputation. But "reputation cannot be sustained without everyday acquiring a greater [one] ... glory is a mistress whom one is never able to neglect."[133] He cautions his heir not to pursue personal glory for its own sake because glory is derived from the state, so one should always act in the "interests of the greatness, the welfare, and the power of the state."[134] So, what it means to act in the name of glory is to act in the state's interests, which in the end serves the ruler's true personal interests. This expresses the dynastic worldview, in which the glory of the ruler and his house underlies the interests of the polity.[135]

In the 1630s, Henri Duke de Rohan, a general in the Huguenot armies during the *Fronde*, documented his life in a celebrated memoir and wrote an influential treatise on the interests of the Christian states. Rohan, like the thinkers of the Italian Renaissance, sought to offer practical advice that would help ensure the survival of the state in turbulent times.[136] Rohan begins his memoir by declaring that "[t]he Princes command the people, and the Interest commands the Princes."[137] Discursively, he places interests outside and "above" princes, so that they are ruled by "true interests." This marks a transition away from the personalized political discourse on which princes themselves have interests, connected to their dynastic condition. Rohan argues that there are two poles in Christendom, Spain and France, "from whence descend the influences of peace and war on other states."[138] States pursue their designs based on "intelligence" and "counsel" and are successful when

---

[131] Richelieu 1961 [1635], 95. See Richelieu 1988 [1625], 5.
[132] Richelieu 1988 [1625], 5.
[133] Louis XIV 1924 [1666–1679], 59.
[134] Louis XIV 1924 [1666–1679], 169. On French state formation in this period, see Mukerji 1997. It is important to note, with Mukerji, that even though Louis XIV consciously thought of state interests in ideational, relative terms as "glory," he and Colbert did more than anyone else to displace this view with a materialist understanding of the state. Their territorial ambitions and courtly displays introduced a whole new discourse of state power centered on the cultivation of the land. This provides an important link in the story between the orientation to interests and the orientation to the balance of power in territorial terms that dominated at Vienna. See also Branch 2013; Carvalho 2016.
[135] Blanning 2007, 541, 547.
[136] Keene 2005, 107.
[137] Rohan 1640, 5.
[138] Rohan 1640, 6.

these produce "good maxims."[139] Rohan's own counsel is for states to follow their "true interests" as constituted by the geographic, religious, monetary, and military composition of their states. Spain, for example, must manage its monks and preachers, diplomatic relations, money, arms, and reputation. France "by nature" is "carried to make a counterpoise" to Spain. Here the analysis of interests points directly to France's interest in balancing behaviour. The term counterpoise refers to one side of a device for determining weight. This is not strictly speaking a term drawn from natural philosophy, but by the mid-seventeenth century, the counterpoise was an important experimental instrument.[140] The metaphor of a balance rooted in weights and counterweights gathered increased significance as the Copernican ontology of mechanical motion rose in prominence.

So, in Rohan, which exemplifies a broader seventeeth-century tradition, balance of power thinking emerges from an examination of state interests.[141] The image of the balance introduced a "stabilization principle … to reassure states that they are unlikely to be extinguished in the near future."[142] The new balancing discourse drew not only on the rationalist episteme, but on the new mechanical and materialist ontology. By the middle of the sixteenth century, political philosophers no longer took for granted the Aristotelian ontology in which every entity has an essence or nature that guides its development and actions. Instead, they portrayed the world in mechanical and quantitative terms as matter in motion. Thus, political communities were no longer organisms or natural bodies oriented to divinely derived natural laws. Instead, states were bodies defined in mechanical and quantitative terms.[143] As such, their interests were to be derived from an understanding of their environment, not abstract philosophizing. From here, as we shall see, it was a natural step to the idea that state interests could be calculated.

The calculability of interests within the science of states made possible new articulations of balance of power thinking.[144] States would understand their political environment by measuring and assessing the relative power of other states. As Bartelson puts it, the construction of a table of interests reinforced "a principle of ordered opposition" now understood in quantitative terms as units of power.[145] That is, the construction

---

[139] Rohan 1640, 6.
[140] See Boyle 1666. See OED 2011, "counterpoise."
[141] Tuck 1993, 96.
[142] Keene 2005, 109.
[143] See also Branch 2013.
[144] Bartelson 1995, 181.
[145] Bartelson 1995, 181.

of a balance of power institution depended on shared ideas about what constitutes a state, the elements of state power, and the objective, external interests of a state that were made possible and desirable by the new cosmology. However, the full-fledged reduction of power to quantitative units was not possible in the seventeenth century. First, states did not have the bureaucratic capacity to collect complete data on their populations. Statistics in the seventeenth century amounted to little more than guesswork by individuals.[146] Second, states only absorbed the idea that the power of the state consisted in its land and its people slowly over the course of the seventeenth century.[147] So, the full implications of the analysis of interest were not realized until the eighteenth century when transnational statistics of population, commerce, and so on became available.[148]

European Discourse in the Treaty of Utrecht, 1713

The new discourse of state purpose oriented to interests and the balance of power was institutionalized in international order in the Treaty of Utrecht in 1713. The War of Spanish Succession (1701–1713) proved to be the last of Louis XIV's wars. Louis had won glory on the battlefield and gained territory in the War of Devolution (1667–1668), the Dutch War (1672–1678), the War of the Reunions (1683–1684), and the Nine Years War (1688–1697). These conflicts established France as the pre-eminent power in Europe and expanded French borders eastward to the Rhine and northward into the Spanish Netherlands. Louis had little desire for war in 1701 but could not resist the possibility of bringing the vast and wealthy Spanish inheritance into the House of Bourbon. The problem of the Spanish inheritance was discussed by the European powers from 1697 until the King of Spain's death in 1701. However, all attempts at a peaceful negotiated partition of Spanish possessions foundered on the absence of an acceptable and still living heir to the Spanish throne.[149]

Thus, when the King's will left the whole of the Spanish possessions to Louis XIV's grandson Philip, Duke of Anjou, Louis was tempted to defend the inheritance with intimidation, prestige, and, ultimately, force. In the ensuing conflict, Britain, the Dutch Republic, and the Austrian Habsburgs resumed the alliance from the Nine Years War to force a

---

[146] Hacking 1990, 106–107.
[147] Mukerji 1997; Carvalho 2016.
[148] To my knowledge, these were first made widely available in the Almanac de Gotha beginning in 1763. I thank Benjamin de Carvalho for this information.
[149] Clark 1970, 395.

partition on Louis and Anjou. The war dragged on without a clear victor until war weariness accelerated negotiations.[150] The resultant peace at Utrecht in 1713 divided the Spanish inheritance between the Duke of Anjou and the great powers. Most importantly, the alliance powers obtained an agreement from Louis that his grandson, who would be permitted to remain on the Spanish throne, would never accede to the French Crown.

Whereas the Peace of Westphalia in 1648 emphasized religion, tradition, legality, and loyalty, Utrecht gave primacy to rationality, reciprocity, and the "interests" of monarchs.[151] The treaty's text institutionalized the balance as a core principle of international politics. In the treaty, each of the monarchs of Europe take turns paying homage to the goal of "[p]eace, and securing the Tranquillity of Europe by a Balance of Power," or the "universal Good and Quiet of Europe, by an equal weight of Power."[152]

The collective orientation to the balance was supported by the rationalist postulate that the laws of the balance could be known and should be used to inform political action. For example, in his renunciation of the French throne in the treaty, Philip V of Spain suggests that "right reason does persuade us" to accept the "fundamental and perpetual Maxim of the Balance of Power in Europe."[153] The word "maxim" is important in this context. It means "a self-evident proposition" and its etymological origin is in mathematical reasoning.[154] So to take one's place in the balance of power system meant to use the power of reason to discern and submit to the laws of the balance. The cosmological backdrop here is still fundamentally providential because humans are not in control of the universe. Instead, wise statesmen understand the operations of balance of power through reason and accede to its dictates.[155]

The discourse around the treaty merges the concept of interest with the idea of the balance. Diplomats and politicians supposed that states had "natural interests" which could be discerned by reason.[156] Such interests were to reflect the changing realities of the balance of power. Thus, the purpose of the state in international order was to pursue foreign policies "dictated by the relationships between natural interests and the balance of power."[157] In British and Dutch circles, such policies were

[150] Pitt 1970.
[151] Osiander 1994, 103.
[152] Treaty of Utrecht 1973 [1713], 184, 187.
[153] Treaty of Utrecht 1973 [1713], 47, 187.
[154] OED 2011, "maxim."
[155] Black 1983, 56.
[156] Black 1983, 56. See also Andersen 2016, 71–73.
[157] Black 1983, 56.

used to contest Austrian and French aggrandizement that recalled the old threat of Universal Monarchy.[158] So the balance referred to an equality of power between France and Austria on the continent. This articulation defined the British interest in serving as an offshore balancer to check French or Austrian power.[159] In this sense, the allied powers desired an "even balance."[160] Such a balance was portrayed as in the "public interest" of maintaining European order and was counterposed to "private interests" that threatened to undermine it.[161]

Bolingbroke, who represented the Tory government at Utrecht, presented a forceful defence of this policy. Bolingbroke's articulation of the balance of power provides a bridge between the seventeenth-century discourse, in which interests were defined in terms of glory, and the eighteenth-century discourse, in which interests were defined by the balance. Glory is an important concept for Bolingbroke, but interests are primary: "[the] glory of a nation is to proportion to the ends she proposes, to her interest and her strength; the means she employs to the ends she proposes, and the vigor she exerts to both."[162] In turn, the interests of states are determined by the dynamics of the balance:

The two great powers, that of France and that of Austria, being formed and a rivalship established by consequence between them; it began to be the interest of their neighbours to oppose the strongest and most enterprising of the two, and to be the ally and friend of the weakest.[163]

Bolingbroke uses the authority of nature to illuminate the operation of the balance. His conception recalls the balance-as-counterpoise metaphor, but points in the direction of the Newtonian balance: "the precise point at which the scales of power turn like that of the solstice in either tropic, is imperceptible to common observation."[164] Thus, on his view, one must look behind appearances to the true causes and consequences of the balance, much like a natural philosopher. When Queen Anne set out to defend the peace to the English people she states, "Nothing, however, has moved me from steadily pursuing, in the first place, the true interest of my own Kingdoms."[165]

---

[158] Andersen 2016, 79–87.
[159] Schroeder 1994b, 142. Schroeder goes so far as to say that the "balance" here and elsewhere is really a principle of British hegemony.
[160] Clark 1970, 397.
[161] Anderson 1993, 169; Andersen 2016, 79.
[162] Bolingbroke 1932 [1735–1736], 95.
[163] Bolingbroke 1932 [1735–1736], 17–18.
[164] Bolingbroke 1932 [1735–1736], 32.
[165] Queen Anne 1904 [1712], 51.

The War of Spanish Succession and the Treaty of Utrecht did not rest solely on rationalist balance of power principles.[166] The balance was also legitimated with reference to legal and religious principles rooted in republican thought.[167] King William III of England and Stadtholder in the Dutch Republic aimed to defend "the liberty of Europe," consistent with the republican ideal of securing autonomy from domination.[168] But William also sought to protect the Protestant religion from a Catholic power.[169] Similarly, the Treaty of the Grand Alliance was concerned with French and Spanish capacity to "oppress the liberty of Europe."[170]

Thus, in 1713 the balance only vaguely refers to a systemic arrangement of powers on the continent. The central idea of the balance at Utrecht is of two powers or sides equally poised as between two scales. States are aided in finding and establishing this balance by maxims of reason and natural law. In Butterfield's terms, the balance was not yet conceptualized in Newtonian terms as unfolding within a field of forces.[171] We can see here the lag between cosmological shifts in scientific discourses and their effects on international discourses. Although it had been twenty-five years since the publication of Newton's *Principia*, his central ontological, epistemic, and cosmological ideas were not yet dominant in political discourses. Instead, international discourses reflected the earlier, more general claims that the world could be rationally known and that leaders should obey maxims derived from reason and observation. The displacement and reconfiguration of European discourses was a slow process.

Naturalizing the Balance of Power in the Eighteenth Century

Over the course of the eighteenth century, the principle of the balance was defended and naturalized in Newtonian terms. The Newtonian model of orderly, law-governed systems permitted an understanding of the balance of power as a lawlike regularity produced by a deeper order. The ontological and epistemic shifts introduced by Newtonian

---

[166] Schroeder (1994b) concludes, for example, that while seventeenth-century states did sometimes exhibit balancing behaviour, they often did so only after hiding and bandwagoning failed. For him, balancing was not a rational outcome or one forced upon states by structural constraints, but rather, was the generic result of "a general free-for-all scramble in international politics" (1994, 140).
[167] Andersen 2016, 74–81. See also Deudney 2007; Haslam 2002; Vagts 1948.
[168] Lossky 1970, 156.
[169] Lossky 1970, 157.
[170] Lossky 1970, 157.
[171] Butterfield 1966.

thinking made possible a new concept of the balance guided by the metaphor of the planets in motion. This metaphor and its associated conceptual framework was not the only set of ideas used to theorize the balance in the eighteenth century, but the arrival of the vocabulary demonstrates the effects of new scientific thinking on European political discourses.

M.S. Anderson's survey of eighteenth-century pamphlets reveals the emergence of Newtonian thinking over the course of the century. As at Utrecht, the balance was initially described as "a Maxim of true Policy" or as a "law."[172] However, by the middle of the eighteenth century, the idea of the balance as a law was based on an analogy with Newtonian laws. As one pamphlet summarized:

> What gravity or attraction, we are told, is to the system of the universe, that the balance of power is to Europe: a thing we cannot just point out to ocular inspection, and see or handle; but which is as real in its existence, and as sensible in its effects, as the weight is in scales.[173]

Just like gravity in Newton's rational system to explain the movement of bodies, balancing was depicted as a "maxim in every rational system of politics."[174]

The new line of thinking reconfigured the basic claims of the discourse on display at Utrecht. Whereas the balance at Utrecht was based on the image of the scales, a more dynamic and systemic understanding of the balance emerged in the middle of the eighteenth century.[175] In Britain, David Hume, Edward Gibbon, and Adam Smith all offered "mechanistic" theories of the balance.[176] For each thinker, interaction and competition between atomistic states produced a general order.[177] The balance embodied the harmony generated by the pursuit of self-interest within a mechanical field.[178]

Balance of power ideas were not confined to British thought. In France, Enlightenment thinkers like Fénelon, Montesqieu, and Pecquet

---

[172] Anonymous pamphlet quoted in Anderson 1993, 164; J.J. Lehman 1716 quoted in Anderson 1993, 165.
[173] Anonymous pamphlet quoted in Anderson 1993, 167–168.
[174] Gentz 1801 quoted in Anderson 1993, 165. Gentz was a clerk for Metternich at Vienna.
[175] Andersen 2016, 144. Andersen posits that the systemic view emerged at Utrecht, but I think the textual evidence for that claim is weak. The image of the scales which dominates the texts is systemic only in a simplified, implicit sense. Diplomats and politicians themselves only really start to use the word "system" later in the eighteenth century, which is consistent with Andersen's history there.
[176] Ashworth 2014, 58.
[177] Ashworth 2014, 55–57.
[178] Ashworth 2014, 57.

articulated the balance in mechanical terms as an "equilibrium."[179] The Swiss jurist Vattel combined a newly secularized and physical understanding of natural law with reason of state discourse into a prescription that states manage the balance of power order.[180] Vergennes, French minister of foreign affairs under Louis XVI, rooted European equilibrium in the shifting patterns of power between the great powers.[181] Frederick the Great of Prussia compared the regular turns of the balance to the workings of a watch.[182] Prince Kaunitz, Austrian foreign minister from 1753 to 1793, believed that the balance of power was rooted in a law of nature and that it operated automatically.[183]

The idea of a Newtonian balance regulated by a hidden order is clearly expressed in the writings of Henry Brougham. Brougham was a polymath with interests in natural philosophy but who pursued a political career. In the 1790s, he intervened in scientific debates on optics, defending Newton's particle-based views against the new wave theory of Thomas Young.[184] By the turn of the century, Brougham had emerged as a prominent Whig politician, intervening in debates on foreign policy issues. He was elected to the House of Commons in 1810 and rose to the position of Lord Chancellor under Prime Minister Grey. In his early writings, he brought a distinctly scientific perspective to balance of power theory. In an 1803 essay, Brougham responded to those who dismissed the doctrine of the balance as mere ideology by legitimating the balance in scientific terms. Brougham compared the development of balance of power theory in the eighteenth century to the process of discovery in which the "planetary motions" were "brought to light."[185] Just as the "law of gravitation" keeps "each body in its place, and preserves the arrangement of the whole system," the balance of power:

[R]egulates the mutual actions of the European nations; subjects each to the influence of others, however remote; connects all together by a common principle; regulates the motions of the whole; and confining within narrow limits whatever deviations may occur in any direction, maintains the order and stability of the vast complicated system. As the newly-discovered planets are found to obey the same law that keeps the rest in their orbits; so the powers, which frequently arise in the European world, immediately fall into their places, and conform to

---

[179] Wright 1975, 81; Sheehan 1996, 2. Andersen suggests that the "equilibrium" discourse emerges after 1815, but it is clearly evident in French thought throughout the eighteenth century.
[180] Devetak 2011.
[181] Wright 1975, 83.
[182] Anderson 1993, 167.
[183] Anderson 1993, 167.
[184] Cantor 1983, 78.
[185] Brougham 1872 [1803], 11.

the same principles that fix the positions and direct the movements of the ancient states.[186]

Brougham explicitly grounded the balance of power in the law-governed universe of Newtonian cosmology.[187] Here it is evident that Newtonianism introduced the ontological theme of the field of forces noted by Butterfield.

At the end of the eighteenth century the Newtonian vision of the balance as an automatic law of nature was consolidated and merged with the practice of calculating interests. As Sofka explains:

[T]he balance of power was an essentially Newtonian construct that assumed that the international system evolved through cycles of peace and war, much as a swinging pendulum moved between fixed poles without ceasing its motion, and was predicated on the idea that international politics operated according to quantifiable principles. By measuring and/or manipulating these variables, such as the size of armies, navies, or financial reserves, a state's "power," or capabilities could be calculated. "Parity" could be obtained by matching these standards and competing with each other for economic and strategic assets.[188]

This discursive formation drew on some themes from the 1550 discourses. Political time, for instance, was conceptualized as "a steady state ... changing in a cyclical manner."[189] Thus, the new Newtonian idea of time as an absolute, open plane was not yet dominant in political discourse. Moreover, the cosmological vestiges of divine providentialism remained in the idea that the balance operated automatically, as if guided by natural law or a deeper order. In the Newtonian worldview, God did not actively intervene in everyday life, but had designed a determinist universe and created the laws of nature according to His will. Newtonian natural providence then, like divine providence, had little room for the active intervention of human agency into history and politics.

As in earlier eras, the mechanical and Newtonian configurations were not the only ways to explicate and justify the balance. Republican moral and legal discourses were also used to represent and legitimate the balance.[190] The view that the balance was not an automatic product of natural law, but needed to be managed by wise statesmen, emerged over the course of the eighteenth century.[191] My survey here is not intended to

---

[186] Brougham 1872 [1803], 11–12.
[187] It is this specific mechanist vision that has inspired comments by Gulick (1955, 26–27), Morgenthau (2006 [1948]), Kissinger (1994, 56–67), and Schweller (2004, 162). But these comments reify a single moment in balance of power history. As this discussion shows, there were multiple visions of the balance.
[188] Sofka 1998, 133.
[189] Rudwick 2005, 136.
[190] Vattel 1758 in Wright 1975, 72; Ashworth 2014, 58; Deudney 2007, 137-152; Devetak 2011; Haslam 2002.
[191] Anderson 1970, 189–190.

show that the Newtonian view was dominant. Rather, I want to suggest that the emergence of a new mechanical ontology attached to a rationalist episteme in which the balance could be discerned by reason made possible new conceptions of state purpose. The Newtonian view of the balance as governed by natural law was an important stepping stone for the construction of a new balance of power idea in the nineteenth century. From the claim that European politics unfolded within a knowable field of forces, only a minor reconfiguration of concepts was necessary to posit that the underlying field itself could be controlled. This formed the basis of the new purpose that would be institutionalized at Vienna.

The Cosmological Shift in Political Economic Thought, 1650–1750

Over the course of the seventeenth century, a cosmology rooted in materialist mechanism transformed European political discourses. The emergence of materialism and mechanism challenged the dualist ontological basis of divine providentialism and reconfigured ideas about blood, marriage, land, and sovereignty that supported aristocratic dynasticism. Moreover, the new cosmology was used to redefine human agency and challenge Newtonian natural providentialism.

Inspired by Copernicus' account of motion, Galileo's experiments overturned the Aristotelian orthodoxy by reconceptualizing motion as the product of a quantitative interaction of forces.[192] The basic idea that bodies only change their speed or direction when acted upon by an external force came to be known as "mechanism." This was premised upon a strict notion of causation in which nothing moves without material impetus.[193] Mechanism provided unity to the disparate strands of natural philosophical investigation in seventeenth-century Europe. Mersenne, Gassendi, Descartes, Boyle, and Hobbes all worked in the mechanist tradition, although there were differences between their systems.

In Descartes' "physico-mathematical" system, corpuscles of various sizes and shapes are the constituents of all matter. Matter is set into motion by external causes and motion is conserved. There is no vacuum or empty space, but rather all bodies are linked to one another by invisible corpuscles. To explain the motion of the planets and other bodies, Descartes argued that "vortices" of moving fluid carried the planets in their orbits.[194] While Descartes explained the motion of lifeless bodies using reductive

---

[192] Gaukroger 2007, 418.
[193] Gaukroger 2007, 253.
[194] Gaukroger 2007, 304.

materialist precepts, his famous dualism posited an immaterial mind. In this and other ways, the Cartesian system left the door open for providence, although Descartes never relied on this tenet explicitly.[195]

Hobbes was more willing than his contemporaries to develop mechanist principles into controversial materialist conjectures.[196] Hobbes read Euclid in 1630, and was exposed to mathematics, astronomy, and mechanics through the various intellectual circles he travelled in. In the 1630s, he travelled to Italy to meet Galileo and to Paris to meet with Mersenne, Gassendi, and Descartes.[197] His exposure to natural philosophy, mathematics, and Epicurean ideas explains his uniquely cosmological political philosophy.[198] Hobbes aimed to build a moral and political philosophy on foundations as certain as knowledge in natural philosophy.[199] Hobbes' "orderly method" for establishing true knowledge in any domain begins with empirical observation but the central steps are proper naming, laying down axioms, and deducing the consequences of these axioms.[200] By following this scientific method of investigation, Hobbes believed his philosophy would establish unquestionable, binding precepts for civic and political life.[201]

The opening pages of *Leviathan* demonstrate Hobbes' commitments to materialist and mechanist precepts. Hobbes begins by describing humans as natural machines. The senses, he argues, are caused "by the pressure, that is, by the motion, of external things upon our eyes, ears and other organs thereunto ordained."[202] The images these motions leave in our mind are "impressions," nothing but "decaying sense."[203] Likewise, human passions are simply "voluntary motions" and decision-making is simply the bubbling up of passions at the moment of action.[204] Hobbes is willing to accept the theological implications of this materialist ontology. If the world is composed of matter in motion, then the soul is not immortal and there is no heaven understood as an "aetherial region" beyond the earth.[205] The determinist implications of this materialism were thought

---

[195] Gaukroger 2007, 338.
[196] Gaukroger 2007, 282.
[197] Rogers 2007.
[198] Connolly (1982, 18) invites us to read Hobbes on an ontological level and I take that up here.
[199] Skinner 1996, 298–300.
[200] Hobbes 1994 [1668], p. 25, §V.17. See also, Skinner 1996, 309–311; Shapin and Schaffer 1985, 87.
[201] Skinner 1996; Evrigenis 2014.
[202] Hobbes 1994 [1668], p. 7, §I.4.
[203] Hobbes 1994 [1668], p. 8, §II.2–4.
[204] Hobbes 1994 [1668], p. 33, §VI.
[205] Hobbes 1994 [1668], pp. 301–305, §38.1–4, p. 312, §38.23. For Strauss, this move is important: "He could not have maintained his thesis that death is the greatest and

to undermine not only belief in God but free will and human responsibility.[206] As such, Hobbes' ideas challenged divine providentialism and he was denounced as an atheistic monster.[207] Nonetheless, materialist determinism, as we shall see, is compatible with other versions of providentialism, in which God creates the laws of nature that govern all phenomena, but leaves them to operate on their own.

Hobbes builds his famous argument in *Leviathan* on the materialist foundation laid out in the opening pages. The mortality and vulnerability of people in the state of nature drives them to create a sovereign which will "keep them all in awe."[208] Epistemic disagreement is one of the central causes of the war of all against all in the state of nature.[209] The differences between people "ariseth partly from the diversity of passions in divers men, and partly from the difference of the knowledge or opinion each one has of the causes which produce the effect desired."[210] So the conflicts that arise in the state of nature are due in part to differences of opinion and the overestimation of one's own "wit" which in turn contribute to distrust and war.[211] Since each person must value their life, and life in the state of nature is precarious, Hobbes contends that people will realize the necessity of forming a covenant with one another to erect a sovereign to create laws that adjudicate these epistemic disputes and keep the peace.

The new mechanical philosophy helped to overturn the dominance of the ancient humours theory in medicine.[212] William Harvey's (1578–1657) experiments on the body discovered the circulation of blood, prompting further experiments. As we saw in the above section on 1550 discourse, blood played an important role in constituting the aristocratic basis of political rule. The "class myth" of the nobility held that "blood of the aristocrat was qualitatively different from that of ordinary mortals," predisposing gentlemen to possess the courage and moral compass necessary

> supreme evil but for the conviction vouched for by his natural science that the soul is not immortal. His criticism of Aristocratic virtue and his denial of any gradation in mankind gains certainty only through his conception of nature, according to which there is no order, no gradation in nature" (1936, 167). For Strauss, the materialist monism of *Leviathan* displaces the dualism of Hobbes' earlier thought: the idea that "man, by virtue of his intelligence, can place himself outside nature, can rebel against nature" (1936, 168). In my terms, Strauss is identifying an unintended consequence of adopting a cosmological discourse in tension with the salutary moral content Strauss finds in early Hobbes.

[206] Rogers 2007, 425.
[207] Rogers 2007, 426.
[208] Hobbes 1994 [1668], p. 76, §XIII.8.
[209] Shapin and Schaffer 1985, 103–107; Skinner 1996, 318.
[210] Hobbes 1994 [1668], pp. 57–58, §XI.1
[211] Hobbes 1994 [1668], pp. 74–75, §XIII.1–3.
[212] Gaukroger 2007, 346–348.

112   Natural Philosophy and the Balance of Power

for the defence of the state.²¹³ This discourse was challenged as medical inquiry in the seventeenth century and redefined how blood was understood. Initially, blood was thought to constitute the essence of a person or animal. For example, in early transfusion experiments at the Royal Society, Boyle hypothesized that if a fierce dog were injected with the blood of a cowardly dog, it would become tame.²¹⁴ The failure of these and other experiments eroded ideas of blood as a life-force. While practices of blood-letting continued as medical practice into the nineteenth century, the idea that blood was a chemical substance like any other was common in Europe by the end of the eighteenth century.²¹⁵ The weakening of the idea of blood as life-force, combined with developments in materialist and mechanical thought, denaturalized ideas that supported dynastic politics.

New materialist and medical ideas were also used to rethink how and to what ends to manage lands and states.²¹⁶ Inspired by findings in medicine, William Petty (1623–1687) and William Temple (1628–1699) redefined the polity as a "body."²¹⁷ This helped contribute to the delimitation of the polity as a territorial unit, pushing political discourse away from the dynastic idea that the polity is the family or line of succession. In the dynastic discourse, quite disconnected lands could be easily thought of as unified under the dynastic rule of one prince.²¹⁸ The Habsburg Empire once included distinct territories in Austria, Hungary, Spain, and the Low Countries. These concepts would become harder to maintain once the polity was reconceptualized, not as the possessions of an individual, but as "a country," or "a nation," with the same coherence and properties as other "natural bodies."²¹⁹ As Branch has shown, these changes interacted with the rise of scientific cartography.²²⁰ This altered primary institutions by contributing to the reconceptualization of sovereignty and the construction of a system of independent, territorially contiguous states. Once represented as coherent bodies, the idea of the

---

[213] Bannister 2000, 38.
[214] Boyle 1666, 368.
[215] Maluf 1954, 70.
[216] Mukerji 1997.
[217] Petty 1691, A8, 14; Temple 1814 [1690], 73, 90. Also relevant here is Hornick (1932 [1684], 230–232), an Austrian political economist who conceptualizes the nation as a body in which money "circulates," "[l]ike the human blood by the power of the heart passed every year to a large extent through the prince's treasury." He describes Austria's problems as an "illness" that is beyond "weak and slow treatment."
[218] Medieval patterns of political authority were also multilayered, producing noncontiguous domains of rule. Ruggie 1993; Branch 2013, 23–29.
[219] Petty 1690, 9.
[220] Branch 2013.

polity could be further reified, and assigned a "public welfare" or "common interest" rooted in enhancing the strength and wealth of the body of state, not the prince himself.[221]

For Petty, the interest of the state, like everything else, was to be understood and expressed in material, mechanical terms:

> I have long aimed to express myself in Terms of *Number, Weight,* or *Measure*; to use only Arguments of Sense, and to consider only such Causes, as have visible Foundations in Nature.[222]

By grounding his analysis in visible, natural causes, Petty sought to build political analysis on the same mechanical foundations as Descartes, Hobbes, and Boyle. But by also demanding "measure" Petty introduced a "quantifying spirit" into the analysis of polity.[223] Transposed into political discourse, the demands of materialism, mechanism, and measure highlighted the role of quantitative bases of power: people, troops, and revenues.[224] However, Petty does not reduce the "wealth and strength" of countries to these visible measures.[225] He argues that small powers with good policies, mercantile trade, and quality land can equal the power of much larger powers.[226]

Although Petty aimed to produce a quantitative science of political analysis, his own figures were often simply estimates or averages. So the entry of measure and statistics into states was slow. Nonetheless, Temple, Petty, and others helped to reconstitute epistemic discourses in the late seventeenth and early eighteenth centuries. On their view, reality could be represented by numbers and calculated by arithmetic. Further, they argued that rulers could control their fates by heeding the advice of political economists. That is, they extended politics as a form of rational calculation. By suggesting that the strength of the state was merely land, money, and people, they extended the conceptual shift introduced by the analysis of interests. State interests were no longer to be equated with intangible glory, but to be measured, calculated, and manipulated.

These early forays into political economy inspired statistical ideas and practices that denaturalized and reconfigured a whole host of concepts. For example, marriage statistics became available in the eighteenth century, and so alongside births and deaths, matrimony was one of the first domains to be investigated with statistical analysis. Whereas marriage had

---
[221] Petty 1690, 5.
[222] Petty 1690, 7.
[223] Frängsmyr *et al.* 1990.
[224] Petty 1690, 44, 62, 64.
[225] Petty 1690, 9, 12.
[226] Petty 1690, 12.

previously been depicted as an ancient legal custom commanded by God ("be fruitful and multiply"), it was increasingly represented as "a necessary consequence of nature and sound reason."[227] Marriage remained an important social institution, but not necessarily one governed by ancient tradition and sacred law. It was naturalized as a command of nature and its value was defined in terms of its contribution to human welfare and the political body.

This marked an important transformation. Marriage was a key component of dynastic political orders. Marriage had shaped the primary and secondary institutions of international order by defining the rules for dynastic succession and territorial acquisition. It was able to have this power because it was backed by cosmological discourses of divine providentialism that naturalized the superiority of aristocratic blood and the sanctity of ancient law. But without this cosmological backing, marriage dropped out of primary and secondary institutions. On the new mechanist and materialist view, territory was to be governed by the laws of the balance of power. Reconceptualized as a constituent element of state power, territory was removed from dynastic tradition and placed in the table of interests. Thus, as the cosmological underpinnings of international order shifted, so too did primary and secondary institutions.

In sum, political economic thought in the mechanist, materialist tradition introduced new cosmological elements into political discourses. Recall that the line of thought from Copernicus to Newton constituted the balance of power as unfolding within a field of forces. These forces had initially been understood in determinist terms and so it was unclear what role human agency was to play in politics. Political economy, by contrast, drew on a materialist ontology and a rationalist episteme to introduce the notion that humans could control the field of forces underlying the balance of power and other political phenomena. It was this idea that made possible a new cosmological discourse of materialist rationalism. Materialist rationalism implies that human reason, informed by scientific knowledge, can be used to manipulate and control the matter that comprises the world. After 1700, materialist rationalism slowly entered political discourses, emerging as the dominant configuration by the time of the Congress of Vienna.

Strategic Deployment: Reforming European States, 1700–1800

New ideas from natural philosophy, natural history, and mathematics had been introduced into seventeenth- and eighteenth-century political

---

[227] Porter 1986, 21.

institutions through a variety of formal and informal connections between the transnational natural philosophical community and the aristocratic political elite. These changes unfolded as a form of horizontal change in which a transnational class of brokers took up similar ideas in similar ways across European states. For example, Prince Metternich, who as we shall see played a central role at the Congress of Vienna, received a classical Enlightenment education. This early education had a profound influence on Metternich. As a young man, he aimed to become a doctor or chemist. As he records in his memoirs, he "diligently attended lectures on Geology, Chemistry, and Physics."[228] Later, he "followed with attention the progress of Medical Science." He remarks, "[m]y particular vocation seemed to me to be the cultivation of knowledge, especially of the exact and physical Sciences, which suited my taste particularly."[229] Although he eventually pursued a law degree, Metternich maintained his interest in chemistry throughout his life, referring to chemical metaphors in his letters and writings. Moreover, he studied law under the rationalist Niklas Vogt where he read Kant's "Idea for a Universal History" and endorsed its central tenets.[230] Thus, through his education and interests, Metternich was constituted as a broker who linked scientific and political communities.

An additional channel was opened in the eighteenth century when the administrative capacities of European states expanded, increasing the number of diplomatic officials and adding new informational and statistical offices. Under intense fiscal pressures in the seventeenth and eighteenth centuries, European states aimed to expand and centralize their administrative apparatuses.[231] At the same time, the foreign offices of all European powers grew.[232] In this, many states emulated the French state:

At Louis's assumption of power in 1661, a single coach would have sufficed to transport the minister who controlled French foreign policy and his handful of assistants. By 1715 Torcy and his retinue would have needed 20 such coaches, so many specialised personnel had been added.[233]

As the foreign ministries grew, so too did the demand for a proper education for would-be state officials. The universities in central Europe were the first to take up the call, offering enlightened training for aspiring statesmen. These universities taught mathematics, natural law, and

[228] Metternich 1880a [1844], 23.
[229] Metternich 1880a [1844], 23. Quoted in Sofka 1998, 117.
[230] Sofka 1998, 119–122.
[231] Scott 2006, 37; Bonney 1999; Silberman 1983; Tilly 1992.
[232] Scott 2006, 133–135. See Soll (2009) and Cole (1939) for the French case. See Brown (2016) for a close examination of bureaucratic practices in the French state.
[233] Scott 2006, 134.

political economy to a new class of state officials.[234] Under the influence of this education, state officials yearned for measurements of population, geography, and commerce that did not fully exist.

William Petty and Gottfried Leibniz suggested the creation of centralized statistical offices in the late seventeenth century.[235] Around the same time, Jakob Friedrich von Bielfeld and Johann Heinrich Gottlieb von Justi proposed a science of political geography founded upon state statistics.[236] Inspired by these ideas, European states and principalities conducted censuses to varying degrees of success throughout the sixteenth, seventeenth, and eighteenth centuries.[237] Bureaucrats, university professors, and amateurs collected data on population, births, deaths, health, commerce, and climate.[238] These efforts were always incomplete and the rationalization of the European state was an aspiration, not a reality.[239]

Nonetheless, the emergence of materialist and rationalist ideas from political economy made possible a new conception of power and with it a new approach to thinking about international politics. In the seventeenth century, power was thought of in terms of honour and glory, but Frederick the Great and Kaunitz drew on Enlightenment ideas to reconceptualize power explicitly in terms of calculable material resources. Kaunitz' political algebra promised to calculate the relative power of states and to predict their behaviour.[240] His thinking was grounded in the mathematical deductive reasoning he received from his education at the hands of Enlightenment philosopher Christian Wolff.[241] The articulation of power in materialist, calculable terms helped define the new concept of a "great power" in the 1760s and 1770s.[242] A great power state was to be conceptualized in terms of superior population and territory, the supposedly material elements prized by political economists.

Data collection efforts intensified over the course of the eighteenth century as government ministries expanded. The first statistical offices were created around 1800. Under Napoleon, France created a bureau de statistique in 1800.[243] However, its first detailed census produced "virtually no usable data" and Napoleon eliminated the office in 1811.[244]

---

[234] Scott 2006, 118.
[235] Hacking 1990, 18.
[236] Scott 2006, 117–118.
[237] Hacking 1990, 17.
[238] Hacking 1990, 22.
[239] Silberman 1983; Scott 1998.
[240] Anderson 1993, 67; Scott 2006, 118.
[241] Scott 2006, 118.
[242] Scott 2006, 119.
[243] Desrosières 1998, 34; Porter 1986, 28.
[244] Porter 1986, 28.

Prussia created a centralized statistical office in 1805 that served the Prussian and German state until 1934.[245] Britain conducted an official census in 1801, but it was deemed unsatisfactory.[246] In the 1830s, Britain created a statistics office for economic data at the Board of Trade and a separate General Register Office was set up to gather social statistics.[247]

The "avalanche of printed numbers" that descended on Europe was combined with new notions of statistical law to subvert determinism.[248] The idea of "chance" was used to reconceptualize suicide, crime, madness, and disease. This gave rise to "the notion that one can improve – control – a deviant subpopulation by enumeration and classification."[249] So the rise of a quantitative view of the world helped to introduce a cosmological shift in ideas about humanity's role in the universe. The idea that chance could be understood and mastered helped to introduce the notion that the world could be controlled by human knowledge and practice.

Thus, by the time of the Napoleonic wars and the Vienna Congress, horizontal associational changes in the European great powers had reoriented ideas of state purpose from God and glory to land and people. Statistical thinking had introduced an episteme of rationalist calculation and an ontology of materialist measurement into international politics, paving the way for the construction of a cosmology in which humans could control politics. Providential or deterministic thinking had not been entirely displaced, as we shall see in the next chapter. Nonetheless, rationalist ideas inspired and legitimated new practices of statistical collection and calculation that were deployed at the Congress of Vienna.

Cosmology and the Congress of Vienna, 1815

During the eighteenth century, aristocratic elites and state organizations had strategically deployed the rationalist episteme and the materialist ontology of the new cosmology. These appropriations produced discursive changes in the associations underlying international order. They reconfigured the automatic, harmonious, Newtonian balance that had emerged over the course of the eighteenth century into a rationalist system in which political officials could harness and control the laws of nature. In particular, materialism and measurement made possible the new practice of finely tuning the balance of power and reoriented state

---

[245] Desrosières 1998, 179.
[246] Porter 1986, 30.
[247] Porter 1986, 31; Desrosières 1998, 167.
[248] Hacking 1990, 3.
[249] Hacking 1990, 3.

purpose to the improvement of lands and populations to secure a better position in the balance. These shifts were embedded in a macro-level balance of power institution by the time the Congress of Vienna convened to reorder Europe after the Napoleonic wars.

Napoleon, like Louis XIV before him, galvanized the other great powers of Europe into a balancing coalition to contain French power. After Napoleon's defeat in Russia during the winter of 1812–1813, Britain, Prussia, Russia, and Austria formed an alliance and fielded a large army in central Europe to push Napoleon west. As the inevitable outcome of the war came into view in early 1814, the coalition initiated cease-fire talks with France. However, Napoleon refused to agree to the coalition's terms and the war continued. Nonetheless, these talks produced the landmark Treaty of Chaumont, which declared allied war aims and first outlined plans for a European concert system in which European affairs would be jointly managed by the great powers in a series of regular conferences.[250] Moreover, these talks made explicit the coalition's view that the aim of any treaty would be to re-establish the balance of power in Europe.[251] After the breakdown of these talks, hostilities between France and the coalition continued until Russia's troops got behind the French line and marched on Paris. The end of the war was settled by the Treaty of Paris, which defined the terms of peace between France and the coalition. France was to be reconstituted as a full member of the great power system, but reduced to its 1792 boundaries. However, central territorial issues still needed to be settled and the great powers declared a congress in Vienna to open in October of 1814.

In Vienna, the victorious powers sought to restore the balance of power as a means of restoring the legitimacy of monarchical rule and the "old regime."[252] The French Revolution had threatened to overturn the dynastic institutions of European order and the conservative powers used Vienna to fuse the balance of power and dynastic law into a powerful new configuration. Britain was represented in Vienna by Foreign Minister Viscount Castlereagh. Castlereagh sought to secure a continental balance of power consistent with British naval supremacy and commercial interests.[253] To this end, Britain wanted to check the rising power of Russia and to ensure that France was reconstituted as a major player in central Europe, but in such a way that it would not threaten the buffer states in the Low Countries. Emperor Alexander himself represented Russia. Russia ended the Napoleonic wars occupying the Polish

[250] Webster 1931, 226–229.
[251] Gulick 1955, 159.
[252] Holbraad 1970; Keene 2002, 14–19; Keene 2005, 158.
[253] Webster 1947, 47.

lands. Alexander aimed to keep these as a buffer state and as a laboratory for his liberal constitutional experiments. The Emperor also aimed to enhance Prussia's position so it could serve as an ally and buffer against France and Austria. Metternich represented the Austrian court. He aimed to check the growth of both Prussian and Russian power. However, the realities of conquest and power limited Metternich's ability to minimize the gains of Austria's enemies in the east. For this reason, Metternich's schemes depended on the support Britain and France, which also aimed to balance against the eastern powers. Prussia's gains in the Napoleonic wars further cemented its status in the first rank of European powers. In Vienna, Prussian Minister Hardenberg led the Prussian delegation, including the statistician Hoffmann, which was intent on expanding Prussian territory. But Prussian expansionism could not rest on the force of occupation alone, since new territorial possessions would not be secure unless recognized by the other European powers.

The Treaties of Chaumont and Paris determined French borders but left a central issue to be settled at the Congress: the Polish-Saxon question. At the end of the war, Russia occupied the Duchy of Warsaw, the rump of the old Polish state that had been dismembered by eighteenth-century partitions. Russia aimed to keep the Polish territory, but Austria and England objected because it would strengthen Russia and disturb the balance in central Europe.[254] In 1815, Prussian and Russian forces occupied the Kingdom of Saxony. Saxony had allied with France during the Napoleonic wars and was one of the last states to abandon the French. Russia and Prussia argued that Saxony should be punished for its alliance with France.[255] Prussia, supported by Russia, hoped to keep all of Saxony. While Castlereagh and Metternich also wanted to strengthen Prussia, they felt this would compromise Austria's defensive position and so they opposed Russia and Prussia's plan for the Polish-Saxon lands.

In the end, the 1815 Final Act signed at Vienna divided Poland between Russia and Prussia while Prussia was granted part of Saxony with the rest remaining independent as the Kingdom of Saxony. The precise distribution of lands was based on population numbers produced by a statistical commission. How and why did the powers distribute territory in this way? It is tempting to conclude that the results of the Congress of Vienna can be explained by the laws of the balance of power, or that the "force of events" compelled the negotiators to finely tune the balance of power in Europe to prevent both the resurgence of French aggression and

[254] Gulick 1955, 190–211.
[255] Talleyrand 1881 [1814], 46.

the emergence of Russian hegemony.[256] Or, with Jervis we might emphasize that the ability and desire of the European states to cooperate can be explained by alterations in the costs and benefits of war in the era after Napoleon.[257] The high domestic costs of war and the necessity of balancing against French national power pressured the coalition states to cooperate. Alternatively, for Ikenberry the concert demonstrates the calculated restraint of the great powers in reconstituting France while checking its power with an open system of rules and communication.[258]

Each of these perspectives helps explain why the great powers aimed to balance power in roughly the way they did. But, these explanations beg the question, why did the European great powers have the concept of the balance they had? How did it become possible for them to divide territory in the way that they did? As Wight, Schroeder, and others have pointed out, "the balance of power" has had many meanings for the European and international leaders.[259] In order to explain how and why power was conceived and balanced in one way rather than another we must examine the ideas of the policy-makers themselves. Rather than assume that power politics and the valuation of territory and population is naturally given or structurally determined, I show how the new cosmology constituted balancing techniques and state purposes at Vienna.

The Discourse of State Purpose at the Congress, 1814–1815

The discourse of state purpose at the Congress of Vienna is dominated by the balance of power (understood in terms of interests and principles) and material resources. This is supported by ontological representations of population and money and a rationalist episteme of calculation that are linked together into a cosmology in which humans can actively control the forces of politics.

In Vienna, the dominant purpose of the state is to act in accordance with the state's "interests" and "principles" to take up a prominent, powerful position in the balance. However, these terms – interests, principles, and the balance – were defined and used in a variety of ways. Interests are variously associated with the balance or equilibrium, material resources like territory and population, security concerns like defensible borders, and commercial interests. States, especially Prussia, also exhibit ambition or

---

[256] Waltz 1979, 173.
[257] Jervis 1985.
[258] Ikenberry 2001.
[259] Wight 1966, 151; Schroeder 1989, 1994a.

concern with influence and rank. An important theme is the opposition of "common interests," usually embodied in the balance of power, and private or separate interests.[260] Representatives frequently argue that private interests are best realized through the common interest.[261] For some actors, "interests" are base and immoral, and for others they have the status of moral ideals, especially if they are "European interests." Interests are also depicted as "natural" in that they are determined by geography and the principles of the balance.[262]

Principles are also defined in various ways. In one sense, the term "principles" is used in the texts to refer to any significant dictum that guides policy. So Metternich denounces Russia and Prussia because they have no principles and act "from the convenience of the moment."[263] In another sense, the term refers to agreed upon rules of conduct, such as the rights of dynastic legitimacy. The central shared rules in the discourse are legitimacy, dynastic rights, public law, and justice. Talleyrand is famously the standard bearer of "principles" at Vienna. Talleyrand aims to end the demonization and exclusion of France by legitimating the restored Bourbon monarchy. But his appeal to the legitimacy of dynastic rights is also a calculated move in the interest of his Bourbon patrons. In Castlereagh's correspondence, principles refer to principles of public law, justice, and also just generically to the ideas that guide other leaders' actions. In the latter sense, principles bleed into interests. So for Talleyrand, French commitment to the balance of power in Europe is a "principle."[264] In one sense, both principles and interests are moral categories, used by each to criticize others' policies. Talleyrand sneers that England's "principles are her interests."[265] Castlereagh reproaches Emperor Alexander by asking, "whether the present Congress is to insure the welfare of the world, or merely to present a scene of disorder and intrigue, an ignoble contest for power at the expense of principles."[266]

Both interests and principles are oriented to the purpose of the balance of power. Castlereagh declares at the outset of the Congress that, "I wish to direct my main efforts to secure an equilibrium in Europe."[267] The agreement of all the powers would be necessary to "found

---

[260] Castlereagh 1947 [1816], 510–511; Metternich 1880b [1814], 586.
[261] E.g. Talleyrand 1881 [1814], 27.
[262] Castlereagh 1853 [1814], Vol. 10, 175.
[263] Metternich 1880a [1801], 9.
[264] Talleyrand 1881 [1815], 136.
[265] Talleyrand 1881 [1814], 37.
[266] Quoted in Talleyrand 1881 [1814], 41.
[267] Castlereagh 1853 [1814], 173.

a satisfactory system of balance in Europe."[268] Metternich's first state paper, from 1801, declares that the central aim of Austrian policy should be "the restoration of the European balance of power."[269] Metternich's chief concern in 1814 is that the actions of Prussia and Russia threaten the "equilibrium and tranquility of Europe."[270] But Metternich does not seek to unduly weaken Prussia as it is "one of the most useful weights in the balance of forces."[271] Talleyrand reports that his instructions express the hope that the Congress would establish "a real and durable equilibrium" and "[t]hat impartial method leads them to enter into the principles of public law recognized by all Europe."[272] Later Talleyrand states that the twin aims of France are the restoration of "the principles of public law, and of a just balance of power in Europe."[273]

The balance of power is more than a state purpose. It is an institution governed by specific rules and practices: compensation, indemnity, and containment. Containment dictates that no power should be given so much territory that they will threaten the security or liberty of the other powers. The principle of compensation stems from the need, on the one hand, to recognize the rights of conquest and, on the other, to create a stable balance of power or equilibrium. Under this principle, a power may be entitled to compensation if they concede territories conquered during war. Possessions given up for the common good, that is, the balance of power, could be exchanged for territorial compensations or *convenances* elsewhere. A power may also be entitled to compensation if their rivals are given territory or have gained territory. In Vienna, Prussia was to be compensated for relinquishing Saxon lands. The principle of indemnity allows for payments in exchange for damage caused by a power in a war.[274]

On the one hand, the balance of power accommodates war and its modes of destruction, as well as ambition and security concerns. On the other hand, the balance of power is closely linked to agreed upon principles. In the balance of power institution there are rules to govern the acquisition of territory and population as well as provisions for the security of large and small powers. The great powers respect and take seriously the principles of the balance of power and accommodate compensation, pay indemnities, and recognize the rights of conquest within

---

[268] Castlereagh 1853 [1814], 173.
[269] Metternich 1880a [1801], 10.
[270] Metternich 1880b [1814], 568.
[271] Metternich 1880b [1814], 589.
[272] Talleyrand 1881 [1814], 6, n. 6.
[273] Talleyrand 1881 [1814], 65.
[274] The French indemnity promised in 1814 for war damages was 700 million francs, but claims of the powers amounted to 1,200 million francs. Webster 1947, 82–83.

limits. A great power can conquer territory during war justly fought, but cannot initiate a war or claim territory on a false pretext and expect the other powers to ratify the acquisition. Since territorial possessions are only as secure in the long run as their guarantors, it behooves ambitious powers to work within the rules. Ambition, aggrandizement, and *amour propre* are negative ends and are denounced.

The interesting innovation in balance of power politics in 1814–1815 is in the idea and practice of balancing itself: to achieve a systemic, designed arrangement of powers. In 1713, to balance was to create two equal dynastic houses. In 1814–1815, to balance power is to create a system of powers in which there are natural allies and enemies to easily form blocs that will contain aggressors. For example, Castlereagh explains that he considered two alternative arrangements in Europe:

> Two alternatives alone presented themselves for consideration – a union of the two great German Powers, supported by Great Britain, and thus combining the minor States of Germany, together with Holland, in an intermediary system between Russia and France – or a union of Austria, France, and the southern States against the northern Powers, with Russia and Prussia in close alliance.[275]

Here, Castlereagh considers two different systems of natural alliances and enmities. To balance is to finely position the distribution of territory so as to structure patterns of alliance. The goal is not to prevent war, but to create natural blocs that check the power of the other. The act of balancing in Vienna is designing a system that provides for the interests of all.

There is some discursive support for a dynastic purpose on which the great powers acted in concert so as to prevent revolution because they feared monarchical legitimacy was being eroded.[276] Castlereagh's postmortem, for example, lauds the alliance for saving Europe "from becoming again prey to revolutionary anarchy and violence."[277] However, this is a secondary theme in the discourse.

Nonetheless, dynastic rights are still important elements of the discourse of state purpose, although they no longer form the taken-for-granted backdrop of European order. International order, after all, is still carried and reproduced by the shared ideas of a transnational class of aristocrats that guard their special privileges. Political

---

[275] Castlereagh 1853 [1814], 173–174.
[276] On this reading of Vienna, see Kissinger 1957; Holbraad 1970.
[277] Castlereagh 1947 [1816], 510.

124   Natural Philosophy and the Balance of Power

legitimacy still rests on rules of succession and the acceptance of "ancient" laws. Older themes rarely disappear, but are reconfigured, adapted to a new age. Here, the old concepts of blood and sacred nature of the nobility have gone away.[278] And, the decreased salience of ancient dynastic rights is evident in the politics of the Congress, as we shall see. Nonetheless, dynastic rights were an important component of Talleyrand's strategy at Vienna.[279] By defending the legitimacy of dynasticism, Talleyrand could bolster the restored French monarchy and defend the Saxon king from being deposed in favour of France's rival Prussia. But dynasticism had long ceased to form the taken-for-granted background of European discourses. Instead, dynastic claims are "principles" or "interests" like any other.

Cosmological Elements at the Congress, 1814–1815

This concept of the balance of power is supported by a cosmological discourse that links together materialist representations of reality and the ideal of rationalist control into a narrative in which humans can control the forces of politics. This is a distinct cosmological configuration from that which dominated Utrecht where leaders sought to follow rational maxims, but not to manipulate reality. Here, the delegates have internalized the rationalist materialist view that the world of matter can be known and that knowledge can be used to control reality.

Ontologically, all the representatives take for granted the importance of military force as a means to ends. Force is represented materially and quantitatively in terms of the number of men under arms and the quality of defensive lines. War is generally accepted as a viable means for accomplishing political ends. The acceptance of war and primacy of force means that states value resources like revenues and population. These are essential means to power, security, status, and holding the balance of power. Alliance here emerges as a conscious means to achieve the balance of power and thus both the interests of state and the interests of Europe.

The episteme includes a new category not present in 1550: the idea that designs, projects, and schemes have a place in power politics. In place of divine providence, there is now a sense that politics can be shaped by planning and execution, so the discourse is filled with

---

[278] As Osiander puts it, "Talleyrand justified the concept of dynastic legitimacy in purely pragmatic terms, and explicitly rejected any admixture of mysticism" (1994, 212).

[279] E.g. Talleyrand 1881 [1814], 20, 24.

"designs," "plans," and "projects."[280] This is evident also in the concept "regulate." This is a general term that means, broadly, to rule or to govern, but carries connotations of design. Metternich's aim is no less than "to regulate the general affairs of Europe."[281] The idea of balancing, connected to regulation, becomes more active. To balance is to govern Europe by creating a system of powers, positioned vis-à-vis one another in such a way as to guarantee the security of all. It is important to note that the idea that the balance is a law or maxim of politics drops out of the discourse between 1713 and 1815. The balance is no longer seen as an entity governed by external laws, somehow beyond the will or control of statesmen.

Designing and regulating are supported epistemically by new modes of calculation and reasoning. Calculation was initially a mathematical concept, derived from the Latin for "count." It was used widely in the sixteenth and seventeenth centuries in astrological or astronomical contexts.[282] By the nineteenth century, it came to mean, more generally, to design or plan. In 1814–1815 political discourse, it carries both this general meaning, to design or plan, and the mathematical application. Castlereagh lauds the recent "repose, without forming calculations that always augment the risks of war."[283] One British official "calculated that a corps of 35,000 [Prussian troops] still remained" in France.[284] The statistical commission was "appointed for the verification of the calculations that are produced."[285] But in some uses, to calculate is to attempt to tell the future, as in the original astrological meaning. Gentz tries to "calculate" how long the Congress will last.[286]

These new discursive elements constitute a cosmological shift because they make possible redefinitions of the nature of the universe and the place of humanity in it. Humanity is no longer merely the subject of divine or natural law, but is able to intervene and manipulate the forces of politics. As we shall see, materialist representations and ideals of rational control made possible new practices of balancing. These practices carefully carved territories and calculated populations to finely tune the arrangement of power conceptualized in material terms as land and people. But they also promoted a new conception of state

---

[280] For example, Castlereagh 1853 [1814], Vol. 10, 143, 187.
[281] Metternich 1880b [1814], 586.
[282] OED 2011, "calculation."
[283] Castlereagh 1853 [1814], Vol. 10, 174.
[284] Castlereagh 1853 [1815], Vol. 10, 272.
[285] Talleyrand 1881 [1814], 120.
[286] Metternich 1880b [1815], 585.

purpose: states should use their power not to obey the rational maxims of the balance but to shape and constitute the balance.

The Statistical Commission and the Final Compromise

The discourse analysis shows that balancing at the Congress was understood as an intentional act to arrange the powers in such a way as to create natural coalitions of alliance and enmity. It seems natural to us today that the great powers would intend to construct a balance, but in 1815 a rationalist discourse in which the world could be controlled had just emerged. This notion was made possible by mechanism, materialism, and measurement, which eroded old discourses and laid the discursive preconditions of the core ideas of balance of power politics. However, ideas from natural philosophy had a more direct effect on the outcomes of the Congress. In this section, I trace the negotiations at Vienna to show how materialism and measurement changed ideas about land and population, making possible the crucial role of the statistical commission in the final outcome of the Congress.

The Congress quickly reached an impasse over the Polish-Saxon question. Castlereagh and Metternich agreed that they could accept only half of Russian and Prussian demands: they could either allow Russia to keep all Polish territories or allow Prussia to keep all of Saxony, but not both.[287] The fear was that this would strengthen the eastern powers beyond the ability of France and Austria to check their combined power. However, Alexander refused to give up any Polish territory, forcing Metternich and Castlereagh to persuade Prussia to reduce their demands in Saxony.[288] Throughout the fall of 1814 Prussia would not yield and by December the coalition threatened to break down into warring parties.[289] By this time, Talleyrand had earned the trust of Metternich and Castlereagh and a Triple Alliance was signed between France, Austria, and England. The threat of war pushed all parties back to the bargaining table in search of a compromise. As Kraehe argues, "the superpowers had no choice but to pursue their goals by other than military means."[290] However, this was no guarantee that a solution acceptable to all parties would be found.

Metternich succeeded in breaking the Russo-Prussian front by promising Russia to recognize its gains in Poland and devising the partition of

---

[287] Gulick 1955, 202.
[288] Webster 1931, 342–355; Gulick 1970, 652.
[289] Gulick 1955, 223–224.
[290] Kraehe 1992, 712.

Polish and Saxon lands. No longer seeking its own aims, Russia was a less than stalwart ally to Prussia and recognized the need to compromise.[291] Compromise would come at the expense of Prussian claims in Saxony. However, the matter was not so simple. If Prussia was to be denied Saxony, which it had earned in the fight against Napoleon, it would have to be compensated with equivalent possessions elsewhere. Moreover, The Russo-Prussian Treaty of Kalisch, signed in 1813 as the coalition chased Napoleon back across Europe, promised to restore Prussia to its 1805 "proportions."[292] So the Congress needed both an estimate of Prussian proportions in 1805 and an estimate of what possessions would be necessary to compensate Prussia in such a way as to restore these proportions. As stated in instructions to Talleyrand:

Convenance is the law. The Allies have, it is said, pledged themselves to replace Prussia in the same condition of power as she was before her fall, that is to say, with ten millions of subjects.[293]

Providing compensation was not merely necessary as a quid pro quo to motivate Prussia, but was required by the shared rules of the balance of power institution.

The solution to the Polish-Saxon question was to divide Saxony between the independent Saxon Kingdom and Prussia, and to divide Polish territory between Prussia and Russia. Russia would be allowed to keep most Polish lands but a few villages and fortresses there would be given to Prussia. Prussia would also be compensated for its concessions in Saxony with territory elsewhere in Germany. But this solution raised questions about where and how Prussia would be compensated. On 10 December, Metternich wrote to Hardenberg and enclosed a table proposing a partition of Saxony, with compensations for Prussia in other principalities.[294] Metternich's aim was to meet the Congress' legal obligations to Prussia, while offering as little population as possible. Prussia disputed the statistics in Metternich's tables and refused to agree to the partition. The impasse led to talk of war over Christmas.[295]

The compromise that settled the Polish and Saxon questions and saved the Congress from devolving into war depended on the calculations of a "statistical commission." The partition of Saxony was proposed as early

---

[291] Gulick 1970, 655.
[292] Gulick 1955, 110.
[293] Talleyrand 1881, xiii.
[294] Metternich 1880b [1814], 586ff.
[295] However, there is reason to doubt that there was a genuine threat of war. Members of the British Cabinet and Parliament made it clear that they refused to go to war and Russia was hardly in a position to fight. Perhaps, then, the episode is one of many classic diplomatic storms before the calm of the settlement. See Webster 1931, 230–235.

as November, but there was no agreement until after threats of war and the formation of the statistical commission.[296] The official charge of the statistical commission was to enumerate the total number of "souls" in a long list of towns and principalities the coalition had retaken from Napoleon.[297] This would provide the necessary information for redistributing all territories according to the principle of compensation. The valuation of each possession in terms of its number of souls reduced, the powers well knew, the value of territory to population. Talleyrand, for example, objected to this:

> I requested it to be added that the population should be estimated not by numbers merely, but by its condition also; for a Polish peasant, without capital, land, or industry, ought not to be placed upon a par with an inhabitant of the left bank of the Rhine, or the richest and most fertile districts of Germany.[298]

Moreover, the defensive position of the territory was also supposed to be taken into account. However, the commission's reports are dominated by tables filled only with numbers of souls, without any mention of the quality of those souls or their lands. There are a few explanations for this. First, the legal obligations of compensation under the Russo-Prussian did not require quality or position to be taken into account. It stated simply that Prussia had to be restored to her 1805 proportions and this was interpreted in narrow terms as population. Second, population statistics proved more legible and credible as tools of diplomacy than arguments over position or quality. In effect, as Talleyrand put it, Prussia had "subordinated their claims upon Saxony" to "a numbering of population."[299]

The commission first met on 24 December 1814 and presented its final report on 19 January 1815.[300] The reports of the commission provided precise measurements of population in Saxony, the German principalities, and the Polish lands. In total, the commission concluded that the allies possessed former Napoleonic territories containing almost thirty-two million people.[301] This total was broken down into long tables detailing the populations of occupied principalities and towns. This quantification of land permitted pieces of land to be conceptualized and traded as parcels of people. This made possible any number of distributions of power and negotiations quickly came to centre on designing a

---

[296] Talleyrand 1881 [1814], 59.
[297] d'Angeberg 1864, 624.
[298] Talleyrand 1881 [1814], 122.
[299] Talleyrand 1881 [1814], 123, 110.
[300] d'Angeberg 1864, 561, 646. Talleyrand's (1881, 120) account dates the first proposals for the commission around 20 December.
[301] d'Angeberg 1864, 647.

compensation offer that would meet the powers' obligations to Prussia while maintaining the balance of power in central Europe.

On 12 January 1815, Hardenberg presented a proposal for the reconstruction of Prussia to the conference of the five powers. The proposal precisely calculated the population Prussia had lost since 1805, the population it had regained and the amount yet to be recovered.[302] The proposal enumerated compensations including the whole of Saxony as well as populations in the Duchy of Warsaw and western German principalities. On 28 January Metternich replied, armed with statistics contained in the final report of the commission (Figure 3.4).[303] He presented a counterproposal for the partition of Saxony and the distribution of other possessions. Although Metternich had proposed similar numbers before, they were not agreed to because they had not been "verified."[304] Metternich's central contention was that Prussia could be compensated and restored to its 1805 population without retaining all of Saxony. Metternich offered 810,268 souls in Polish territories, 1,044,156 along the Rhine, 829,951 in Northern Germany, and 782,249 in Saxony.[305] This amounted to less than half the population of Saxony, but a little more than half the land. Metternich's memo finely tuned the balance of power, recognizing Prussian rights, but offering no more than necessary given the precise calculations of the commission. As Talleyrand states, the commission proved "Prussia does not require Saxony to obtain more than what the treaties had assured her."[306] By demonstrating an epistemic point of fact, the commission made an agreement possible.

Prussia presented another counterproposal on 8 February, claiming 855,305 souls in Saxony. The great powers agreed to this division and the central problem at Vienna was resolved. By the Prussian calculations, Prussia would be compensated in excess of its obligations by 41,630 souls.[307] Their first proposal claimed an excess of 681,914, so negotiations reduced the scale of Prussian gains considerably.[308]

In the case of statistical commission we can clearly see how the emergence of rationalist materialism constituted and altered the fundamental rules of the balance. The precision with which these calculations were executed would have been neither possible nor thinkable two hundred years prior. The balance of power techniques upon which the Congress

---

[302] *British and Foreign State Papers 1814–15*, 602–603.
[303] *British and Foreign State Papers 1814–15*, 606–612.
[304] Talleyrand 1881, 59, 120–123.
[305] *British and Foreign State Papers 1814–15*, 610.
[306] Talleyrand 1891 [1815], 5.
[307] *British and Foreign State Papers 1814–15*, 629.
[308] *British and Foreign State Papers 1814–15*, 603.

(Annexe 4.)—Compensations en Saxe.

LA démarcation Prussienne en Saxe suivroit, en partant des Frontières de la Bohême, la rive droite de la Wittich jusqu'à son embouchure dans la Neisse, et de là la droite de ce fleuve, en laissant Gœrliz et sa banlieue à la Saxe ; plus bas une ligne à tirer depuis Rothenbourg, qui seroit à la Prusse, le Zeissholz, le Gultauer-Heide, entre Kœnigswartha et Wittichenau, sur la Schwarze-Elster, vis-à-vis d'Ortrand ; de là la droite de ce fleuve jusqu'à Elsterwerda, qui seroit à la Prusse, d'où l'on tireroit une ligne jusqu'à l'Elbe, entre Belgern et Torgau. La Prusse auroit la route de Torgau à Eilenbourg, d'où la démarcation suivroit une ligne à tirer par Delitsch, Landsberg, jusqu'au Territoire de Halle. Du côté de la Thuringe, la Frontière seroit tracée par la Saale.

Les Pays Saxons compris dans cette Démarcation sont les suivans :

		Habitans.
a. La Basse-Lusace		143,921
b. De la Haute-Lusace		170,000
c. Du Cercle de Misnie, les Bailliages de		
Senftenberg	5,765	
Finsterwalde	3,218	
Elsterwerda	8,000	
Torgau	22,277	
		39,260
d. Du Cercle de Leipzig :		
A peu près trois-quarts des Bailliages d'Eilenbourg et Duben	14,000	
Zerbig	4,729	
Une Partie du Bailliage de Delitsch	16,000	
		34,729
e. Le Cercle de Wittenberg		110,990
f. Jüterbock et Dahmen		12,998
g. Barby et Gommern		10,309
h. Querfurt et Heldrungen		11,538
i. Partie Saxonne de Mansfeld		28,060
k. Partie du Cercle de Thuringe, savoir :		
Les Bailliages de		
1. Langensalza avec Tennstedt	23,641	
2. Sangershausen	18,860	
3. Sachsenbourg	8,198	
4. Weissensee	16,138	
5. Stolberg	12,552	
6. Enclaves de Schwarzbourg	10,638	
7. Wendelstein	3,054	
8. Ekartsberg	25,475	
9. Freybourg	21,199	
10. Partie de Mersebourg, sur la rive gauche de la Saale	17,000	
		156,755
l. Le Cercle de Neustadt, pour être échangé avec Weimar		38,949
m. Le Comté de Henneberg		24,740
Total		**782,249**

La surface de ces Districts comprend 281 lieues carrées, et par conséquent au-delà de la moitié de la surface totale du Royaume de Saxe.

Figure 3.4 Excerpt from Metternich's proposal for the reconstruction of Prussia (source: *British and Foreign State Papers 1814–15*, 611–12)

of Vienna depended were the result of changes in the deepest levels of political discourse over the course of the eighteenth century. A materialist concept of power measurable in terms of population and revenue, an apparatus of calculation, and a mechanical view of the universe as subject to rational human control made balancing power at Vienna possible and desirable. Moreover, as a result of these changes, the principle of compensation came to be understood in reductive, quantitative terms as an exchange of souls. Thus, at Vienna, to balance meant not merely to oppose the strongest state or potential hegemon, but to arrange populations and parcels of land in such a way as to produce two natural blocs of enmity and alliance.

The settlement of the Polish-Saxon question hinged on the ability of the great powers to precisely measure the balance and distribute territories on the basis of population figures. Although this satisfied the interests of the great powers, those interests were structured by the rise of materialist ideas in natural philosophy and political economy. But more specifically, the statistical commission helped the powers execute a key rule in the balance of power institution: that states should be compensated for losses and for participating in victorious expeditions. The statistical commission made that possible by producing a detailed, precise partition. In this way, the statistical commission helped the great powers simultaneously fulfil their obligations and their interests. But the reductive demands of materialist and quantitative ideas meant that the principle of compensations was interpreted in a specific way, privileging population numbers over the quality of the land or its defensive position.

In short, the discourse of state purpose that dominated the Congress was oriented to a materialist understanding of the balance of power, interpreted in statistical terms. This discourse was constituted by a scientific cosmology that defined the world in terms of matter and motion. While this cosmology was originally grafted onto the providential terms of early modern religious discourses, by 1815 materialism and mechanism were articulated to the view that natural laws governing power could be manipulated by rational human control. So between 1550 and 1815, the emergence of balance of power politics depended on a cosmological shift that weakened the determinist ideas of divine providentialism and bolstered faith in materialist rationalism.

Explaining Balance of Power Politics

In contrast to the view that the balance of power has always dominated international politics, or that the balance of power had to emerge because of the pressures on early modern states, my account provides an ideational history of the balance. The emergence of a shared cosmological discourse helps explain why the European system was built on a balance of power purpose in the first place. As Finnemore says, there is no functional reason that the balance of power had to emerge. After all, "[o]ne can easily have material multipolarity without a concomitant notion of balance of power."[309] A shared conception of the balance supported by consistent rules is necessary to maintain a rigorous system of counter-hegemonic balancing.[310] Otherwise, states are just as likely to bandwagon, hide, or form collective security arrangements. So the emergence of a balance of power order depends both on military-economic dynamics governing the rise and fall of relative power *and* on the construction of rules and discourses grounded in a cosmological vision to legitimate and naturalize certain patterns of behaviour.[311] It is not surprising then, that other international systems looked very different. The succession of East Asian orders centred on China, for example, matched cosmology to purposes and rules in a distinct manner.[312]

Scientific discourses not only framed macro-level understandings of order. They directly shaped the meso-level dynamics of order-building. At Vienna, the mechanical, materialist, and quantitative conception of the balance culminated in the precise calculation of territorial interest in terms of the number of souls on parcels of land. Statistics helped the Congress finely tune the balance while respecting core principles of great power politics. This argument builds on Gulick's contention that the leaders at Vienna intended to produce an equilibrium based on natural blocs of alliance and emnity while maintaining their commitments to common principles, such as the principle of compensation. The great powers initiated a complex system of consciously designing the balance collectively.[313] Mitzen demonstrates that the Congress and the form of politics it produced was more than mere governance by a coalition. It represented the first time that the great powers formed collective intentions about the management of international politics.[314] Indeed, Castlereagh himself saw that the real value of the Congress was in "combining these great Powers in one common system of policy."[315] In line with Mitzen's

---

[309] Finnemore 2003, 99.
[310] Finnemore 2003, 105.
[311] This ideational theory also helps explain the variation between balance of power orders institutionalized between 1550 and 1815. It also helps to explain Wight (1966) and Schroeder's (1989) observation that European leaders understood the balance in many different ways.
[312] Phillips 2011.
[313] Gulick 1955, 306–308.
[314] Mitzen 2013.
[315] Castlereagh 1947 [1816], 509. See also, Webster 1947, 56–57.

argument, I have shown some important epistemic and ontological preconditions for the emergence of the idea that international politics itself can be managed and controlled.

Siding with Gulick here is not a fashionable move. Schroeder argues that Gulick's account cannot be right because the "balance" achieved at Vienna rested on hegemony.[316] However, this misconstrues Gulick's argument. It does not follow that since the settlement rested on a hegemonic or bipolar material distribution, it therefore could not be shaped by ideas about the appropriate way to balance power. On my reading, Gulick shows that the leaders worked within the rules and goals of a shared balance of power institution. Thus, their shared vocabulary and intersubjective world was equilibrist. Whether their agreement was supported by a bipolar or unipolar distribution of military power is a separate concern. They nonetheless *intended* to construct a balanced system rooted in the underlying power realities. Kraehe agrees with Schroeder, but states the argument in a different way.[317] For him, Schroeder is right to reject Gulick's thesis because the "so-called equilibrists were pursuing hegemony, however much they preached balance."[318] However, this ignores the fact that while Russia and Great Britain may have dreamed of achieving hegemony, they had to work within the discourse of state purpose embedded in international order. International order did not legitimate the pursuit of hegemony so it channelled states towards constructing and taking up a position within the balance of power. Moreover, the great powers had to more or less abide by the shared rules and practices that governed international order and this meant engaging in the collective task of creating a balanced system, albeit one that served the most powerful states.[319]

Osiander takes the shared ideas of the great powers seriously, but not the idea of the balance. For him, the balance of power is underdetermined. Many different policies could be justified by use of these terms, and so it could not have dictated behaviour. Specifically, in the case of Saxony, Osiander argues that the great powers denied all of Saxony to Prussia not on balance of power terms, but to punish Prussia for its intransigence.[320] Osiander denies that partitioning Saxony was good for the overall balance, although clearly Metternich thought so and convinced

---

[316] Schroeder 1992, 1994b.
[317] Kraehe 1992.
[318] Kraehe 1992, 709.
[319] In Mitzen's (2013) terms, the force of hypocrisy drives states towards collective intentions and shared principles. In my terms, the institution of the balance channelled private interests in some directions rather than others.
[320] Osiander 1994, 196.

Castlereagh it did, and so he suggests that great power motivation had to lie elsewhere.[321] However, this ignores the extent to which all the powers, including France, thought they were bound to provide Prussia with a set number of souls as compensation due to it by the Treaty of Kalisch. It was this obligation, rooted in the principles of the balance, which gave the statistical commission a central role in Vienna. In the end, Osiander falls into an unnecessary dichotomy. On one side, a material concept of interest drives the balance. On the other, ideational principles of legitimacy motivate states. The fact is that the leaders themselves did not think in these dichotomous terms and so our explanation should not be cast in that way. For the great powers, finely tuning the material balance fulfilled their obligations to Prussia under international law. Territorial interests were bound up with principles. The rules of the balance of power were principles of international law. So it does not make sense to argue, as Osiander does, that principles of legitimacy were more important or more central to explaining the outcomes at Vienna than the balance.

The statistical commission enabled the great powers to balance power while fulfilling their commitments to Prussia. In this way, the precise calculation of the balance played an important role in the agreement. This interpretation goes further than Gulick, who states that Vienna's long statistical tables have "no great value for us today."[322] Indeed, one could dismiss the role of the statistical commission as unimportant to the overall negotiations because the precise division of territory is secondary. However, the statistical tables produced by the commission helped the great powers reach a deal by allowing the precise division of territory. The numbers revealed a mutually agreeable outcome that negotiators did not previously see. The commission thereby opened up the bargaining space and provided a legitimate accounting of territorial units to be disbursed in accordance with the principle of compensations. Moreover, it allowed the territory to be finely divisible. As Goddard points out, ideas about territory and legitimacy can often hinder the ability of states to divide land.[323] In this case, the precise divisions of the statistical commission expanded the number of bargaining outcomes so that all the great powers could agree on one.[324]

Morgenthau went slightly further. For him, the statistical commission is one of many events in world politics when leaders tried to reduce power politics to the "rational calculation" of "objective" factors.[325] In

---

[321] Osiander 1994, 227.
[322] Gulick 1955, 251.
[323] Goddard 2010.
[324] On bargaining and divisibility, see Fearon 1995.
[325] Morgenthau 2006 [1948], 44–45.

the hands of Metternich, "[t]he delimitation of territory thus became a kind of mathematical exercise." One could respond to Morgenthau that it is hyperbolic to suggest that the outcome can be reduced to calculation alone. Certainly, the power position of each of the states was felt in the negotiations and constrained the range of acceptable alternatives. Nonetheless, in the end all the powers agreed to be bound by an authoritative calculation produced by the commission. The case is suggestive of delegation to an expert body and so provides an early case of expert authority in international politics.[326]

Conclusion

The role of the statistical commission at Vienna is symbolic of the radical reconfiguration of European discourses between the sixteenth century and the nineteenth century. First, the representation of the world in mechanical, materialist, and measurable terms eroded the conceptual bases of divine providentialism and aristocratic dynasticism. In addition, they provided the discursive elements for a new balance of power discourse that combined a constellation of new ideas: the doctrine of calculable, knowable state interests; the existence of discoverable natural laws governing political economy; materialist concepts of territory and population; an image of states embedded in a field of forces; statistical techniques for determining compensations; and the whole notion that politics can be controlled. This cosmological constellation placed humans in a rational, knowable, physical universe amenable to human influence. The recursive institutionalization of this constellation in European states and the core sites of international order reoriented the discourse of state purpose from the pursuit of religious dynasticism to the pursuit of material capabilities. Moreover, it created a moral, legitimate consensus that supported the balance of power order consolidated in eighteenth- and nineteenth-century Europe. Cosmological changes reconfigured discourses of state purpose, which in turn altered the institutions of international order at Utrecht and ultimately Vienna.

The desire to control international politics was still inchoate in 1815. Although the ideas of regulation, design, and administration had appeared over the course of the eighteenth century, their institutionalization at the macro-level was weak. Similarly, the associated idea of "improvement" had emerged, but was not yet a prominent state purpose. These ideas first emerged in Britain where improvement had served throughout the eighteenth century as a general term for economic, technological,

---

[326] On expert authority and delegation, see Barnett and Finnemore 2004.

and moral progress at home and in the colonies.[327] In India, the notion of improvement served as a tool of political persuasion to convince the indigenous elite to ally with the colonial regime.[328] Thus, it is no surprise that improvement appears in Castlereagh's papers most often in a colonial context. For example, one memo from an officer contends that British influence is the author of "improvements, agricultural, commercial, and moral" in Haiti and of "beneficent improvements" in West Africa.[329]

By the time of the Vienna Congress, the emergence of a new, more general idea of improvement had been made possible. The statistical arguments underlying the Congress settlement bestowed on the European powers a new understanding of what power consisted in. Power flowed from population, its number and its quality. The next logical step was for European states to intervene in and improve the quality of population. As Talleyrand put it:

> We have at last learned – rather later, perhaps – that for states, as for individuals, real wealth is acquired not by conquering and invading foreign countries, but on the contrary by improving one's own.[330]

For Talleyrand, power politics no longer consisted solely in seizing territory and maintaining a strong defensive line. Now politics was about improving the lives of the people as a means to international strength and standing.

The new purpose was institutionalized, although weakly, in a reference to "improvement" in the 1815 Final Act. In the article by which Russia took possession of the Duchy of Warsaw, a clause notes that "His Imperial Majesty reserves to himself to give this State, enjoying a distinct Administration, the interior improvement which he shall judge proper."[331] Improvement here is related to another emerging concept, "civilization." Talleyrand notes that Poland "should be given to institutions calculated to implant and cultivate all the principles of civilization."[332] Elsewhere he refers to the "advance of civilization."[333] The emergence of improvement surely owed something to the influence of liberal ideas on Russian conduct at Vienna.[334] But it also reflected the

---

[327] Guha 1997, 31–33.
[328] Guha 1997, 33.
[329] Castlereagh 1853 [1818], Vol. 12, 16, 24.
[330] Talleyrand 1881 [1792], 294. Quoted in Osiander 1994, 166.
[331] Final Act of the Vienna Congress, 76.
[332] Talleyrand 1881 [1814], 30.
[333] Talleyrand 1881 [1814], 31.
[334] See, for example, Czartoryski's statement to Talleyrand that "his principle was the welfare of the people" (Talleyrand 1881 [1814], 123).

Conclusion

power of scientific ideas. Newton's concept of absolute time had been used to underwrite linear narratives of progress and the political economists had brought the control of land and peoples into political discourse.

The arrival of improvement at Vienna foreshadows a broader discursive reconfiguration that would unfold over the course of the nineteenth century. After Vienna, political elites became more confident in the idea that politics could be controlled and shaped by human action. This was expressed, internationally, in the intent to construct the balance and, domestically, in the drive to improve population, revenue, and land via administration or "government." This is consistent with Foucault's history, which traces the origins of "governmentality" to Europe in the middle of the eighteenth century. For Foucault, it arises out of the mercantilist imperative to increase the constituents of power. Governmentality emerges when this is translated by political economists into the doctrine that states must improve land and population through administration rooted in knowledge. For Foucault, the shift from "the people" to "population" is a central moment in this story.[335] In nineteenth-century government discourse, population is no longer a collection of subjects, but a set of variables to be governed.[336] Furthermore, the idea of population reorients political economy and the whole field of governing to the management of the human species. Population, Foucault argues, is the "operator of transformation" from natural history to biology, from the analysis of wealth to political economy, from grammar to the history of ideas.[337] So the idea that the population can be improved is central to the constitution of nineteenth-century governmentality, which is structured by the historical sciences of biology and geology.[338]

However, the inchoate nature of the concepts improvement and progress in Vienna shows that the full force of Enlightenment ideas had not yet entered European political discourse by 1815. But as we shall see in the next chapter, new ideas from the emerging historical sciences conveyed improvement and progress into the centre of political discourse. From this vantage point, Francis Bacon (1561–1626), thus far missing from the story, takes on the character of a tragic figure. Bacon now has a place in the pantheon of the history of science, but in his lifetime Bacon was not particularly scientifically influential. When his natural

---

[335] Foucault 2007, 67–79.
[336] Foucault 2007, 74, 77.
[337] Foucault 2007, 78.
[338] Collingwood 1945.

philosophic works did become influential, they were interpreted as promoting an "empiricist" doctrine that emphasized inductive experience over deductive logic. However, Bacon himself advocated working by a "regular system of operations" guided by logic.[339] But more importantly, this conception of Baconianism did not quite capture the modernist core of Bacon's project. For Bacon, the goal of knowledge was "human utility": "the benefit and use of life."[340] He represented the pursuit of knowledge as a voyage, analogous to the discovery of the New World, into the "remoter and more hidden parts of nature."[341] Bacon's idea of science as a voyage depended on and promoted a linear conception of time, in which knowledge allows humankind to see and control the future.[342]

In short, Bacon aimed to use knowledge to cultivate human life and society. He was the first public advocate of the idea that knowledge can be used to control the world. However, neither Europe nor science was ready for this notion in the early seventeenth century. Europe was not ready because it was dominated by providentialist ideas that downplayed human agency in a world of divine and natural laws. Science was not ready because knowledge simply was not very useful in the seventeenth century.[343] But, as we shall see in the next chapter, over the course of the nineteenth century British thinkers and colonial officials used these ideas to transform political discourses and international orders.

---

[339] Bacon 2012 [1620], 17. Reiss 1982, 199–202; Gaukroger 2001, 138–148.
[340] Bacon 2012 [1620], 21.
[341] Bacon 2012 [1620], 19.
[342] Bacon 2012 [1620], 12. See Reiss 1982, 199–200.
[343] See Gaukroger 2007, 472; Wuthnow 1979, 222.

4 Darwin, Social Knowledge, and Development in the British Colonial Office and the League of Nations, 1850–1945

> The inorganic has one final comprehensive law, GRAVITATION. The organic, the other great department of mundane things, rests in like manner on one law and that is, DEVELOPMENT.
>
> – Robert Chambers[1]

Introduction

After 1800, the natural sciences of geology, botany, and zoology produced new ideas about the universe, time, and humanity. They combined the growing fossil record, geologic surveys, theories of organismic development, and the Newtonian idea of absolute time into a cosmic vision of a dynamic, historical earth in which mountains were constructed from millions of everyday events. At first, the new cosmological elements reconfigured divine providentialist ideas into a secular, naturalist variant of determinist thinking. On this view, all entities evolved through stages towards progressive ends according to the laws of nature. Meanwhile, changes in the mathematics of statistics led to a debate about whether these events were determined by God, natural law, or whether they were contingent and subject to human manipulation. The idea that events were "probable" rather than determined was used to challenge determinist narratives of progress.[2] It was in this context that the emerging social sciences articulated a new vision of developmental progress as knowable, predictable, and controllable.

The cosmological shift in the historical and statistical sciences made possible and desirable a series of changes in state purpose. As we saw in the last chapter, the notion of improvement was inchoately institutionalized in the 1815 Vienna Final Act. Over the course of the nineteenth century, improvement became an important element of colonial ideology.[3] In the British Colonial Office, improvement was combined

---

[1] Chambers 1844. Cited in Cowen and Shenton 1996, 2.
[2] Hacking 1975; Hacking 1990; Porter 1986.
[3] Guha 1997.

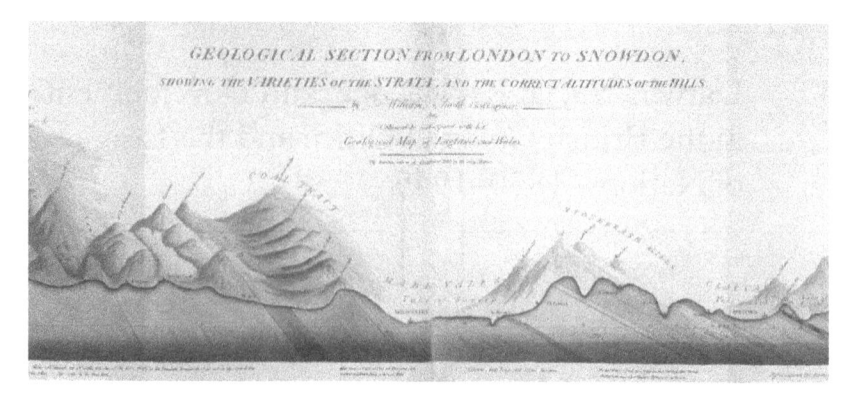

Figure 4.1 Geological section from London to Snowdon, William Smith, 1819

with historical, evolutionary ideas and transformed into the concept of development. Development was initially conceptualized as a deterministic process driven by natural laws. Colonial officials posited that native societies would move naturally through the stages of development from barbarism to civilization. European empires were merely to facilitate this process as trustees.[4] This view of development was recursively institutionalized in international order throughout the nineteenth century. It was spread amongst the European states by a transnational network of natural and social scientists and embedded in multilateral treaties at Berlin (1885) and Versailles (1919).[5] While various versions of this view circulated in European discourses, it was the British hegemon's variant that was institutionalized in the core sites of international order.[6] Thus, while the rise of development was led by the hegemon, it did not need to be imposed on other European states.

However, throughout the 1920s colonized peoples led widespread resistance to British rule.[7] Resistance to development called into question the view that native societies would develop naturally towards Western civilization. This created a crisis in colonial thought and administration.

---

[4] Bain 2003; Bowden 2009.
[5] On the transnational scientific networks spreading this discourse, see Neill 2014; Rosenberg 2012; Stutchey 2005; Tilley and Gordon 2007; Tilley 2011. On the long history of European science and colonialism, see Adas 1989.
[6] On the French and German civilizing missions see, Conklin 1997; Neill 2014; Zimmerman 2010. On imperial ideologies, see Adas 1989; Pagden 1995.
[7] On anti-colonial resistance, see Young 2001, 161–181. Indigenous resistance also challenged colonial ideology throughout the nineteenth century, notably in the Indian Rebellion of 1857 and first Italo-Ethiopian war in the 1890s. On the effects of the 1857 Rebellion on British ideology, see Mantena 2007b, 2010.

Introduction 141

The British response was to harness a new cosmological discourse of epistemic modernism to the project of colonial development. On the new doctrine, an alliance of expert knowledge and a rational state bureaucracy would drive development. This new ideology of late colonial development did not alleviate normative pressure for decolonization. Nonetheless, the core idea of late colonial development, that all states should pursue moral and material progress via the application of scientific knowledge to social and political problems, transformed the discourse of state purpose underlying international order.

Buzan and Lawson offer a multifactor explanation for the emergence of development as a state purpose in the nineteenth and twentieth centuries.[8] First, at the macro-level, the material changes in the mode of production and the movement of capital introduced by the industrial revolution introduced rapid and successful changes in societies throughout Europe. These changes rewarded countries that industrialized and generated pressure for economic expansion. Second, at the meso-level of the British state, narratives of progress and scientific racism reoriented British imperialism towards the spread of British civilization.[9] While these are essential elements of any explanation of the emergence of development, Buzan and Lawson do not provide a theoretical or empirical account of where these ideas come from. By contrast, my account illustrates the meso-level mechanisms by which these ideas entered and transformed the ideational and institutional structure of international order. Moreover, by situating it in the nineteenth century, Buzan and Lawson miss that the rise of development is part of a broader cosmological shift grounded in scientific knowledge production. In what follows, I show that the rise of development as a state purpose must be placed in the context of a long-run process of cosmological change in which scientific ideas have transformed the discourses of international order.

At the meso-level, I show how associational changes in the British Colonial Office produced the discursive configurations that were later institutionalized in the core sites of international order from the 1880s through the 1940s. In particular, I try to explain why the Colonial Office moved to a policy of developing the colonies.[10] In the nineteenth century,

---

[8] Buzan and Lawson 2015.
[9] See also Chakrabarty 2000; Mantena 2010; Pitts 2005.
[10] The meso-level argument is based on a discourse analysis of primary documents drawn from Colonial Office minutes and authoritative texts written by colonial officials and academics. Kirk-Greene (in Hailey 1979 [1940–1942]) reports that works by F.D. Lugard (1922), Raymond Buell (1928), Margery Perham (1937), and Lord Hailey (1938) were the four most "authoritative" texts on African development in colonial policy. I focus on African policy because the concept of development that was institutionalized in the

British colonialism operated on a strict "self-sufficiency" principle: colonies had to pay for infrastructure and bureaucratic costs with local revenue. Moreover, since development through the stages of society was thought to proceed automatically in accordance with natural law, there was no need for active government intervention to advance civilization in the colonies. However, between 1929 and 1945 the policy of self-sufficiency was dismantled and the British Treasury set up a series of funds for state-led colonial development. These initiatives produced new ideas and practices of development that later formed the basis of postwar global public policy in the World Bank, Food and Agricultural Organization, International Labour Organization, and UN Educational, Scientific, and Cultural Organization.[11]

Why did British colonial policy radically change between 1850 and 1945? One possibility is that the shift from liberalism to interventionism simply reflects the domestic politics and interests of the British state. On this view, colonial development was intended to bolster the legitimacy of the imperial enterprise, increase British exports, and restore the balance of payments.[12] These were important factors, but as explanations they leave many questions unanswered. First, the funds allocated were quite small and unlikely to change colonial life or spur economic recovery on their own. Moreover, these explanations beg the question of how it became thinkable that Britain could intervene in and shape the process of development in the colonies. That is, state-led development could not be used to solve domestic political problems until the idea and its concomitant modes of intervention had been invented.

This chapter traces the history of colonial development discourse to the eighteenth century when Enlightenment philosophers articulated a new historical and progressive vision of human development. British thinkers combined these ideas with geological and biological ideas to produce a new vision of human progress as guided by natural laws of growth. However, the failures of colonial development contradicted this vision, inspiring a new generation of anthropologists and social scientists to offer a more complex theory. Experts in political economy, labour, nutrition, and public health created new ways of representing and measuring social reality that made state-led development possible.

post-Second World War settlement emerged from academic and policy projects in the British colonies there. Moreover, British colonies in West and East Africa (including Nigeria, the Gold Coast, Sierra Leone, Kenya, Rhodesia, and others) were overseen directly by the Colonial Office, which provides an opportunity for a focused case study of associational change.

[11] Staples 2006; Finnemore 1996b; Mazower 2012.
[12] Abbott 1971; Constantine 1984; Cowen and Shenton 1996.

These new experts marshalled the authority and techniques of the natural sciences to build faith in the application of knowledge to social and political problems. They thereby displaced the liberal and evolutionary views that had dominated British policy in the nineteenth century. The new expert discourse explained individual behaviour and societal outcomes as a function of social forces such as class dynamics and cultural norms. This form of analysis was extended to various social scientific "objects" that could be controlled via interventions in social and economic life.[13] This reconceptualization displaced "races" and "nations" as the central analytical units for societies and permitted the design of policy interventions meant to alter social processes and forces. The reconceptualization of society and the economy as objects also made possible the idea that they were self-contained units that could progress and develop. In short, modernist social knowledge displaced evolutionary theories of human nature and society, making possible and desirable a new doctrine of state-led development.

The Cosmological Shift in the Historical Sciences, 1750–1850

In the first half of the eighteenth century, the synthesis of Newtonian natural philosophy with Christian theology encouraged the study of nature as God's design. The search for knowledge was a quest for the eternal laws that regulated both human and natural spheres, holding them in timeless, harmonious order. However, in the latter half of the eighteenth century a new historical perspective emerged. Historical thinking in the human and natural sciences promoted the cosmological themes of process, change, and development.[14] First, ontologically the new historical sciences depicted a universe of changing, developing entities. Epistemically, they foregrounded a view of knowledge as tracing cause and effect chains through time. They challenged the old medieval depiction of cyclical time, offering a conception of time as linear development. Finally, they redefined the origins of the universe and humanity in secular terms. This represented the universe in a naturalist cosmogony without reference to divine providence.

Histories of humankind had of course always existed, but these histories were usually ordered genealogically, as in the Bible, and emphasized

---

[13] On scientific objects see Daston 2000; Mitchell 2002.
[14] Collingwood 1945; Stocking 1987, 13, 17; Cowen and Shenton 1996, 14. We cannot say "introduce," as these themes could be seen in the work of Lucretius which had been rediscovered in Italy during the fifteenth century.

themes of decline and degeneration.[15] Enlightenment thinkers presented a new vision of history as a narrative of progress driven by the advance of scientific knowledge.[16] In Turgot's 1751 lectures, he ordered human history into a series of "stages." For him, the central driver of historical progress was the advance of reason.[17] Reason manifested itself in the genius of scientific men, culminating in Newton. At the end of the century, Condorcet (1744–1795) argued that just as the faculties of an individual develop, "according to the same general laws" the "mass of individuals" also develops, providing a "picture of the progress of human intellect."[18] For Condorcet, this was to be a "historical" analysis, in which early nomadic tribes, the Greeks, and the Europeans all existed on the same plane of progress. The progress of the human mind was equated with the overthrow of inquiry from authority and the rise of the "sciences." These narratives gathered together themes inchoate in earlier works that provided a linear, progressive history of humankind.[19]

At the same time, Scottish Enlightenment thinkers conceptualized civilizational progress as an increase in societal complexity.[20] Also working from an analogy to individual development, they argued that society advanced through stages from childhood to maturity.[21] For Adam Smith (1723–1790), this historical work was no mere exercise in conjecture, but the necessary precondition for establishing a Newtonian science of political economy.[22] Smith's goal was to reveal the "laws of motion of the social world" to build an ordered system of liberty that drew on his natural history of human society.[23] Smith argued that human society advances through four stages, driven by increases in the division

---

[15] Stocking 1987, 12. For a sixteenth-century history of humankind, see Lanquet 1548 and my discussion in Chapter 3. For an account of unsuccessful declinist cosmogonies in the seventeenth century, see Porter 1979, 106–108.
[16] On the origins of historicism, see Deudney 2007, 195–198.
[17] Bowden 2009, 60–61; Collingwood 1945, 10; Deudney 2007, 195; Stocking 1987, 14–15. Adam Smith's lectures from the same year on the same theme are lost.
[18] Condorcet 1795, 3. See also Bowden 2009, 63.
[19] Similarly, in Germany, a historical, romantic school of thought articulated progress along both cultural and utilitarian lines (Stocking 1987, 20–25). However, in the German romantic tradition, the progress of culture was often threatened by the progress of mechanist, utilitarian civilization. See, for example, the ambivalent thought of Johan Gottfried Herder.
[20] Deudney 2007, 198–200.
[21] Berry 1997, 64–65.
[22] Redman 1997, 207–215. Mirowski (1989) places Smith in a tradition that borrowed its "substance theory of value" from pre-energy physics concepts. This tradition sought to "reduce economic value to a conserved substance in motion, and thus consequently to elevate moral philosophy and political economy to the status of a natural science" (Mirowski 1989, 186).
[23] Redman 1997, 211; Amadae 2003, 205–207.

of labour.[24] The division of labour makes possible the rise of jurisprudence and the increase in production that constitutes civilization. Smith thought that legitimate laws protecting individual rights would "construct a framework for the achievement of material prosperity, given the universal principle of industriousness."[25] Smith developed a linear understanding of progress reminiscent of the Enlightenment definition of progress as scientific and intellectual development.

In building their stadial theories of human progress, Smith and his contemporaries retained an element of earlier providential cosmology. In divine providentialism, God's will determined human events. On the Enlightenment view, the laws of nature, not divine will, determined the course of history. In Smith, universal human sentiment takes the place of God in driving human events to their predestined, harmonious conclusion.[26] In Kant, "natural law" and "nature" drive inexorable progress towards cosmopolitanism and perpetual peace.[27] Similarly, J.S. Mill (1806–1873) drew on natural providentialist ideas to posit a stage theory of societal "improvement" that he used to justify liberal imperialism.[28]

Mill's vision was inspired by Francois Guizot's (1787–1874) sensational lectures *The History of Civilization* and Auguste Comte's (1798–1857) general theory of societal progress.[29] Guizot conceptualized "civilization" as material progress to be understood in terms of "prosperity" and "well-being." But he also gave an important role to individual mental development. Mill combined this with Comte's linear stadial theory, which posited progress towards civilization defined in terms of material and moral improvement. For Mill, following Smith and Cobden, commerce could be the engine of civilization because it produced the conditions for peace and prosperity in accordance with natural, almost automatically operating laws. In articulating this vision, he explicitly rejected the idea that the balance of power was the sole end of international politics.[30] Instead, progress towards civilization itself deserved to be the central end of states.[31]

---

[24] Berry 1997, 93.
[25] Amadae 2003, 207.
[26] Smith 1976 [1759], 1976 [1776]. See Amadae 2003; Pitts 2005.
[27] Kant 1991 [1795].
[28] Pitts 2005, 138–146; Mantena 2007b, 117–118.
[29] Keene 2005, 162–168; Pitts 2005, 141.
[30] For Keene (2005, 166–167), Mill challenges the conservative, monarchical discourse that emerged at the time of the Concert, on which the balance of power, embodying the achievements of European civilization, was counterposed to French Revolutionary notions of rational progress. However Talleyrand himself, a defender of monarchical discourse and the ancient law at Vienna, had no problem with a progressive concept like "improvement" (e.g. Talleyrand 1881, 283), so we should not strictly counterpose these discourses.
[31] While I am drawing Smith and Mill into the same natural providentialist cosmological frame, see Pitts (2005, 138–146) for important differences in their underlying philosophical anthropologies and social theories.

Enlightenment ideas about human history contributed to the historicization of the natural sciences. In the eighteenth century, geology was a branch of natural history. Natural history in the wake of Newton was less an attempt to reconstruct the chronological history of the earth than a search for the "lawlike" regularities that produced diverse forms of valleys, tributaries, and rock formations.[32] Geologic time was not only restricted by the Newtonian model but by the biblical creation story and the many attempts to establish a date for creation by tracing biblical genealogies.[33] But natural history was transformed when concepts and methods from human history were transposed onto the study of the earth. The earth came to be seen not as "essentially stable and bound by unchanging 'laws of nature'" since the days of creation, but as a "contingent" product of its own history.[34] Together, the human and natural sciences "burst the limits of time."[35]

Geology progressed rapidly in the eighteenth century. The first successes were achieved by Linnaeus (1707–1778) and Buffon (1707–1788) who built the foundations of geological classification and explanation. In Britain eighteenth-century histories of the earth were dominated by natural theology, which attempted to reconcile the fossil record with biblical chronology. In 1795 James Hutton's *Theory of the Earth* broke from this tradition. Hutton refused to rely on the flood or other moments of divine intervention to explain geological phenomena. Fossils and rock strata were to be explained only by "analogy to processes which we are able to observe occurring in the present and by the evidence of the resulting formations."[36] Hutton's history thus posited an ancient earth. This could have brought Hutton into conflict with the religious orthodoxy that dominated the universities.[37] However, many readers generously concluded an ancient earth was consistent with a non-literal reading of "days" in Genesis. Moreover, Hutton could be read, like Newton, as revealing how God's laws work, without pretending to explain why they work. The idea that geology merely revealed God's will left room for God's mystery and providence.[38] Nonetheless, for many the geological record contradicted Christian cosmology.[39]

---

[32] Rudwick 2005, 103–104.
[33] Rudwick 2005, 115–118.
[34] Rudwick 2005, 6.
[35] Rudwick 2005.
[36] Gillispie 1951, 47.
[37] On the dominance of Anglicanism in the universities, see Heyck 1982, 62–70.
[38] Gillispie 1951, 76–78.
[39] Rudwick 2005, 116.

Amongst the British elite, geology slowly displaced the image of a young earth, replacing it with a new conception of the earth as changing slowly over time. Over the course of the nineteenth century, the geological vision of a changing earth merged with Enlightenment histories of human society into a vision of progress. Like the earth, human society developed over time into more complex and civilized forms. It was this basic idea, social evolutionism, rather than the notion that the human species was evolving, that was taken up into the British Colonial Office. This idea emerged from a broad tradition of historical, evolutionary thinking that was revolutionized by Charles Darwin (1809–1882). Darwin did not invent historical or evolutionary ideas, but the cosmological implications of his work raised their profile. By leading a cosmological shift in European thought, Darwinism made possible widespread changes in political discourses.

Darwin was neither the first to bring the historical perspective into biology nor the first to argue for a history of species change. His grandfather, Erasmus Darwin (1731–1802), articulated a theory of evolution that proposed all mammals might have descended from a single organism.[40] Cuvier (1769–1832), "the Napoleon of fossil bones," amassed a database of fossils and employed the techniques of comparative anatomy to argue that the history of life on earth unfolded as a series of "revolutions" not unlike the one he witnessed.[41] Lamarck (1744–1829) argued that organisms were constantly changing within a universe of "continual flux" that created "an endless cycle of environmental change."[42]

Theories of social evolution also pre-dated Darwin. Turgot, Smith, Condorcet, Comte, Herbert Spencer and others had all articulated the view that societies advanced through stages, guided by natural law.[43] Ontologies of change were in the scientific and political atmosphere of early nineteenth-century Europe. In 1844, during the early, heady days of Victorian Britain, Robert Chambers caused a sensation with a best-selling account of evolutionary science.[44] Chambers likened the unfolding of biological development to a natural law and argued that a divine mechanism was not necessary to explain natural history. Chambers' work was controversial, but it reflected a shift from a view of time and

---

[40] Herbert 2011, 19–20. In fact, Charles read Erasmus, likely in 1837, while formulating his theory of natural selection.
[41] Rudwick 2005, 349–388.
[42] Rudwick 2005, 389, 390.
[43] Smith 1997, 478–479; McDonald 1993, 243; Stocking 1987, 14–35.
[44] Cowen and Shenton 1996, 9–12. Chambers' *Vestiges of the Natural History of Creation* was mass-produced for the middle-class reading public. It went through eleven editions between 1844 and 1860, selling over 100,000 copies. See Herbert 2011, 26.

nature as the "life-cycle of all living things" to the notion that all things grow, mature, and progress.[45] But Chambers went much further: by arguing that development was a natural law, akin to gravitation, he secularized linear progress, naturalizing it in biological terms.

The beauty and clarity of Darwin's theory of natural selection presented in *On the Origin of Species* (1859) provided a scientific mechanism to explain evolutionary development.[46] The book built on the zoological and geological traditions of Cuvier, Lamarck, Hutton, and Lyell. These sciences had established the record of historical change, and Darwin offered the explanation. Prior to Darwin, the close fit between species and their environments was thought to be a sign of design.[47] The scientific analysis of species therefore proved God's existence.[48] Darwin's theory was controversial and influential, not because it introduced the ideas of evolution or development, but because he provided a scientific, secular mechanism to explain the fit between species and their environment. From this vantage point, the line of thought from Hutton to Darwin can be seen as gradually circumscribing the role of divine providence in nature.[49] Darwin, in the end, brought the creation of species into nature.[50] In so doing, he provided a central piece of a materialist, secular cosmological configuration.

This cosmology constituted and motivated visions of creative progress in both natural and social domains. The core of the *Origin* is an account of how species produce variation and how certain variants of species are selected. Darwin first argues that species diversify as they move through the "stages of development" (Figure 4.2).[51] Then, he suggests that this process of producing variation leaves species better off: "the modified descendants of any one species will succeed so much the better as they become more diversified in structure."[52] This is because, "the more diversified in structure ... the more places they will be enabled to seize on."[53] This argument picked up an important theme from Adam Smith: progress is driven by increases in the division of labour. So Darwin's theory

---

[45] Cowen and Shenton 1996, 11.
[46] Nonetheless, there was resistance to Darwinian ideas and this delayed their effects. Darwinian ideas spread quickly in academia, but only entered political discourse indirectly, via intellectual development in anthropology, psychology, history, economics, and so on. See Heyck 1982, 114–122.
[47] Gillispie 1951, 219.
[48] Gillispie 1951, 222–226.
[49] Gillispie 1951, 220.
[50] Collingwood 1945, 134.
[51] Darwin 1861, 122.
[52] Darwin 1861, 122.
[53] Darwin 1861, 125.

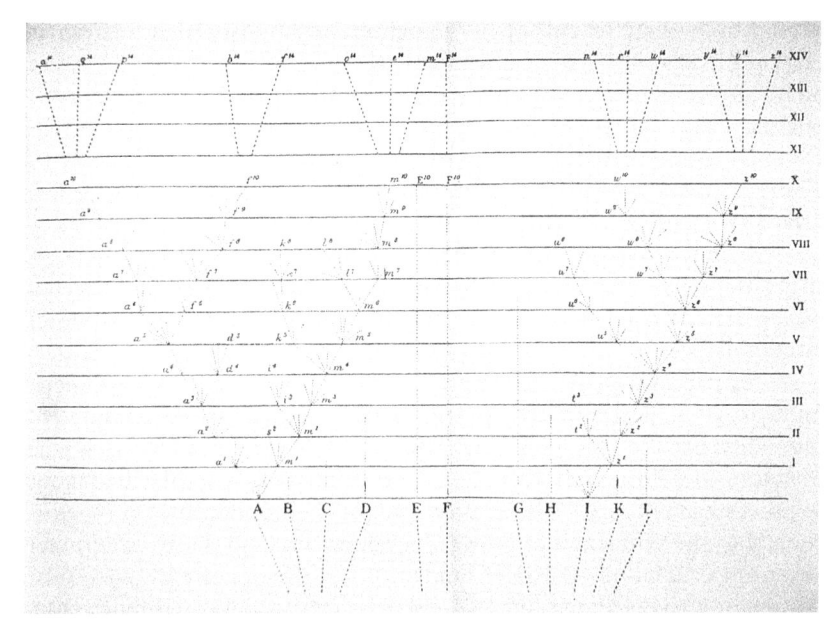

Figure 4.2 Diagram illustrating variation within and between species through the stages of development, Charles Darwin (Darwin 1861, 122–123)

of evolution naturalized the idea that variation and complexity were signs of progress through the stages of development. This, more than biological determinism, structured thinking about colonial development in the British Colonial Office.

Darwin is often linked with the rise of scientific racism and, in particular, with biological determinist thought.[54] Darwin's ideas were certainly used to strengthen already existing beliefs in the superiority of European civilization.[55] Moreover, the scientific authority of Darwin's theory legitimized new racial theories and emboldened generations of biological determinists.[56] However, there were other important variants of Darwinian thought. Indeed, Victorian Britain was the site of a vigorous scientific and moral debate about racism and hierarchy.[57] A central axis of debate was whether the backwardness of savage societies was

---

[54] See Crook 1994; Deudney 2007, 202.
[55] Smith 1997, 481.
[56] Crook 1994, 25–47.
[57] Crook 1994, 29–62. Buzan and Lawson (2015) and others miss this debate between scientific ideas about race.

caused by biology or culture.[58] Biological determinists such as Francis Galton (1822–1911) and his student Karl Pearson (1857–1936) were politically influential in debates about war and eugenicist social reforms, but they were largely ignored in the realm of colonial policy.[59] In the Colonial Office, sociocultural theories of evolution were more influential. These theories argued that progress through evolutionary stages was driven by economic, cultural, and intellectual developments. These theories, which formed the basis of the emerging discipline of anthropology, suggested that humans were still subject to "natural laws" but rejected biological determinism in favour of a more complex understanding of human nature.

Biological and geological ideas thereby promoted a new cosmology built on geological concepts of time and biological understandings of humanity. On the emerging view, "humans were to be treated as a part of nature, and human developments over time were presumed analogous to processes in nature."[60] Physiology and psychology depicted humans as subject to the same laws of growth, development, and decay as all other organisms.[61] The new ideas of humanity and time finally displaced the dominant ancient and medieval view of history as governed by the lifecycle of all living things as a succession of birth, growth, degeneration, and death.[62] In its place rose the idea of progress as "the linear unfolding of the universal potential for human improvement that need not be recurrent, finite or reversible."[63] In other words, Darwinism was used to produce a new biological and historical discursive configuration I call evolutionary developmentalism. Evolutionary developmentalism borrowed the central tenets of earlier forms of divine and natural providentialism. But rather than obeying the timeless laws of God, natural beings were now thought to progress through evolutionary stages guided by the

---

[58] Stocking 1987, 230–236.
[59] McDonald 1993, 243. But as Crook (1994) shows, biological and genetic determinism could be linked to both sides of most issues. For example, Darwin was linked both to militarism via the idea that humanity was a warring animal and to pacifism on the grounds that war was evolutionarily counterproductive because it sent the bravest and most fit out to die. At root there is a tension in Darwinian thought between competition and harmony. In one sense, Darwin challenged the providential view of nature as a harmonious order guaranteed by a benevolent God. By drawing on ideas from political economy and Malthusian natural theology, Darwin elevated competition and strife to the level of scientific principle (see Bowler 1976). Yet, Darwin also presented the image of the "tangled bank" in which many different species harmoniously co-exist. For an alternative reading of Darwin on these points, see Connolly 2017, Ch. 2.
[60] Lorimer 2009, 186.
[61] Cooter 1979, 78.
[62] Cowen and Shenton 1996, 14.
[63] Cowen and Shenton 1996, 14.

laws of organicist development. Natural law no longer held the world in ordered harmony, as in Newton, but drove entities through a series of progressive stages.

Laissez-Faire Liberalism and International Colonial Discourse, 1850–1885

Britain emerged from Vienna in a position of unrivalled strength. It was strengthened further by the long peace maintained under the auspices of the Concert of Europe.[64] The long peace provided Britain with the opportunity to expand its empire and oversee the construction of a global colonial economy.[65] Despite the success of these efforts, in the middle of the nineteenth century the British Empire remained a patchwork, lacking both a clear administrative centre and a unifying ideology. It consisted of possessions in India, the West Indies, Egypt, West Africa, and East Africa. The colonies were administered by an ad hoc collection of government departments. India was ruled out of the India Office, Egypt was governed largely by the Foreign Office, and the Colonial Office administered the rest of the colonial empire. Since there was little in the way of top-down direction from the prime minister or Cabinet, these offices were free to solve problems as they saw fit.

The absence of a coherent ideology or centralized administration did not result in incoherent policy. Rather, colonial policy was unified by the "common sense" of British culture and the curriculum of the Oxford- and Cambridge-educated elite.[66] These institutions absorbed and reproduced the ideologies of liberal imperialism. As a result, foreign and colonial policy in the first half of the nineteenth century was based on the liberal ideas of Smith and Mill.[67] Despite the diversity and sophistication of liberal and scientific thought in the middle of the nineteenth century, the cosmological ideas that ended up shaping British foreign policy were rather simple and reductive.[68] In short, it was the determinist idea that societies moved automatically through stages of development that transformed state purpose.[69] By the middle of the century, British visions of international order were premised on the idea that commerce

---

[64] See Jervis 1985; Mitzen 2013.
[65] Buzan and Lawson 2015; Darwin 2009. The construction of a new colonial order was already underway in 1815. See, Castlereagh 1853 [1818]; Keene 2007.
[66] Ehrich 1973, 650–651. See also Parkinson 1945; Constantine 1984.
[67] Bell 2016; Bell and Sylvest 2006; Cowen and Shenton 1996; Hodge 2007; Howe 2007.
[68] For a range of these views, see Bell 2007, 2016; Mantena 2010; Muthu 2003; Pitts 2005. On the importance of scientific rationalism to these ideologies, see Morgenthau 1946, 13–29; Morgenthau 2006 [1948], 41–49.
[69] Bain 2003, 17–20; Howe 2007, 27.

would create international peace amongst the great powers and civilize native populations in colonies. So whereas in 1815, Britain primarily advocated and promoted the balance of power in international forums, between 1850 and 1900 Britain also began to promote the purpose of civilizational development. It did not abandon balance of power principles, but by 1860 it was clear that the British conception of its purpose in international order had expanded.

In 1860, Britain had just finished establishing a series of tariff reductions meant to shift European order from a basis in balance of power mercantilism to one rooted in laissez-faire liberalism.[70] The 1846 repeal of the Corn Laws was in effect a unilateral shift towards free trade.[71] This was the first in a series of moves designed to build a liberal international order.[72] The 1860 Anglo-French free trade agreement served as a "pacemaker," inspiring agreements with Italy, Belgium, Prussia, and others.[73] The underlying vision motivating this bid for a free trade order was "of a global order based on free trade, peace, and progress in civilization."[74]

This commercial order between European sovereigns was built alongside a colonial order in which Europeans denied subjectivity and sovereignty to indigenous peoples throughout the world.[75] Throughout the nineteenth century, Britain and France competed for access to trade networks and natural resources in the global south. Disagreements were usually settled easily with small territorial concessions and border adjustments. However, as Germany, Belgium, and other European powers sought to expand their holdings and influence, the informal system Britain and France had worked out broke down.[76] The great powers did not want colonial competition in Africa to disrupt the balance of power in Europe. They wanted to preserve the European trading order and continue vigorous geopolitical competition in the colonial world. With these

---

[70] My account draws on a tradition of British historiography that revised the political economic, often Marxist, consensus that dominated accounts of British imperialism from 1900 to 1960. The older view maintained that British imperialism could be explained by economic factors like the power of domestic firms and the necessity of overseas markets to British economic development. In the 1960s and 1970s, Roger Louis placed British economic policy in geopolitical context and gave pride of place to the beliefs of leaders. The subsequent literature still recognizes economic factors, but there is lots of evidence that trade and economic policy served geopolitical conceptions of British interests, not the other way around. See Hyam 2010, 73; Eldridge 1973. See also Earle 1986, 222–225; Baldwin 1985, 71–86.
[71] Howe 2007, 26.
[72] Howe 2007, 33–34.
[73] Howe 2007, 34.
[74] Howe 2007, 27. See Eldridge 1973.
[75] Anghie 2004; Crawford 2002; Grovogui 1996, 2002; Keene 2002.
[76] Newburg 1999, 636–640.

goals in mind, they met in Berlin in 1885 to outline the rules of engagement for African colonization.

British Colonial Discourse, 1850–1885

In order to clearly show changes in the discourse of state purpose, in this section and the next I present a discourse analysis of British foreign policy documents and the proceedings of the 1885 Berlin Conference. Comparing these two discourses reveals the importance of the British hegemon in leading the institutionalization of new purposes in the middle of the nineteenth century. While other European powers also held imperialist doctrines premised upon civilizational development, the version that was institutionalized at Berlin drew heavily on British ideas.[77] Thus, recursive institutionalization took a hybrid form. On the one hand, the core associations carrying and reproducing international order already shared important cosmological and purposive ideas. On the other hand, the role of the hegemon in uniting and institutionalizing these ideas cannot be ignored.

British colonial discourse in the mid-nineteenth century was dominated by laissez-faire liberalism. Circa 1860, the central goal of the British Colonial Office is to advance the interests of the Empire and the world at large by facilitating trade. British interests are equated with the advance of "free and unrestricted commerce" which would ameliorate "the conditions of life of the European residents and the civilization and improvement of the native population."[78]

Traditional economic interests aimed at generating revenue for the good of the empire are balanced alongside humanitarian goals such as the suppression of slavery, the advancement of "civilized rule," and the "improvement" of native societies.[79] In one memo, an official declares that the central aims of the British government are "the abolition of slavery and the civilization of Africa by the extension of legitimate commerce."[80] In the 1870s, the Liberal Colonial Secretary argue that withdrawal from the Gold Coast would "destroy all hope of improvement of the natives ... and all prospect would be lost of opening the interior to commerce."[81] Leading up to the Berlin Conference in 1884–1885, one official argues that the British should press for rules that permitted

---

[77] For the French and German cases, see Conklin 1997; Neill 2014; Zimmerman 2010.
[78] Partridge and Gillard 1996, 5, 20.
[79] Partridge and Gillard 1996, 82–101.
[80] Partridge and Gillard 1996, 101.
[81] CO 96/104 quoted in Eldridge 1973, 157.

"legitimate commerce" which "will confer the advantages of civilization on the natives, and extinguish such evils as the internal Slave Trade, by which their progress is at present retarded."[82] Progress is barred by moral and political evils that the British must remove on behalf of the barbarians.

In short, British officials believed that free commerce unleashed a natural process of improvement from primitive to agricultural to industrial society.[83] Suppressing the slave trade would encourage the natural inclinations of free men to sell their labour for wages. As individuals entered a free labour market, their natural inclination to industriousness and desire for well-being would increase welfare and fuel commercial development. Guided by universal natural law, communication and commerce would stimulate primitive peoples to take up productive labour, generate desires for baubles and trinkets, and thereby lead inexorably to a material and scientific civilization.[84] Neither trusteeship nor direct intervention was necessary for progress. This conception of colonial purpose was backstopped by the cosmological configuration of natural providentialism and determinist evolutionism.

International Colonial Discourse, 1850–1885

The discourse of the Berlin Conference reflects that fact that between 1860 and 1885 the British hegemon drove processes of associational change that created a constituency for the institutionalization of laissez-faire liberalism and new ideas about human welfare. The final treaty signed at Berlin affirms the central ends of states are to establish "free navigation" and "freedom of trade" in the basin of the Congo. These terms mark the central compromise of Berlin: colonial competition was not to spill over and disrupt the balance of power amongst the European powers. In its general content then, the discourse of state purpose here includes both the balance of power and the advance of liberal commerce. However, Berlin is important because it shows that a new configuration of evolutionary development had entered the discourse of state purpose, constituting new practices and interests.

---

[82] Partridge and Gillard 1996, 176.
[83] This is not without some basis in a careful reading of Smith. As Amadae argues, Smith thought that legal protection for individual rights would "construct a framework for the achievement of material prosperity, given the universal principle of industriousness" (2003, 207).
[84] Cooper argues that this rested on a free labour ideology: "the belief that a labor market unconstrained by bonds of personal servitude and governmental coercion provides the best means to achieve a just wage, just working conditions, and social progress" (1989, 745). For Cowen and Shenton, the British believed that the spread of free commerce

The emergence of development in the international discourse of state purpose is evident in the European powers' justification of the colonial enterprise in humanitarian terms. Free commerce in the Congo Basin would contribute to the "development of trade and civilization" of the continent. It would improve the "moral and material well-being of the native populations."[85] The General Act aims to "regulate" commerce and competition amongst the great powers. These regulations will underwrite "improvement" now understood, not in the exclusive balance of power terms of improving the populations and lands, but more broadly as advancing the welfare of people.

The proceedings of the conference reveal the extent to which British laissez-faire liberalism and its associated cosmology was accepted in other countries. Bismarck, serving in his capacity as President of the Congress, announces:

> [A]ll the invited governments share the wish to bring the natives within the pale of civilization by opening up the interior of that continent to commerce, by giving its inhabitants the means of instructing themselves, by encouraging missions and enterprises calculated to spread useful knowledge, and by preparing the way for the suppression of slavery.[86]

The spread of education and knowledge, along with the end of the slave trade, are means to the ends of civilizational progress. But in Bismarck's comments, it is also clear that his central goals are the "natural development of commerce" and "restraining commercial rivalries within the bounds of legitimate competition."[87] Similarly, the Portuguese delegate refers primarily to the "liberty of commerce and navigation" but also aims to deliver the "progress of civilization" to the natives.[88]

At the cosmological level, this is supported by the belief that nature obeyed deterministic natural laws that drove the unfolding of human nature in linear, progressive time. This discursive formation sat alongside the old balance of power discourse. But now, the economic imperatives expressed in laissez-faire principles are independent from security concerns. At Vienna, improvement was embedded with the balance: economic ends served the ends of enhancing power. But here, economic ends have their own justification: to advance and civilize human welfare. This has been regarded as a liberal idea in international order.[89] What

would unleash "the potentially unlimited capacity for improvement through the human effort of labor" (1996, 14). See Conklin (1997) for the French view.
[85] Gavin 1973, 288.
[86] Bismarck 1884 in Gavin 1973, 129.
[87] Bismarck 1884 in Gavin 1973, 130, 129.
[88] Penafiel 1884 in Gavin 1973, 135.
[89] Reus-Smit 1999, 124–126.

has not been demonstrated before is that this liberal idea depended on a complex concatenation of scientific ideas that had been accumulating in international order for three hundred years.

The Cosmological Shift in Social Knowledge, 1885–1945

By 1885, the discourse of state purpose reflected the rise and power of the historical sciences in the wake of Darwin. This cosmological shift introduced by the historical sciences united two central ideas: a naturalist ontology of the world and a linear conception of time in which natural entities develop. Together, these ideas suggested a universe in which all entities progressively develop through evolutionary stages in accordance with natural laws. Throughout the latter half of the nineteenth century new forms of statistical, natural, and social knowledge challenged the determinist elements of this configuration. These forms of knowledge promised to provide states with the expertise necessary to control the complex social and political problems governments faced in the nineteenth century.

The rapid bureaucratization of European states in the nineteenth century created demand for new forms of expertise and opened a new channel by which scientific ideas could enter political discourses.[90] In the seventeenth and eighteenth centuries, scientific ideas had primarily entered international politics via aristocratic brokers. However, starting in the latter half of the eighteenth century, newly formed government offices to manage foreign policy, commerce, health, labour, and so on provided a more direct route for knowledge to enter state discourses. These state institutions emerged as key brokers. Meanwhile, the professionalization of the natural and social sciences created a broader landscape of disciplinary knowledge.[91] These new disciplines provided the specialized knowledge demanded by the growing ranks of political associations, corporations, and state bureaucracies.[92]

These developments created a new class of experts that promised to "replace aristocracy with a professionalized meritocracy, to push aside the well-connected amateurs and bring in new cadres of educated and rational elites."[93] As the social and political influence of the aristocracy waned and that of state bureaucracies, businesses, and political associations rose, scientific knowledge was increasingly channelled into politics directly from specialized disciplines and semischolarly political

[90] de Swaan 1998; Rueschemeyer and Skocpol 1996; Silberman 1983.
[91] Ross 1991; Wittrock and Wagner 1996; Wagner 2003a.
[92] Wagner 2003a, 543.
[93] Mazower 2012, 95.

associations.[94] This allowed more complex, technical, and practical forms of knowledge to move into the associational basis of international politics. In 1815, European aristocratic elites had just begun to believe that knowledge from political economy, agriculture, and public health could be used to enhance their position in the balance of power by improving lands and peoples. But by the early twentieth century, the use of expertise in European government was ubiquitous.

Two important forms of social and political knowledge emerged at this time. First, the new discipline of anthropology emerged out of ethnology, comparative philology, and history.[95] Anthropological thought initially strengthened providentialist thinking about linear progress, but transposed it from a commercial to a Darwinian context. However, twentieth-century anthropologists challenged the determinist beliefs of their nineteenth-century predecessors, offering a more nuanced image of the historical development of society. Second, there arose forms of social knowledge constructed around new types of social objects and problem-centred expertise. These forms of social knowledge made possible new active, interventionist ways of thinking about development that would later displace the providentialist assumptions of earlier colonial discourse.

The Invention of British Anthropology

Darwin inspired a sociocultural interpretation of human development that formed the basis of the new discipline of anthropology that emerged in the 1870s and 1880s.[96] Writers like John Lubbock, John McLennan, and Edward Burnett Tylor took up the central problematic of the Darwinian age, the origins of humankind, and sought to address it with a comparative analysis of human development. These thinkers were self-consciously "scientific" and modelled their work on astronomy and geology, the two most visible and successful sciences in nineteenth-century Britain.[97]

It was the sociocultural evolutionism of Lubbock and Tylor, rather than the scientific racism of Galton and Pearson, that directly shaped British colonial policy from the 1880s through the 1920s. The core tenet of evolutionism was that individuals and societies were autonomous entities that

---

[94] Wagner 2003a, 545–546; Wagner 2003b, 592.
[95] On the influence of anthropology and ethnography on colonial policy, see Asad 1973; Kuklick 1978; Mantena 2010; Povinelli 2002; Steinmetz 2007; Stocking 1987; Tilley 2011; Tilley and Gordon 2007.
[96] Stocking 1987, 146.
[97] See, e.g., Tylor 1871, 29.

158   Darwin, Social Knowledge, and Development

developed "along a linear evolutionary sequence" from childhood to maturity or savagery to civilization.[98] Unlike its biological determinist relatives, the sociocultural theory of evolution could be easily linked to the liberal, paternalist view that the task of colonialism was to guide primitive peoples through these stages. Evolutionist theory merged the stage theory of Smith and Condorcet with a geological model of change. Stages were analogous to strata and the task of the anthropologist was to excavate evidence of the process of development from its sedimentation in surviving forms and relics.[99] Sociocultural evolutionism did not lock primitive peoples into a strict racial hierarchy but it nonetheless legitimized the brutal exploitation of peoples. Under its influence, the British liberal elite had full confidence both in their own racial superiority and "belief in the progress of the 'barbarian' and the 'savage' towards a civilised ideal over time."[100]

Karuna Mantena has recently argued that British anthropology in the tradition of Henry Maine helped shift British colonial ideology from a universalist to a culturalist ethos in the late nineteenth and early twentieth centuries. For her, whereas the liberal, reformist imperialism of the early and mid-nineteenth century advocated the radical reconstruction of native societies along Western lines, Maine and others made possible a culturalist ethos that "recognized and respected the cultural specificity of native society."[101] My story here also tracks the rise of anthropological ideas about society and culture that posited differences between Western and non-Western societies. However, while I concur with Mantena that these ideas made officials in the Colonial Office more attentive to native institutions, they did not eliminate the underlying universalist, unidirectional presumptions of imperial ideology. Instead, evolutionary ideas from anthropology strengthened the idea that all societies would progress towards a Western civilizational and economic ideal underwritten by scientific and technological progress.[102]

Mantena is right that the idea of native society as an autonomous sphere had a profound influence on colonial discourse. It made possible

---

[98] Kuklick 1978, 98. Mantena (2010, 87–88) downplays the connection between anthropological thought and the stadial theory of progress, but as I show below, evolutionism was still a major premise of this body of thought and one of the key ideas that was taken up into the British Colonial Office.
[99] Kuklick 1978, 98; Stocking 1987, 163.
[100] Lorimer 2009, 194.
[101] Mantena 2010, 6. For a similar line of thought, see Chakrabarty 2000, 7–11.
[102] Some of the differences between my argument and Mantena's account may be a result of the fact that her analysis focuses on India, where Maine had a direct influence on policy, and my analysis centres on the Colonial Office, which mainly governed African colonies. That said, Mantena's (2010, 171–177) claims about the application of ideas about native society to African colonialism are missing some key elements, which I outline below.

The Cosmological Shift in Social Knowledge 159

the notion that there was a crisis in native society that imperialism had to deal with. However, for officials in the Colonial Office the solution was not, as Mantena maintains, the defence of static native institutions. Rather, at first, the protection of native institutions through indirect rule was justified because it would enable societies to go through the natural stages of sociocultural development.[103] Later, after 1930, when the idea of a natural process of sociocultural development seemed untenable, the crisis of native society was used to legitimate state-led development through interventions in political-economy, labour, and education. Thus, the common backdrop of British colonial policy from the 1850s through the 1940s was the belief that British imperialism was helping to advance the progress and civilization of colonized peoples. This backdrop was drawn, as we shall see, from anthropological thinking in the Darwinian tradition led by Tylor and others.

E.B. Tylor took up the first post as reader in anthropology at Oxford in 1884. Tylor made his name in the 1860s and 1870s with works that clearly articulated the evolutionist vision.[104] In doing so, he equated progress with "movement along a measured line from grade to grade of actual savagery, barbarism, and civilization."[105] His method was to treat existing indigenous societies as the embodiment of "the savage state" which Europeans had already passed through: "the savage state in some measure represents an early condition of mankind, out of which the higher culture has gradually been developed or evolved, by processes still in regular operation."[106] Thus, just as Hutton had rebuffed providential and Vulcanist theories by clearly showing evidence of ongoing geological processes that could explain the long history of the earth, Tylor offered up present societies and processes as evidence of a long history of human development. For Tylor, these laws were sociocultural, but no less amenable to scientific discovery than the laws of astronomy or geology.

Later diffusionist and functionalist strands of anthropology were created in opposition to the evolutionist approach of Tylor. Diffusionism was the dominant approach to British anthropology around 1910, but it was soon displaced functionalist anthropology led by Bronisław Malinowski

---

[103] Some evidence for this view appears in Mantena's own readings. For example, she cites Perham's statement that "the preservation of native law and custom is ... a transitional stage" (Mantena 2010, 176). So native societies were not static; they were developing towards a civilized end. However, Perham and others were arguing that the process of civilization needed to unfold at a natural pace. This argument was, as we shall see, influential in the Colonial Office.
[104] See Stocking 1987, 157–165.
[105] Tylor 1871, 28.
[106] Tylor 1871, 28.

and A.R. Radcliffe-Brown.[107] At first, functionalists were criticized for focusing on the synchronic analysis of culture at the expense of a historical, diachronic analysis of human development privileged in evolutionism and diffusionism.[108] However, over time the functionalists took an interest in cultural contact and change. In doing so, Malinowski and his followers were not unsympathetic to evolutionism, but their model modified it along functional lines.[109] Rather than assume the process of societal evolution was deterministic, automatic, linear, or universal, functional anthropology argued that the primacy of beliefs and sociocultural factors meant that the process of development was contingent and unique. Thus, functionalist anthropology challenged the epistemic and ontological underpinnings of evolutionism.

The challenge to evolutionism followed from the basic assumptions of the functional approach. Malinowski argued that the customs and institutions of primitive groups were embedded in the social bonds of society.[110] He privileged in-depth fieldwork and argued that social anthropology needed to closely study "native behaviour" to understand the origins of social and economic problems in the colonies. Further, functional anthropology produced an image of society as a system of institutions (familial, cultural, juridical, or economic) that served basic human needs. The task of functional anthropology was to achieve the "integration of all the details observed" so as to depict the overall structure of society.[111]

Starting in the late 1920s and early 1930s, Malinowski and his students turned to address the criticism that they ignored history. In so doing, they called for an expansion of theories of development and thus a change in anthropological knowledge itself. This turn followed Malinowski's rise to prominence in British society and colonial policy circles from his position at the London School of Economics (LSE). During the interwar period, Malinowski established himself as an advisor to the Colonial Office.[112] He was involved in the creation of the International African Institute of Languages and Cultures that grew out

[107] Kuklick 1978, 99. A brief note on diffusionism: "Diffusionists argued that modern civilization grew out of a culture complex called the 'archaic civilization,' which originated in Egypt and was carried throughout the world by a hardy band called the 'children of the sun.' Doing diffusionist analysis meant tracing the migrations of this group" (Kuklick 1978, 99). Among functionalists, Radcliffe-Brown rose to prominence first, but left Britain to train colonial officers in South Africa and Australia, leaving Malinowski as the most influential anthropologist in Britain.
[108] Kuklick 1978, 96; Leach 1966, 560–561.
[109] Stocking 1987, 292–293; Stocking 1995, 248–249.
[110] Malinowski 1961 [1922]. See Stocking 1987, 321.
[111] Malinowski 1961 [1922], 84.
[112] Kuklick 1978, 95.

of a Colonial Office advisory committee. The role of the African Institute was to foster conversations between colonial officers, anthropologists, businessmen, missionaries, and Africans. The Institute linked academic and political institutions, offering Malinowski a position as broker between the two spheres. Government officials attended Institute lectures and Malinowski's seminars at LSE.[113] This support helped Malinowski and his functionalist students dominate British anthropology. Colonial governments were directed to sponsor only functionalist anthropologists and Malinowski influenced the dissemination of Institute funds provided by the Rockefeller Foundation.[114]

In this context, Malinowski moved to apply functionalist anthropology to colonial problems. In a 1929 paper delivered at the Institute Malinowski defended a "practical" social anthropology concerned with "the changing African" and contract between "European culture and primitive tribal life."[115] He argued that regardless of the preferred governance strategy (though Malinowski certainly had his preferences), "a full knowledge of indigenous culture in the special subjects is indispensable."[116] But to do this, anthropology had to move away from its focus on "the explanation of customs which appear to us strange, quaint, incomprehensible."[117] This was an essential piece of Malinowski's overall view: anthropologists should seek to explain the logic and purpose behind apparently strange customs by subsuming themselves within the world of the indigenous people they were studying. Anthropology was only useful if it revealed the complex structure of rules and beliefs that governed land tenure, justice, art, and so on.

Once inside the world of the native, schemes could be devised that fit with the logic and norms of native societies. For example, Malinowski was confident that anthropological analysis could help improve labour output in the colonies:

Forced labour, conscription or voluntary labour contracts, and the difficulties of obtaining sufficient numbers – all these form another type of practical difficulties in the colonies. The chief trouble in all this is to entice the Native or persuade him to keep him satisfied while he works for the white man ... anthropological generalization teaches that satisfactory conditions of work are obtained by reproducing those conditions under which the native works within his own culture.[118]

---

[113] Kuklick 1978, 96.
[114] Kuklick 1978, 96–97; Stocking 1995, 399–426.
[115] Malinowski 1929, 22.
[116] Malinowski 1929, 24.
[117] Malinowski 1929, 27.
[118] Malinowski 1929, 35.

For Malinowksi, anthropology was important because the analysis of native culture, anthropology could be used to design effective schemes that motivated natives to participate in the colonial economy.

However, Malinowski, writing in 1929, was aware that the natives themselves were not operating within a closed world of indigenous culture. Contact with the Europeans was disrupting and altering the practices and beliefs of the native. Thus, to be useful anthropology would have to create a new branch: "the anthropology of the changing Native."[119] The central question of how "European influence is being diffused into native communities" was vastly understudied.[120] Here, Malinowski was responding to the claim that functionalist anthropology could not do diachronic analysis. For one, the diffusionists were not closely studying these changes anyhow, so it was up to the functionalists. More importantly, Malinowski believed that functionalist theory was well suited to studying change. The careful methods of the functionalists would be able "to study the savage as he is, that is, influenced by European culture, and then to eliminate those new influences and reconstruct the pre-European status."[121] Such a study would be necessary if anthropology were to serve a practical purpose.

However, Malinowski pointed out that anthropology, even of the functionalist variety, was not yet ready to serve. For example, "[t]he honest anthropologist will have to confess at once that as subject-matter primitive economics has been neglected."[122] But economic, demographic, and medical questions would be essential from the practical point of view. Thus, Malinowski called for "studies of primitive economics, primitive jurisprudence, questions of land tenure, of indigenous financial systems and taxation, a correct understanding of the principles of African indigenous education, as well as wider problems of population, hygiene and changing outlook."[123]

The shift in the dominant ideas of anthropology from evolutionism to functionalism tracks the shift in colonial policy from faith in automative, laissez-faire liberalism to a more interventionist form of developmentalism. Anthropological knowledge was not solely responsible for this broader shift, but it provided the discursive architecture within which colonial problems were conceptualized and grappled with. The close ties between colonial officials and anthropologists meant that anthropology

---

[119] Malinowski 1929, 36. See also Mair 1936.
[120] Malinowski 1929, 36.
[121] Malinowski 1929, 28.
[122] Malinowski 1929, 33.
[123] Malinowski 1929, 23. See also Mair 1936.

had a major influence on how the Colonial Office understood the problems of colonial development. As we shall see, the increasingly practical orientation of social anthropology in the 1930s opened up a space for the incorporation of other forms of social knowledge such as public health and political economy into colonial policy.

Social Objects and Expertise

A second strand of British social thought grew out of the problems of the 1830s, the decade of the "social question": "the complex of poverty, class conflict, and racial and ethnic diversity created by industrialization and its social dislocations."[124] Urban squalor, crime, and disease created demand for new forms of expertise to help improve the welfare of citizens in industrial Britain. The problems of this era increased demand for new forms of what I call social knowledge. Social knowledge drew on both the natural sciences and forms of expertise emerging from what we can think of as the proto-social sciences. These forms of social knowledge were centred on a series of distinct objects of analysis like labour, public health, agriculture and education.

A couple of trends in knowledge production converged to constitute and shape social knowledge in the late nineteenth and early twentieth centuries. First, over the period 1870 through 1945 the rise of colonial science transformed the landscape of European knowledge. The sciences of geography, geology, medicine, agriculture, forestry, anthropology, and political economy all benefited from the late colonial scramble for Africa. They treated African colonies as an "imperial laboratory" and used the data to rapidly expand both basic and applied knowledge.[125] The development of these sciences was a transnational phenomenon. Scientific societies and associations in Britain, France, Germany, Portugal, and Belgium attended one another's conferences, exchanged materials, and developed common modes of discourse and practice.[126] These sciences formed around a series of epistemic objects increasingly defined in mathematical, statistical terms: land, public health, nutrition, forests, labour, the economy, and so on.

Meanwhile, the social sciences emerged from the complex amalgam of political economy, statistics, and moral philosophy that had previously formed the background knowledge of Europe's liberal educated

---

[124] Ross 2003, 209. Other versions of the same problematic appeared in France, Prussia, and other European polities. See Porter 2003, 26; Wagner 2003a, 540.
[125] Tilley 2011.
[126] Neill 2014; Rosenberg 2012; Stutchey 2005; Tilley and Gordon 2007; Tilley 2011.

elite.[127] The forerunners and founders of the social sciences "embraced the Enlightenment ideal of modernity as a progressive and culminating stage in human history, grounded in individual liberty and guided by scientific social knowledge."[128] Although not all social scientists agreed with the Comtean positivist creed that the social sciences should be modelled on the natural sciences, many borrowed authority and concepts from biology, geology, physics, and cartography.[129]

These new knowledge forms drew upon a new episteme and ontology centred on objects like society, culture, public health, and classes.[130] These objects could not be seen with the naked eye and so they had to be created and rendered legible by expert knowledge. The basic idea was that the properties and dynamics of these entities were to be explained by placing them in a field of forces. This was an old idea in the natural sciences. It had emerged with mechanism and been raised to the level of first principles in the Newtonian era. But advances in experimental sciences such as chemistry reinvigorated the image of entities in fields of forces.

In the context of the social sciences, this experimental vision suggested that individual behaviour and social outcomes were not to be explained in terms of human nature or race or national character or individual morality, but by the "composition of the social forces."[131] The new social scientific discourse displaced the naturalism and evolutionism of earlier social knowledge and permitted a host of new ideas about how states might intervene in the operations of these objects. If the behaviour of individuals and phenomena were produced by "the composition of the social forces," then experts could isolate and control those forces, improving welfare.

The rise of statistics also played an important role in these changes. When they first emerged in public discourse, statistical regularities such as the ratio of male to female births or the suicide rate undermined the idea that government action could affect social progress. Statistical regularities were interpreted as reflecting eternal realities of human nature. This left little room for either free will or government to shape the course of history.[132] This interpretation of statistical laws was consistent with the determinist and evolutionary cosmology that dominated intellectual

---

[127] McDonald 1993, 216–261; Porter 2003, 16–18.
[128] Ross 2003, 208–209.
[129] Ross 2003, 215.
[130] On the objects of development, see Mitchell 2002. On scientific objects, see Daston 2000; Allan 2017a.
[131] Observations on the International Statistical Congress, London 1860, quoted in Mazower 2012, 95.
[132] Porter 1986, 64.

thought in Britain and France up until the mid-nineteenth century. The view that government could not and should not intervene was rooted in the idea that "society" was an ontologically distinct entity from "the state." Society was an aggregate of individuals, and individual behaviour was determined by human nature. The aggregation of human nature in society produced social laws that could be recovered by statistics. Thus, society was "governed naturally by statistical laws" that obeyed their own historical, evolutionary logic.[133] For early statistical thinkers like Quetelet, secular progress was driven by increases in societal knowledge and learning slowly over time. Therefore, the "wise legislator would not try to impose his will on the social system" but rather work within the constraints imposed by "secular social evolution."[134] In 1850s Britain, this bolstered laissez-faire liberalism, which aimed to reduce the role of government in social and commercial life.[135]

However, a change in theories of statistical law made possible a reconfiguration of these ideas. In the 1860s and 1870s, "determinism became untenable precisely when social thinkers who used numbers became unwilling to overlook the diversity of the component individuals in society."[136] The diversity of human behaviours gave lie to the idea that statistical laws expressed natural laws. In effect, the new forms of social knowledge discredited naturalist views of statistics, foreshadowing and contributing to the rise of ontological indeterminism that revolutionized physics in the quantum era.[137] But the short-run effect was a shift in the conception of government. The new forms of natural and scientific knowledge that emerged in the late nineteenth century posited a world of objects that could be controlled by knowledge.

These trends in colonial science, the nascent social sciences, and statistical thinking transformed the cosmological basis of colonial discourses. First, they oriented colonialism to a new episteme and ontology centred on objects. Second, more deeply, they changed cosmological ideas about the role and possibilities of human agency in the universe. Taken together, these new forms of knowledge produced a late colonial purpose on which the forces of societies could be mapped and controlled by knowledge. I call this discursive configuration, modifying Scott, epistemic modernism. It was epistemic in the sense that it was premised on a faith in knowledge. It was modernist because it was premised upon the

---

[133] Porter 1986, 56.
[134] Porter 1986, 56.
[135] Porter 1986, 57.
[136] Porter 1986, 151.
[137] See Porter 1986, 193–227.

idea that knowledge could help states master nature and design rational social and political orders.[138]

Particularly important in the colonial context was the constitution of society, public health, labour, and the economy as abstract objects. In eighteenth-century European thought, society referred to an "aggregation of human beings that have come together for a purpose," as in a geographical or scientific society.[139] Only in the late nineteenth century did the term society come to refer to the space between the household and the state.[140] Like the state and the household, society came to be seen as a single, self-contained entity analogous to the "nation" as a source of the state. In the nineteenth century, society was variously conceptualized as culture, race, or collections of statistical aggregations. But in the early twentieth century Durkheimian sociology and functionalist anthropology introduced a new image of "society" as a set of shared social ties linked to integrated institutions.[141] Functionalist anthropology challenged the naturalist and organicist assumptions of earlier forms of anthropology, conceptualizing society as a collection of "social forces" that shape human behaviour.[142] These social forces were no longer conceived as rooted in laws of human nature and as such they could be influenced by government policy.

In the early nineteenth century, public health issues like hunger and disease were, like labour, part of a divinely ordained Malthusian moral universe in which ill-health was a natural fact caused by immoral behaviour and lack of virtue. Over the course of the nineteenth century, public health research and legislation had transformed "disease" from a moral condition to a scientific object that could be controlled by sanitation and medicine.[143] Underlying this shift was the scientific discovery of bacteria and viruses. The discovery of bacteria made it possible to see that epidemics were the result of the social and economic forces that produced the crowded, dirty conditions that facilitated the spread of disease. Scientific concepts and techniques were used to abstract disease from its local and cultural context. Once it was constituted as an abstract object caused by social forces, disease was amenable to state interventions.

---

[138] This is based on, Scott 1998, 4–5. Whereas Scott focuses on high modernism, especially in the post-Second World War period, in this period modernism is, if not low, then simply standard modernism.
[139] Wagner 2000, 133.
[140] Wagner 2000, 134.
[141] On the history of this idea, see Mantena 2010, 56–88.
[142] E.g. Malinowski 1961 [1922], 157–159.
[143] Vernon 2007, 2.

In a parallel development, pressure for a government food policy grew out of the social analysis of poverty by Seebohm Rowntree. Rowntree made important contributions to the statistical representation of abstract phenomena like poverty. His statistical tables and graphs (e.g. Figure 4.3) used data to illustrate the differences between classes and defined poverty in quantitative terms. Rowntree also broke with nineteenth-century moral determinism by arguing that hunger was caused by social factors.[144] This was made possible by a reconceptualization of what hunger was. Just as disease had to be abstracted from its local and cultural context, hunger was also constituted as a scientific object before it entered government policy. Rowntree conceived of the human body as a motor and food as necessary fuel. This fuel was composed primarily of calories and protein. This reductive conception of food permitted the creation of an abstract, quantitative standard of nutrition.

Rowntree argued that calories and protein could be combined into a "man-value" that indicated the amount of food necessary to maintain a labouring man. The discovery of vitamins contributed to this reduction and quantification of nutrition.[145] In this way Rowntree and others "developed a range of techniques that appeared to allow objective, standardized and universal ways of defining and measuring hunger."[146] Hunger was then redefined as "failure to reach a minimum nutritional standard."[147] The constitution of hunger as an object was necessary for the doctrine institutionalized in the League of Nations and other international organizations that "poverty and hunger, in Britain and all over the world, could be eradicated if the market were disciplined by scientific planning."[148]

The idea that free labour inexorably created progress was central to British colonial policy in the late nineteenth century. However, throughout the early twentieth century, colonized peoples contradicted the idealized expectations of free labour ideology.[149] Colonized peoples did not automatically enter the labour market and British and French colonies experienced chronic labour shortages.[150] Colonial governors turned to forced labour campaigns in West Africa. While many governors denied using forced labour, plenty admitted it, offering strained justifications.[151]

---

[144] Rowntree 1901.
[145] Vernon 2007, 101.
[146] Vernon 2007, 83.
[147] Vernon 2007, 86.
[148] Vernon 2007, 135.
[149] Cooper 1989, 746–747; Cooper 1997.
[150] Doty 1996, 52.
[151] Buell 1928, 657. See Conklin (1997) for the French case.

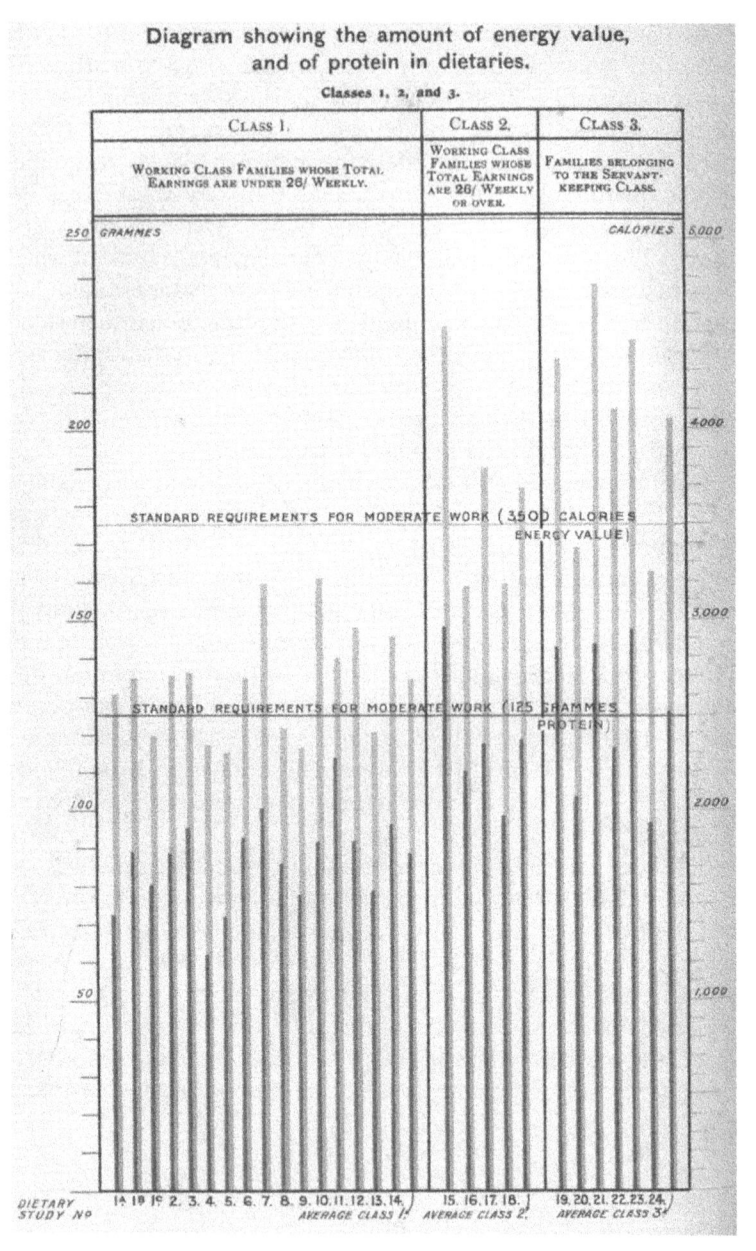

Figure 4.3 Statistical graph showing class differences in caloric and protein intake, Seebohm Rowntree (Rowntree 1901, 254)

Many colonized peoples who did enter the labour force refused to accept the deplorable conditions and petty wages. In the 1920s and 1930s rising trade union activity and labour "disturbances" in African and West Indian colonies forced the Colonial Office to grapple with the question of labour rights. The "labour problem" demanded new forms of expertise to explain these behaviours and devise solutions. So "labour" too emerged as an object of social analysis, institutionalized in the expert committees of the International Labour Office under the auspices of the League of Nations. League experts constructed an abstract, universal notion of labour and explored "whether natives were in fact capable of work and whether the reduction in native populations was due to disease or work."[152]

Finally, the object of the "economy" made it possible to reorient government policy to economic development. Mitchell has recently argued that there was no such thing as "the economy" before the 1930s. Prior to that, the term "economy" was synonymous with the "prudent management" of revenues, goods, and trade. Adam Smith's "political economy" referred to something akin to household management, as in the original Greek meaning of the term economics.[153] In the early twentieth century, the British government conceived of "economic development" narrowly as increases in exports, and thus revenue, and thus profit for the metropole.[154] At this time, however, economists were busy consolidating their discipline and in so doing they constituted the economy as an abstract object independent of society and culture.[155] Mitchell argues that the emergence of the economy was not simply the application of a label to an existing set of processes. Rather, he argues that economists and policy-makers actually *made* the economy.[156] This process was neither solely cultural nor material: it meant creating new ideational and material realities and then yoking them together into a new social sphere.

The constitution of the economy grew out of the colonial perspective. The colonial position of the Western powers provided an "outsider" perspective that facilitated a view of the colonized society as a separate object.[157] While this colonial perspective made the idea of the economy

---

[152] Anghie 2004, 164.
[153] This differs from the history of political economy in Polanyi and Foucault. See Mitchell 2005, 127.
[154] CAB 24/158, 1923.
[155] Mitchell 2002, 2005.
[156] Mitchell 2002, 5–7, 82.
[157] Mitchell 2002, 100. See Stocking 1987, 233.

more likely, the concept was slow to enter the British government. For example, in 1917 wartime shortages necessitated "economy in the uses of wheat and flour for purposes other than human food."[158] A 1931 cabinet-level "Economy Committee" dealt primarily with spending cuts.[159] So the term still meant prudent management into the 1930s. Only in the 1930s, when Keynes himself formulated the concept of "economic society" or "economic system," did the inchoate concept of a bounded entity come into view.[160] Even then, it was not until the 1940s that government documents gave life to the economy with its "circulation" and "liquid form."[161] An "Economic Survey" report from 1949 celebrates economic growth as progress, but warns that the "rapid expansion in many parts of the economy is drawing to a close. The process of growth has changed."[162] Namely, whereas economic growth in the early postwar years relied on returning soldiers adding bodies to the labour force, "[f]uture progress" would depend on new capital equipment and greater productivity via "improved organisation" and the "application of our scientific knowledge."[163] So it is only in the 1930s and 1940s that the concept of economic development really became possible.

In sum, two forms of social knowledge altered cosmological discourses in the late nineteenth and early twentieth centuries. First, anthropology bolstered evolutionary thinking about time and provided an influential voice for the new ideas of epistemic modernism on which anthropology and other forms of knowledge could serve the state. Second, a variety of social knowledges redefined social and political ontology in terms of a series of objects. These new forms of knowledge displaced the determinism of providentialism and evolutionary developmentalism. Instead, they cultivated a belief that governments could harness the power of knowledge and scientific ideas to advance social progress. Changes in the concept of time had opened up the possibility of unending social progress. Now, the improvement of social conditions over time was to be achieved by the application of expert knowledge to the control of social forces. These changes altered British thinking about state purpose in two waves. Up until 1930, evolutionist ideas dominated. After 1930, functionalism and social knowledge helped constitute a new purpose of state-led development.

---

[158] CAB 24/7 1917, 3.
[159] CAB 24/223 1931,1.
[160] See Mitchell 2008.
[161] T 236/4090 1943 (Ashton and Stockwell 1996 Pt. I, 153).
[162] CAB 129/32 1949, 4.
[163] CAB 129/32 1949, 4, 35.

Strategic Deployment: Reforming the Colonial Office, 1895–1930

As these new forms of knowledge were being created, the British Empire entered a crisis that stimulated the strategic deployment of scientific ideas. By the end of the nineteenth century, it was clear that the presuppositions of laissez-faire liberalism were naive. Independent states and colonized peoples alike failed to act as liberal ideology predicted, undermining the commercial stage theory.[164] Development towards "civilization" did not seem to be forthcoming and anti-colonial resistance grew. In established, independent states such as the Ethiopian Empire, the Kingdom of Asante, and the Shona Kingdom, indigenous leaders led armed resistance to colonial conquest.[165] In Britain, the horrors of the Boer War led to questions about the value and validity of the imperial project. Then, throughout the first decades of the twentieth century, anti-colonial resistance spread.

In the 1910s and 1920s, a series of spontaneous peasant rebellions, nationalist campaigns for self-determination, and revolutionary political movements rocked the empire.[166] Labour leaders organized resistance to unfair working conditions in mines and on plantations.[167] Marxists inspired by the Bolshevik Revolution promised freedom and prosperity.[168] Religious revivalism including Islamic and Hindu nationalist campaigns as well as messianic movements led opposition in North Africa, India, and Southeast Asia.[169] In addition, the First World War intensified pressure for change by tightening the financial constraints on the great powers and bolstering the political consciousness of black conscripts. Wilson's Fourteen Points recognized the right of self-determination and challenged the legitimacy of empire. Nationalism and Pan-Africanism was further fuelled by the constitution of a transnational, intercontinental group of black intellectuals including Edward Blyden, Alexander Crummell, John Mensah Sarbah, W.E.B. Du Bois, Marcus Garvey, and others.[170]

So from the 1880s through the 1920s there was sustained pressure to reform the administration of the colonies and legitimate the imperial

---

[164] On the importance of an earlier set of challenges for nineteenth-century imperial ideology, see Mantena 2007a, 2007b, 2010.
[165] Ranger 1977; Lewis 1987.
[166] Isaacman and Isaacman 1976; Young 2001, 161–181; Anghie 2004, 138–139.
[167] Cooper 1989, 1997.
[168] Grovogui 1996, 113.
[169] Young 2001, 163; Johnson 1977.
[170] Young 2001, 218–219; Hanchard 2003.

project. These pressures created the impetus for meso-level associational changes in the British state that redefined colonial purpose. The resulting debates and processes of knowledge production had wide-ranging, if indirect, effects on the recursive institutionalization of international order after 1900. New ideas arising from colonial debates made possible new ways of thinking about the role of the state in processes of economic development. Alongside the rise of the welfare state in European polities, the emergence of colonial development transformed the discourse of state purpose underlying international order.

In this context, Joseph Chamberlain, Secretary of State for the Colonies between 1885 and 1903, sought to reinvigorate colonialism with a doctrine of "constructive imperialism." He advocated for an early form of "trusteeship" in which the British government would fund "improvement" by expanding modern communications, railways, medical research, and agricultural training.[171] Chamberlain is often credited with turning British colonial policy towards development.[172] However, for Chamberlain, development merely meant mobilizing the untapped resources in the colonies to increase British revenues. In this sense, Chamberlain's project was closer to the original, agricultural meaning of the term improvement than to the postwar concept of development, which had a broader and more humanitarian meaning. Chamberlain's vision gave little thought to the desires, rights, or well-being of the natives. He argued that it would be necessary to "shed some blood" in order to "exercise control over barbarous countries."[173] A combination of violence and technocratic management would bring the barbarians into civilization.

Chamberlain's project was not an unqualified success. He could not secure all the funds he desired, nor could he convince Parliament or the Treasury to abandon the principle of colonial self-sufficiency which barred the use of English money for colonial development projects. His ideas and plans faced resistance from district officials, whom the Colonial Office lacked the capacity or inclination to override. Even where Chamberlainite ideas were tried, they gave the lie to the optimistic visions of automatic development. Indigenous societies organized armed resistance that forced Chamberlain to field multiple bloody expeditions in West Africa. Finally, the disastrous intervention in the Boer War galvanized opposition to increased colonial interventionism and spending.[174]

---

[171] Hodge 2007, 22–23.
[172] Abbott 1971, 68.
[173] Quoted in Hodge 2007, 47.
[174] Hodge 2007, 45–53.

Nonetheless, Chamberlain's project initiated the strategic deployment of experts and scientific advisors to devise plans and guide the disbursement of funds for colonial improvement. Before 1900, the officials staffing the Colonial Office were drawn from students of Oxford and Cambridge.[175] Few of the officials had special knowledge and though they took examinations like other civil service staffers, the tests were not technical.[176] Rather, the examinations reflected the generalist education offered at Oxford and Cambridge. Chamberlain began the process of hiring experts and forming scientific advisory committees that would undertake research and submit findings to the Colonial Office, but it was only after his departure that the formalization of training and the mass hiring of experts accelerated.[177]

In 1929, Parliament authorized the 1929 Colonial Development Act. The Act made available £1,000,000 for investment in the colonies. This initiative was small and did not revise the doctrine of colonial self-sufficiency. However, successive bills in 1940 and 1945 eliminated the self-sufficiency condition and provided significant funds for colonial development. As with Chamberlain's initiatives, domestic economic concerns were central to these bills. First, it was thought that colonial development would help alleviate problems at home brought on by the Great Depression in the Western economies.[178] If the economic situation in the colonies improved, this would bolster exports and employment in Britain. Second, government officials hoped that economic interventions would raise revenues and legitimate the colonial enterprise in an era of self-determination. In 1938, the Secretary of State suggested that the legitimacy of the British Empire hinged on its behaviour in the colonies: "In the future, criticism of Great Britain would be directed against her management of the Colonial Empire, and it was essential to provide as little basis as possible for such criticism."[179] The threat of hypocrisy pressed the Colonial Office into passing development bills that would help colonial officials treat native labourers better than they might otherwise have.

While domestic motivations were important in changing British policy, they cannot explain the origin or effects of the development discourse. In the terms introduced in Chapter 2, domestic economic concerns can explain the strategic deployment of scientific knowledge, but not the subsequent discursive reconfigurations they set off. The effects of the new policies outstripped their instrumental purposes. By hiring natural and

---

[175] Hodge 2007, 44–45; Kirk-Greene 1999.
[176] Parkinson 1945.
[177] Hodge 2007, 44–54.
[178] Abbott 1971; Constantine 1984.
[179] CO 852/190/10, no 12 1938 (Ashton and Stockwell 1996, 65).

social scientists, colonial officials after Chamberlain set off discursive reconfigurations of ideas about state purpose in the Colonial Office and ultimately the broader British government.

Importing Expertise

The desire to incorporate scientific knowledge and expertise into the Colonial Office crossed party lines and was advanced by Conservative, Labour, and Liberal Colonial Secretaries between 1900 and 1945. A series of reforms created formal institutional channels between the natural and social sciences and colonial policy. First, the office of Government Anthropologist was created in 1908 to facilitate research and provide knowledge for colonial administration. Second, the Colonial Office formalized the training and recruitment of African colonial officers. After 1909, district and colonial officers were required to take two months of vocational training on language, tropical hygiene, and colonial history.[180] The demand for expert knowledge intensified in the 1920s as anti-colonial resistance spread. In 1926, Colonial Secretary Leo Amery formalized colonial officer training by establishing joint "Tropical African Service" (TAS) Committees at Oxford and Cambridge.[181] These committees were to set up a new academic, not vocational, curriculum for prospective officers.[182] In the 1940s, the TAS programme was extended to include London University. These TAS courses were taught by leading British anthropologists such as C.G. Seligman and A.R. Radcliffe-Brown. Henceforth, students seeking to take the Colonial Service exam received instruction in anthropology, colonial history, forestry, and agriculture.[183]

Third, the Colonial Office hired experts and constituted scientific advisory bodies that were to conduct research and guide colonial policy. Recruitment records show that 39 per cent of hires between 1913 and 1952 were natural scientists and a further 20 per cent were experts drawn from technical and social sciences.[184] By 1920, the Colonial Office was hiring more experts than administrative personnel (Table 4.1).

In addition, new committees were formed to advise colonial officers (Table 4.2). Early committees, such as the Veterinarian Committee,

---

[180] Kirk-Greene 1999, 17; Dimier 2006, 349–350.
[181] Heussler 1963, 44; Dimier 2006, 339. The TAS was renamed the Colonial Administrative Service (CAS) in 1932 and further developed into the "Devonshire course" in 1945 (Kirk-Greene 2006).
[182] Kirk-Greene 2006, 44.
[183] Heussler 1963, 126.
[184] Hodge 2007, 11. Based on Kirk-Greene 1999.

Table 4.1 *Colonial Office recruitment, 1913–1952 (Hodge 2007, 11. Based on Kirk-Greene 1999)*

	1913–1919	1920–1929	1930–1939	1940–1944	1945–1952	Total	%
Administrative personnel	190	983	610	166	1,934	3,883	22%
Natural science experts	167	1,720	928	311	3,724	6,850	39%
Other experts	32	514	161	141	2,689	3,537	20%
Other appointments	154	725	548	154	1,801	3,382	19%
Total	543	3,942	2,247	772	10,148	17,652	

Table 4.2 *Growth of Colonial Office advisory committees, 1900–1961 (Hodge 2007, 10)*

	1900–1909	1910–1919	1920–1929	1930–1939	1940–1949	1950–1959	1960–
Total committees	2	3	4	4	10	9	8

the Geological and Mineral Survey, and the Medical and Sanitary Committee, drew on expertise from the natural sciences. It was in the 1930s that advisory bodies were struck to contribute assistance on social and economic issues such as labour and education.

Cosmo Parkinson's memoir provides a window into the changing role of expertise in the policy-making process of the Colonial Office in the first half of the twentieth century. Parkinson served as a clerk in the Colonial Office for many years before moving up to the rank of Under-Secretary of State for the Colonies. In 1909, when the young Parkinson arrived at the Colonial Office, it had "no specialist advisers."[185] The structure of the office made the empowerment of outside experts unlikely. Departments were organized by region and clerks worked for one of the regional offices (West Indies, the East, Nigeria, West Africa and the Mediterranean, and East Africa). Each department had an independent staff and a degree of autonomy.[186] By 1945,

---

[185] Parkinson 1945, 34.
[186] Within this regional structure, policy-making was quite informal. New papers were received from the Colonial Service (officers working in the colonial territories) and then "minuted" or commented on by the second-class clerks of the regional department.

Cosmo Parkinson was astounded by the changes in the Colonial Office. The structure of the Office had been modified ad hoc, adding "subject" departments like commerce and social services to the existing regional ones "as the complexity of the work grew."[187] By 1945 there was a "veritable galaxy of advisers" on development planning, law, medicine, agriculture, forestry, fisheries, labour, education, business, animal health, food supplies, engineering, marketing, demography, etc.[188] These reforms acted as a constitutive mechanism that channeled new ideas from anthropology, public health, labour studies, education, and economics into the Colonial Office where they set off discursive reconfigurations in ideas about state purpose.

Discursive Reconfiguration: Evolutionary Developmentalism in the British Colonial Office, 1900–1935

The rise of experts in the British Colonial Office made possible new ideas about state purpose that would be institutionalized in international order in the interwar period. Associational change in Colonial Office was not the only discursive reconfiguration that drove the broader institutionalization. Similar developments took place in agencies of French and German states. Nonetheless, it was the discourse that emerged in the British Colonial Office in the 1910s and 1920s that became embedded in the League of Nations.

As experts entered the Colonial Office in the 1910s and 1920s, they oriented policy to a new purpose of evolutionary developmentalism. The central idea was that "indigenous societies adapt themselves instinctively of their own volition, to changing conditions – and therefore the task of government is merely to guide, not create development."[189] Colonial peoples and societies would develop along a predictable linear path governed by the laws of sociocultural evolution.[190] This policy of "development along native lines" displaced laissez-faire liberalism while leaving in place its providentialist foundations.

However, evolutionary developmentalism also drew on a new image of the native as embedded in anthropological models of tribal society.

Then the paper would "pass up the ladder until it reached the Secretary of State himself." On the way, officials would simply comment on the paper as they saw fit, unless someone felt they had the authority to dispose of the memo by putting it aside or giving some action order. See Parkinson 1945, 29; Hyam 2010, 217.

[187] Parkinson 1945, 55.
[188] Parkinson 1945, 56. See Tilley 2011.
[189] CO 847/35/9 1947 (Hyam 1992, 149).
[190] Kuklick 1978, 46.

Discursive Reconfiguration

In the early twentieth century, experts argued that British colonialism was ripping the native from tribal life and that this was causing social dislocation and unrest across the empire. The response was the doctrine of "development along native lines." This doctrine retained the linear view of progress but suggested that the evolution of primitive societies towards Western standards required guidance and trusteeship by the European colonial powers. Thus, evolutionary developmentalism acted as a bridge between the natural providentialism of liberal colonial ideology and the rise of more modernist, interventionist ideas in the 1930s and 1940s.

This discursive configuration had its roots in evolutionist anthropology. Evolutionism was the most influential anthropological doctrine from its academic rise in the 1880s through 1910.[191] When the Colonial Office created the post of Government Anthropologist in 1908, it consulted E.B. Tylor, J.G. Frazer, and other prominent evolutionists. Tylor recommended N.W. Thomas, a professional anthropologist, but not an accomplished academic for the new post.[192] Thomas served in Sierra Leone, where more often than not he clashed with the colonial governors.[193] But Thomas' evolutionary analyses were well-received. His examination of crime and punishment in Nigeria concluded that civilized laws would come to replace barbaric ones, but that this would take time.[194] The underlying premise was that Nigerian society was going through a natural, inevitable process of development.

R.S. Rattray, Government Anthropologist in the Gold Coast during the 1920s, was a professed follower of Tylor and Frazer. He argued that indigenous societies should be preserved against the corroding influence of European contact.[195] Thus, he advocated for colonial policy that encouraged indigenous traditions, homogenization, and tribal centralization.[196] The restoration and preservation of tribal life would enable

---

[191] Diffusionism, led by William Rivers and Elliot Smith, was the dominant school from approximately 1910 through 1930, when Malinowskian functionalism became ascendant. Diffusionism never had much of an effect on colonial policy, so I largely ignore it here. See Kuklick 1978, 96–97; Stocking 1995, 179–232.
[192] Kuklick 1978, 102; Stocking 1995, 377–378.
[193] For example, in Sierra Leone he was sent "to investigate the 'Human Leopard Society,' allegedly a cannibalistic secret society whose practices had grown more barbaric as a result of colonial contact; but he refused to accept the government's requirement that as its employee he was bound to report the murderers he interviewed, since this conflicted with the professional code of respect for informants' confidences" (Kuklick 1978, 103).
[194] Kuklick 1978, 103. Nonetheless, his boss in Nigeria, Governor Lugard, recommendation his termination. Lugard argued that political officers would be better suited to collect information on native customs. See Basu 2015, 88.
[195] von Laue 1976, 35.
[196] Kuklick 1978, 105.

African tribes to resume the process of sociocultural evolution towards civilization. As Kuklick explains, colonial anthropologists like Thomas and Rattray "took from evolutionism a preoccupation with the necessity for maintaining the integrity of each evolutionary stage of development, if further healthy growth was to be possible."[197]

Evolutionism provided the impetus and justification for the consolidation of the British policy of "indirect rule."[198] On this view, the British would rule through "traditional" or "tribal" rulers. Various forms of indirect rule had been worked out in the nineteenth century because they were convenient and efficient. But the policy was unevenly applied and never articulated explicitly until the 1920s. Indirect rule was codified as more or less official policy after the publication of F.D. Lugard's influential text, *The Dual Mandate* (1922). Lugard served as the governor of Nigeria from 1914 to 1919. Lugard argued that the British Empire had a "dual mandate." First, it had "moral obligations" to train the natives and create a system of free labour. Second, it had "material obligations" to advance the "mutual benefit of the people and of mankind in general."[199] But with liberal, commercial policies discredited, Lugard articulated a new means to fulfil the mandate: the British should govern indirectly, through "native institutions." This would be cheaper, of course, but it also promised to guide the progress of primitive tribes towards civilization.

Lugard selectively engaged with evolutionist anthropology, but his thinking reflected the "racialist evolutionism of the later nineteenth century."[200] Lugard had equal confidence in the natural law of evolutionary progress and the necessity of British trusteeship to advance civilization in the colonies. On the one hand, without the British, natives would remain mired in a state of "inter-tribal war" in which "extermination and slavery were practiced by African tribes upon each other."[201] On the other hand, he maintains that "[e]volution and progress are a law of nature" and the "ascent of man to a higher plane of intelligence, self-control, and responsibility" is an inevitable, if painful, process.[202] Governmental institutions, like man's intelligence, followed a "natural process of evolution" through "stages."[203] Likewise, regarding material conditions, there is a "natural evolution of industrial progress."[204] But these processes could

---

[197] Kuklick 1978, 102.
[198] von Laue 1976; Kuklick 1978; Mantena 2010.
[199] Lugard 1929 [1922], 58.
[200] Stocking 1995, 383–384.
[201] Lugard 1929 [1922], 5.
[202] Lugard 1929 [1922], 91.
[203] Lugard 1929 [1922], 65, 97, 282, 285.
[204] Lugard 1929 [1922], 509.

be more or less painful and it was incumbent upon the British to "see to it that the process is accompanied by as much benefit and as little injury to the natives as may be."[205]

Another influential voice in colonial policy, Harvard Professor Raymond Buell, supported the policy of indirect rule in his book *The Native Problem in Africa* (1928). Buell travelled extensively in Africa to gather material for the book, which was widely read and influential. He argued that supporters of indirect rule believed that "European standards and methods must be introduced in the form and measure in which they can profitably be grafted onto the pre-existing stock."[206] That is, while supporters of indirect rule believed that change was "inevitable and in fact desirable" in the long run, too much change in the short run would be detrimental: "if the traditional group life of the native disappears without a new group life being put in its place, the continent of Africa will disintegrate."[207] For Buell, native conduct was controlled by an "intricate" social system that must be preserved.[208] The well-being of Africans depended on maintaining their traditional social relations: "If the continent of Africa is to be saved from Anarchy, these bonds must not be cut, but rather annealed."[209]

As D.C. Cameron, an influential Governor of Nigeria and Tanganyika, put it, indirect rule would protect "the natural evolution of tribes in a larger system."[210] On this view, the British were there to guide development along native lines, but fundamentally primitive peoples must move naturally through the stages of sociocultural evolution on their own terms.[211] The solution was European trusteeship of native societies, not independence and isolation.

Evolutionary developmentalism appeared in various Colonial Office practices and debates through the 1920s and into the 1930s. For example, in the 1920s riots by oppressed labourers forced a debate on the rights of colonized peoples. Did primitive peoples have the same rights to organize as British citizens? A central question in the debate was whether the labour associations were a "natural" or "artificial" development in Africa and the West Indies. Everyone agreed that "the laws of people should be a natural growth developing to meet the conditions

---

[205] Lugard 1929 [1922], 91.
[206] Buell 1928, 717.
[207] Buell 1928, 717.
[208] Buell 1928, 720.
[209] Buell 1928, 720.
[210] CO 323/1077/12, no 11 1932 (Ashton and Stockwell 1996, 235).
[211] This is the central point that distinguishes my understanding of colonial anthropology and indirect rule from Mantena's (2010, 173–178). For me, native societies had to be preserved as an element in a larger theory of development, not as part of a "practice of cultural tolerance and cosmopolitan pluralism" (Mantena 2010, 178).

obtaining."[212] The question, as D.C. Cameron put it, was whether the organization of labour was "a natural growth warmly esteemed by the people themselves, or merely a somewhat artificial system of administration which owes its existence to our presence."[213]

In 1930, Secretary of State for the Colonies Lord Passfield (also known as Sydney Webb) intervened in this debate by invoking the principles of trusteeship:

> I regard the formation of [trade unions] in the Colonial Dependencies as a natural and legitimate consequence of social and industrial progress, but I recognise that there is a danger that, without sympathetic supervision and guidance, organizations of labourers without experience of combination for any social or economic purposes may fall under the domination of disaffected persons ... I accordingly feel it is the duty of Colonial Governments to take such steps as may be possible to smooth the passage of such organizations, as they emerge, into constitutional channels.[214]

Trusteeship was designed to guide, channel, and smooth the natural and inevitable process of development from primitive life to European civilization. Since labour was believed to be subject to natural evolution that could only be guided by politics, all colonial officials could do was protect free labour and labour rights with legislation. At this time, colonial governors were ordered to implement labour legislation along the lines of British law.[215]

Throughout the period, officials forged links between development and narratives of scientific and technological progress. Take, for example, Lugard's defence of colonialism in *The Dual Mandate*:

> Though we may at times entertain a lingering regret for the passing of the picturesque methods of the past, we must admit that the locomotive is a substantial improvement on head-borne transport, and the motor-van is more efficient than the camel. The advent of Europeans has brought the mind and methods of Europe to bear on the native of Africa for good or for ill, and the seclusion of ages must perforce give place to modern ideas. Material development is accompanied by education and progress.[216]

In this argument, Lugard measures the pre-colonial life of Africans in two ways. First, he measures it against the technological standards of Europe, which are asserted as superior.[217] Second, he measures it against

[212] CO 323/1077/12, no 11 1932 (Ashton and Stockwell 1996, 235).
[213] CO 323/1077/12, no 11 1932 (Ashton and Stockwell 1996, 234).
[214] CO 323/1077/10, no 3 (Ashton and Stockwell 1996, 225). For an excellent discussion of Webb's socialism and his evolutionism, see Bevir 2002.
[215] CO 323/1071/12, no 2a 1930 (Ashton and Stockwell 1996, 226).
[216] Lugard 1929 [1922], 5.
[217] On this, see Adas 1989.

the standard of "efficiency," which itself is drawn from utilitarian scientific discourses. Together these standards lend credibility and solidity to the desirability of "material improvement." In this way, scientific and technological concepts structure the ends of colonial development. Science is not merely a means to progress, but a marker of progress and an end in its own right. Economic and scientific progress go hand in hand. This powerful discursive configuration links Smith, Darwin, and Lugard in a postcolonial vision of development as an increase in complexity. As we shall see in the next chapter, a scientific and technological conceptualization of development slowly displaced the older "civilizational" conception that included a diverse group of artistic, religious, political, and scientific standards.

Evolutionary Developmentalism in International Order, 1919–1935

In Chapter 2, I outlined two pathways of recursive institutionalization: hegemonic imposition and horizontal change. The rise of ideas about civilization and evolutionary development generally took the form of a hybrid pathway in which the British hegemon led, but did not impose, the institutionalization of new purposes in the core sites of international order. Britain did not need to impose or actively spread these ideas because a transnational network of natural and social scientists had already done so. Thus, in the late nineteenth century notions of civilizational progress and development already structured the purposes of the great powers.[218]

However, the discursive configurations that took hold differed from country to country. In France, for example, the colonial enterprise centred on the *mission civilisatrice* which was understood in moral, republican terms.[219] In Britain, the standard of civilization was interpreted differently because it was rooted in colonial anthropology and social knowledge.[220] But in hybrid pathways it is the specific interpretation of the dominant state that is likely to become embedded in international order. Thus in 1885 and again in 1919, it was the British understanding of civilization and human progress, rather than the French or German variants, that became embedded in international order. Thus, the same discursive configurations that led the British Colonial Office to a policy of evolutionary developmentalism were institutionalized in the League of Nations.

---

[218] Bain 2003, 19–20.
[219] Conklin 1997.
[220] Cowen and Shenton 1996.

182   Darwin, Social Knowledge, and Development

The League is usually portrayed as the first permanent multilateral organization and the first expression of liberal principles like collective security.[221] In contrast, my reading builds on the work of Morgenthau and others who claim that the League of Nations was also significant because it marked the "age of the scientific approach to international affairs."[222] On this view, the League was an expression of a broader movement of international societies and conferences that sought "to cure the ills of humanity in a scientific way."[223] For Morgenthau, the great hubris of this era was the belief that the realities of power politics could be reduced to a scientific formula and abolished by reason. Scientific ideology led policy-makers to substitute "supposedly scientific standards for genuine political evaluations" rooted in an appreciation of power.[224] Morgenthau interpreted the doctrine of collective security, the promotion of international courts, the Kellogg–Briand Pact, and all the other elements of the League of Nations as the transposition of scientifically inspired domestic liberal institutions to the international level.[225] Morgenthau criticized this move because it did not recognize that liberal institutions depend on social and economic conditions not present at the international level.

While tendentious, Morgenthau's reading has been supported by the recent historiography on the period.[226] In the nineteenth century, the Saint-Simonian idea that scientific and technological associations could unite and transform the world had inspired an "explosion of meetings, conferences, and international networking."[227] International conferences were organized to coordinate public health, telegraphs, penal policy, measurement, and time.[228] These early experiments in international association laid the foundation for the League of Nations. Moreover, the League was primarily designed by British officials who modelled it on their own system of government. They envisioned a parliament, the "council of nations," attached to a technocratic bureaucracy.[229] This technocratic bureaucracy was to both solve problems of interstate cooperation and to supervise the administration of colonial development under the Mandate

---

[221] Reus-Smit 1999; Ikenberry 2001.
[222] Morgenthau 2006 [1948], 44.
[223] Morgenthau 2006 [1948], 44.
[224] Morgenthau 2006 [1948], 48.
[225] Morgethenthau 1946, 75–104; Morgenthau 2006 [1948], 48.
[226] For reviews see Pedersen 2007; Gray 2007; Mazower 2012, 103–150; Gorman 2012, 55–57.
[227] Mazower 2012, 103. See also, Rosenberg 2012; Tilley 2011.
[228] Mazower 2012, 102–103.
[229] Mazower 2012, 120–136; Sending 2015. Mazower says these designers "abandoned the legalist paradigm." But, as we shall see, this is not quite right: the legalist paradigm and the technocratic one co-existed in the League.

System.[230] While the League was certainly a diplomatic failure because the council did not prevent fascist aggression, "its technical services took the organization of international humanitarian cooperation further than anyone had imagined possible."[231] Cooperation on international labour, drug trafficking, the slave trade, and international health was so impressive that its basic organs and experts were incorporated into the post-Second World War order.[232]

Along these lines, in this section I present evidence that while peace, disarmament, and collective security were central ends of the League, these goals were suffused with a scientific spirit. Moreover, my analysis of primary documents reveals that an orientation towards civilizational progress and evolutionary development had emerged as central ends in the discourse of state purpose of the 1920s and 1930s. Both sets of ideas about state purpose drew on the cosmological shift arising from the historical natural sciences and modernist social knowledge. In this way, the League embodied the idea of a world organized and controlled by scientific knowledge.[233]

Discourse of State Purpose

My analysis of League documents reveals two central discursive configurations. First, there is a legalist formation oriented to "maintaining peace and promoting security" via the peaceful settlement of disputes. Second, there is a welfarist formation oriented to the "well-being and development of all peoples" to be advanced by the trusteeship of colonies and technical bureaucratic committees at the League. Both of these formations were underwritten by the purposes of civilizational progress and evolutionary development. In relations between the European colonial powers, this meant using the standard of civilization to discipline and constrain states to accede to the principles and practices of collective security.[234] In relations between the colonial powers and their colonies, this meant using international institutions to guide the development of backward peoples and nations towards civilization.

The Covenant of the League of Nations itself emphasizes the legalist formation. The preamble states that the purpose of the League is to "promote international cooperation and to achieve international peace

[230] Anghie 2004, 121, 140.
[231] Mazower 2012, 143. It is here that the recent historiography departs from Morgenthau's line of argument.
[232] Mazower 2012, 148–149, 192–193.
[233] Mazower 2012, 95.
[234] Bowden 2009.

and security." The covenant outlines a number of means by which peace and security is to be achieved. First, open relations and public discussion were meant to promote agreement and advance international law. This in turn, would support the various legal measures for the peaceful settlement of disputes. Members of the League were to submit disputes to judicial settlement or arbitration under the requirements of articles 12 through 15. The security dilemma was to be ameliorated by coordinated reductions in armaments. The intent to reduce arms was indicated in article 8, but negotiations towards an arms regime faltered at every step. All of this was to be backstopped by the collective security measures specified in articles 10, 11, and 16.

The welfarist discourse appears in articles 22 through 25 of the Covenant which outline the Mandate System, cooperation on economic and social issues, the centralization of existing international offices under the League, and support for the International Red Cross. Here the evolutionary developmentalism we saw in the British Colonial Office appears at the core of international order. Article 22 states that those colonies formerly governed by the defeated powers "not yet able to stand by themselves under the strenuous conditions of the modern world" are to become mandatories governed by victorious powers under League supervision. The principle underlying the Mandate System is that "the well-being and development of such peoples form a sacred trust of civilisation." On this scheme, colonies were placed in different classes depending on "the stage of the development of the people." Former Ottoman territories "had reached a stage where their exercise as independent actors can be provisionally recognized." However, other "peoples" such as those in Central Africa, "are at such a stage that the Mandatory must be responsible for the administration of the territory." Others still, due to their small and sparse populations or their "remoteness from the centres of civilisation," are best "administered under the laws of the Mandatory as integral portions of its territory."[235]

Article 23 lays out a series of economic and social problems that the League was also charged with addressing: labour conditions, just treatment of "native inhabitants," the drug trade, the arms trade, communications and commerce, and disease control. Article 24 more broadly states that "all commissions for the regulation of matters of international interest" shall be brought under the direction of the League. On the basis of these articles, economic, social, and other committees were formed under the League. They prepared reports on numerous talks and presented proposals to the League Council.

---

[235] For the classes in action, see e.g., League of Nations 1925, 208ff. See also Wright 1930.

Records from the early League are dominated by the welfarist discourse and faith in the technical and bureaucratic work of the commissions and sections. As Lord Curzon puts it, the League will not only facilitate "mutual confidence" but will "affect many branches of human life and welfare."[236] The League exists "for the ideal of universal brotherhood of Governments and peoples, for social peace and for progress, security and well-being of States and their citizens."[237] Welfare is to be "advanced" by a constellation of technical and expert committees. Expert committees are formed to propose arms control regulations, simplify and advance international law, coordinate transportation, design the permanent court of justice, promote international health, create labour regulations, study financial and economic problems, collect statistics, and so on.[238] Here, the League drew on and reproduced epistemic modernist ideas about the power of knowledge.

What constitutes and motivates welfarist ends in the League? It seems obvious that political elites would want to advance welfare, but the idea was new to international politics. Moreover, we should not assume that "well-being" meant the same thing to League delegates as it does to us. First, the documents reveal a desire for the "mitigation of suffering."[239] This motivates the promotion of health and measures to address disease outbreaks. Second, there is the will to provide a utilitarian "benefit" to humanity.[240] Third, there is a broad ethos of "improvement."[241] Health, labour conditions, education, and economic stability can all be "improved." The alleviation of suffering and improvement of life is thought to go hand-in-hand with the advance of civilization.

Welfare is broadly conceptualized and is not reduced to economic ends. Indeed, although economic concerns are important in the discourse, the principal goal of economic policy in the League remained currency stability.[242] Reports and discussion on the postwar financial reconstruction of Austria are focused on achieving "budgetary equilibrium" and resolving the "banking crisis" by curbing inflation and concomitantly "speculation." Production is discussed, but it is not considered a core component of the "consolidation of economic life."[243] All of this shows that ideas

---

[236] League of Nations 1920a, 22.
[237] League of Nations 1920a, 22. Given the rebellions against colonial rule, the League had to justify the Mandate System in terms of the welfare of native peoples. See Anghie 2004, 138–140.
[238] E.g. League of Nations 1920b; League of Nations 1920c; League of Nations 1925.
[239] League of Nations 1920c, 68; League of Nations 1935, 561.
[240] League of Nations 1925, 118.
[241] League of Nations 1920c, 68; League of Nations 1925, 119.
[242] League of Nations 1925, 122.
[243] League of Nations 1925, 122.

about welfare and increased production had yet to be integrated into a single goal like economic growth.[244]

Cosmological Elements

What is new and widespread in League discourse is an underlying cosmology oriented to the future. As one member of the council states, the founding of the League "will go down to history" as "the birth of a new world" in which all nations chose to "substitute right for might."[245] For another, "all here must feel [that] we stand at a turning point in history, at the dawn of a new era in human life."[246] The goals of the League are the "task of the future" to be achieved by a "sure method."[247] Looking to the past reveals "terrible disasters, which have imperilled civilisation and drenched the world in blood."[248] Instead, "[w]ith eyes fixed on the distant future, but with our feet on the solid ground of political and social realities, we will create a world."[249]

The orientation to the future rests on the idea that history is the story of changes in international politics and that these changes can be controlled. Delegates describe the League itself as a "sure method" or a "saner method" for "regulating the affairs of mankind."[250] We saw variations of these ideas in 1815, but the nineteenth century's scientific optimism amplified the epistemic modernist notion that associations and organizations can apply knowledge or embody methods that can alter the world. Indeed, by the twentieth century, what it means to govern is to create a whole world, rather than merely construct a balance or improve land.

The idea that the League could control the future of international politics rests on a new epistemic foundation. In 1815, an episteme that recognized the power of human knowledge was implicit only in the decision to form and rely on a statistical commission that precisely calculated and finely tuned the balance. The agreement at Vienna depended only on a thin, implied rationalism. A more ambitious epistemic modernism underwrites techno-politics in the League. The epistemic foundation of the League is the Baconian and Enlightenment tenet that knowledge and

---

[244] Still, the object that would unify these elements, the "national economic system," appears in the discourse (League of Nations 1925, 123).
[245] League of Nations 1920a, 18.
[246] League of Nations 1920a, 24.
[247] League of Nations 1920a, 19–20.
[248] League of Nations 1920a, 20.
[249] League of Nations 1920a, 20.
[250] League of Nations 1920a, 20.

understanding are the drivers of progress. Open relations built on discussion in the League Council will create mutual understanding of issues and eliminate the tensions and confusions that lead to arms races and war. More deeply, "peace depends on the progress of science which, by alleviating distress, will remove one of the permanent causes of discontent and conflict."[251] Increased scientific knowledge and information sharing will solve social, economic, and medical problems. The report calling for the creation of what would become the League of Nations Health Office argues explicitly that such an organ is necessary because neither the Council, the Assembly, nor the Secretariat of the League "possesses the requisite knowledge for the necessary technical research, which is scientific as well as social."[252] A permanent organization is necessary to collect statistics and circulate the discoveries of scientific research.[253]

The modernist faith in information and knowledge to transform the world is illustrated by one of the less well-known League initiatives, the International Committee for Intellectual Cooperation. The impetus was the scientific internationalist ideal that "[a] more intimate and active interchange of ideas, impressions, scientific discoveries, moral improvements and literary and scientific publications" will produce a "moral union." This moral union is the necessary preliminary to "an agreement of interests" that would "give to the work of the League of Nations the soundest guarantees of permanency and power."[254] The "ultimate object" of "universal peace" will be built upon "a freer and more extensive circulation of ... knowledge and ideas."[255]

Moreover, the day-to-day work of the League would simply not have been possible without a taken-for-granted modernist faith in the authority and competence of "experts."[256] There are many instances of delegation to committees to conduct an "enquiry" or a "general survey."[257] Members of the Council seek to "obtain impartial and authoritative information."[258] But the basis of expertise is not true knowledge, as in early modern Europe or the Enlightenment, but "qualification."[259] An expert has been educated or trained in a mode of thought or practice, but does not apply a formula or deliver truth.

---

[251] League of Nations 1920d, 446.
[252] League of Nations 1920b, 43.
[253] League of Nations 1920b, 43.
[254] League of Nations 1920d, 445.
[255] League of Nations 1920d, 446.
[256] League of Nations 1920b, 36, 40.
[257] E.g., League of Nations 1920c, 62, 65.
[258] League of Nations 1920c, 64.
[259] League of Nations 1925, 131.

188   Darwin, Social Knowledge, and Development

Delegation to experts created demand for the production of more information and statistics. This is especially evident in the documents of the Mandates Commission. The Mandates Commission was created to supervise the administration of the mandated territories. The colonial powers were to present reports and statistics to be then reviewed by the Commission. In 1925, the Commission was generally satisfied with the colonial administration. However, it made many requests for better information on the education, health, and labour conditions of the peoples in the mandated territories.[260]

Contestation in League Discourse

While present throughout the 1920s, the legalist security formation rose in significance over the course of the 1930s as it was challenged by Japan, Italy, and Germany.[261] A 1925 debate about the potential problems with investigations to verify German arms reductions reveals that the League delegates themselves understood the challenges of collective security. French Minister Briand, for example, wondered aloud how the right of investigation was to be enforced:

> The representatives of the League must be able to move freely and to give proof of their official capacity. This was a simple matter. When, however, they wished to enter a factory they might encounter resistance. Here the difficulties were of a practical character, in regard to which the jurists could not give any useful advice to the Council.[262]

In the last words here Briand shows that he understands that both law and expertise are impotent if countries refused to recognize the authority of the League. And yet, the faith in law and rational solutions pressed the delegates to look for a legal "formula" that would ensure the authority and effectiveness of inspectors.[263]

Of course, the League never got the chance to grapple with the practical realities that concerned Briand. Instead, the states did not agree on a disarmament regime and Germany withdrew from the League just as it violated the terms of Versailles by reinstating conscription in 1935. An extraordinary session of the League Council was called to discuss the matter. What is striking about the records, compared

---

[260] League of Nations 1925, 211–215. Anghie (2004, 183–184) has shown that this produced an avalanche of information that made possible a new colonial science of administration.

[261] For early articulations of the legalist discourse and the ends of peace, disarmament and so on, see, e.g., League of Nations 1920a, 18; League of Nations 1925, 128–130.

[262] League of Nations 1925, 138. See also League of Nations 1925, 141.

[263] League of Nations 1925, 130–132.

with the diplomatic documents in 1815, is the complete absence of balance of power discourse. Instead, the central goal invoked is the "organisation of security in Europe."[264] As in 1815, the collective goal is a European one, but this is not equated with or understood in terms of the balance. German action is not portrayed as a threat to the balance but as a violation of international law. The French proposal for a resolution begins by restating the old principle of *pacta sunt servanda* and declaring that Germany has failed in its duty to uphold its agreements.[265] The legal means available to the Council was to apply sanctions.[266] But, as one delegate pointed out, it was not clear what the meaning and efficacy of such sanctions could possibly be "in a period when import and export restrictions, quotas, licenses and many other similar measures are applied indiscriminately to friendly States."[267] Another delegate conceded that the League and its Council "wield only moral forces."[268]

The members of the League must have had all this clearly in their minds when they, days earlier, had refused the Ethiopian request for European intervention on their behalf. Conflict with Italy had been simmering and with Italian troops amassing on their border, the Ethiopian delegate urgently appealed to the League to enforce its own laws and compel Italy into "a procedure of conciliation and arbitration."[269] The Ethiopian delegate received only assurances that the Council took the Italian government's statement that it was in conformance with legal procedure at face value.[270]

The fundamental problem with the League, it was understood, was the principle of sovereign equality. The principle of perfect equality, upon which the League was built, guaranteed that states could arm and seek security regardless of what the law and other states said.[271] Indeed, delegates were eager to defend "the equality of nations and their incontrovertible right to security."[272] As one delegate put it, "[u]nfortunately, the progress of the world still leaves the use of material forces almost entirely to the sovereignty of the respective nations."[273] An incomplete collective

---

[264] League of Nations 1935, 551.
[265] League of Nations 1935, 551.
[266] League of Nations 1935, 552–553.
[267] League of Nations 1935, 557.
[268] League of Nations 1935, 559.
[269] League of Nations 1935, 547.
[270] League of Nations 1935, 548.
[271] League of Nations 1925, 231.
[272] League of Nations 1935, 556.
[273] League of Nations 1935, 559.

security system, League members and designers surely realized, always contained the germ of its own dissolution.

The Emergence of State-Led Development in the British Colonial Office

I began the empirical analysis of this chapter by showing how the historical sciences constituted British colonial discourses and international order between 1860 and 1885. Then, I demonstrated how the rise of anthropology constituted and naturalized evolutionary developmentalism in the British Colonial Office and the League of Nations. In both cases, the discourse of state purpose in the 1920s and 1930s featured a strong orientation to civilizational progress. This purpose was supported by a cosmology of natural providence on which natural laws guided the progressive development of societies towards civilization. Throughout the period 1900–1930, epistemic modernist ideas challenged the determinist presuppositions of evolutionary developmentalism. This was evident in the League, which was premised upon the idea that knowledge could be used to solve social, economic, and political problems.

In this final section, I show how epistemic modernism forged a new purpose in the British Colonial Office as the deployment of anthropological and social knowledge continued to reconfigure discourses through the 1930s and 1940s.[274] By 1945, the dominant discursive formation in the Colonial Office was state-led developmentalism. While the ontology of linear progress was retained from evolutionism, the goal shifted from civilization to welfare conceptualized in social scientific terms. The deeper transformation here was from determinism to the idea that state intervention by governments was necessary to drive progress. In short, the purpose of state-led development drew upon epistemic modernist ideas that made it possible and natural to think that progress could be directed or controlled by Western experts. Cosmologically, the new discourse naturalized development not as the inevitable outcome of natural law, but as the result of humans harnessing scientific and technological progress. Although this purpose was not directly institutionalized in international order by the British hegemon, it did have a lasting effect on international discourses as it was carried into the post-Second World War order by transnational expert networks.

---

[274] The same developments also played out in the League Mandates Commission, on which Lord Lugard served. See Wright 1930.

Objects and Interventions, 1925–1938

The change in British thinking about development can be traced back to the 1920s. Starting in 1925, the evolutionist consensus in the Colonial Office was disrupted by the influx of experts in public health, labour, and education. These experts were not necessarily anthropologists and did not necessarily share the presumption that societies passed through a process of automatic evolution. Nonetheless, anthropological thinking still structured Colonial Office policy. So, a wholesale change in colonial policy had to be worked out within the intellectual framework provided by functionalist anthropology. This rethinking was initiated by Malinowski in the late 1920s and culminated in Lord Hailey's influential 1938 report, *An African Survey*.

An ontology centred on social objects emerged with the new influx of experts after 1920. As we saw above, this was a reaction to widespread resistance throughout the empire in the 1910s and 1920s. The British response to anti-colonial resistance was to treat it as a technical problem that could be solved by a combination of repressive violence and scientific knowledge.[275] Of the personnel recruited between 1920 and 1939, 43 per cent were experts drawn from the natural sciences (including geology, forestry, medicine, and veterinary medicine) and 11 per cent were experts from the civil and social sciences (including education, labour policy, economics, town planning, aviation, and dentistry).[276] New advisory committees on education, agriculture, labour, economics, and other topics introduced colonial officials to the latest research in these fields. The new experts gave colonial officials the impression that discrete elements of colonial societies could be manipulated by knowledgeable policy.

New forms of expertise permitted a shift from passive, preventive medical policy to active interventions in "public health."[277] At first, health and disease were explained by environmental and racial factors that were beyond control. As the public health movement expanded into the colonial sphere, officials began to see medical problems as manipulable. Addressing public health problems became part of the trusteeship project. The earliest sustained attention to medical problems came in 1925, after labour disruptions in East Africa prompted a rethink of policy there.[278] The report of the East Africa Commission argues that "medical

---

[275] On depoliticization and development, see Ferguson 1990; Cooper 1997. On the police and the army in the colonies, see Killingray 1986.
[276] Hodge 2007, 11.
[277] CAB 24/250 1934, 9.
[278] CAB 24/173 1925.

services" must be re-evaluated because of "the necessity for providing the means to be taken to conserve the labour supply and to ensure its efficiency."[279] Thus "care of the natives" emerges as a way to facilitate the "exploitation of the natural resources."[280] The East African report focuses on cultural causes and social forces highlighted by anthropology. It suggests that the central causes of infant mortality and the prevalence of worm disease in Africa are "superstition" and "ignorant native customs."[281] However, the ability to expand medical services is limited, the report argues, by the lack of "satisfactory statistics" and properly trained personnel.[282]

At first, medical policy aimed to build hospitals and treat individuals in a preventive manner. But by the late 1930s, the Colonial Office advocated addressing "the health problems of the people as a whole."[283] Policy thereby shifted from medical policy to "public health." In the latter discourse, "public health" is constituted as an object or "problem" that can be solved with expert knowledge.[284] This presupposes the existence of a statistically identifiable "general standard of health" that can be improved by medical intervention.[285] This shift in emphasis led to calls from within the Colonial Office for a change from an urban, hospital-centred policy to the use of "agricultural officers and sanitary inspectors who could go into the villages and teach and advise the natives on questions relating to crops, elementary sanitation and the protection of water supplies."[286]

The same ideas that transformed the management of health were applied to the management of food. By the beginning of the twentieth century, natural scientists began to revise the Malthusian theory that hunger is caused by immorality and indolence. The success of public health interventions was held up as a model for other domains of colonial policy. Government action could effect "improvement in the physique and general health of the people comparable in extent to that which in the nineteenth century followed the introduction of cleanliness and sanitation."[287] What was required was "the acceptance by the state of a national feeding policy based on scientific knowledge ... [to] bring the greatest benefits to all sections of people."[288]

[279] CAB 24/173 1925, 53.
[280] CAB 24/173 1925, 53.
[281] CAB 24/173 1925, 54–55.
[282] CAB 24/173 1925, 54, 57.
[283] CO 847/16/9, no 2 1939 (Ashton and Stockwell 1996, 280).
[284] CO 950/1 1938 (Ashton and Stockwell 1996, 269).
[285] CO 859/65/8, no 14 1943 (Ashton and Stockwell 1996, 340).
[286] CO 847/16/9, no 2 1939 (Ashton and Stockwell 1996, 282).
[287] CAB 24/250 1934, 9.
[288] CAB 24/250 1934, 9.

We saw above that evolutionist ideas influenced colonial labour policy up through 1930. Over the course of the 1930s, the evolutionary view gave way to the functionalist and modernist idea that experts could manipulate and manage the labour supply directly. As one official argued, "[t]he Northern Rhodesia trouble is an example of the kind of thing we may get more of unless expert knowledge at the Colonial Office, as well as locally, is available."[289] These experts constructed labour as an object of policy that could be changed with various social and economic programmes. They advocated measuring labour in statistical terms and called for experts to advise colonial governments on how to deal with the "labour problem."[290] If labour conditions would not evolve of their own accord, state interventions could push that evolution along.

The transition from an evolutionary to an interventionist, state-led model is also evident in education policy. In the 1920s, British education policy was cautious, fearing that education led to African nationalism and the degeneration of Western civilization.[291] One British official warned against the French assimilationist education doctrine: "The French aim at creating a new race of black Frenchmen and I cannot wish them success in their attempt which in my opinion will hasten the decline and fall of Western civilization."[292] As late as 1934, colonial officials argued that education, like labour, must evolve slowly, since it is "a plant of very slow growth."[293] Buell wrote a dissent to this policy in *The Native Problem in Africa*. He argued that to achieve progress Africans had to acquire "knowledge, elementary though it may be, of the principles upon which modern machinery and medicine and other apparatus of the Western world are based. In other words, they must be given a scientific and a technical education."[294] On his view, the natives had exercised considerable skill in the "intricate devices of European administration" but "they experience greater difficulty in performing duties such as the construction of public works and the improvement of public welfare which require some applied knowledge of European science."[295] Here Buell initiated a line of argument on which education was justified as a necessary means to scientific and technological progress.

This line of thinking shows that the transition from autonomous development to agentic, state-led development was premised upon epistemic

---

[289] CO 866/29/1166/1936, no 1 1936 (Ashton and Stockwell 1996, 249).
[290] CO 866/29/1166/1936, no 1 1936 (Ashton and Stockwell 1996, 254).
[291] CO 554/74/4, no 1 1926 (Ashton and Stockwell 1996, 223).
[292] CO 554/74/4, no 1 1926 (Ashton and Stockwell 1996, 223).
[293] CO/847/3/15 1934 (Ashton and Stockwell 1996, 242).
[294] Buell 1928, 727.
[295] Buell 1928, 727.

presuppositions about the capacity and value of knowledge. That is, for Europeans and Africans alike, development was a process of harnessing science and technology. However, whereas Europeans had mastered this, the natives needed to be educated before they could themselves take up the reins. This backdrop naturalized development as scientific and technological progress. But noting this reveals a paradox that itself legitimated European rule. To improve their societies in scientific terms, the Africans needed to have already achieved the scientific progress necessary to drive development. Thus, colonial rule, or, rather, trusteeship was necessary to lead Africans along the developmental path.

By the late 1930s, Buell's view had won out in the Colonial Office. A major commission on higher education argued that scientific education was central to the management of changes that were now inevitable. The British had, after all, already irrevocably "impinged upon the old tribal organization" and it was no longer possible to protect traditional ways of life. The authors suggested that "[t]he African has been taught that European ways of life are superior to his. He sees that European methods and education give control over the forces of nature and the circumstances of life" and thus "demands education as a right."[296] Here, the importance of education is linked to the narrative of scientific progress, which offers control over nature. The report's authors took it for granted that control over nature was a desirable end and this justified increased Westernization and education of the natives. This, in turn, justified the trusteeship model of colonialism: "if the concept of trusteeship, if the method of Indirect Rule, are to be anything more than glib evasions of responsibility they must assert that the African shall in due course reach full maturity and take his place among the peoples of the world."[297]

Moreover, officials argued, education was necessary to create the indigenous expertise to guide development. The colonies faced an "increasing need for conserving the fertility of the land and encouraging more scientific methods of production."[298] The problem was finding the necessary staff, especially since "European medical officers, agricultural experts, engineers, and technicians are necessarily very expensive." Thus, "the only remedy lies in the employment of trained African personnel."[299]

It is significant that calls for education of the Africans were justified with scientific and technological ideologies. Alternatively, colonial officials could have justified education on religious, moral, or liberal

---

[296] CO 822/83/11 1937 (Ashton and Stockwell 1996, 256).
[297] CO 822/83/11 1937 (Ashton and Stockwell 1996, 256).
[298] CO 822/83/11, no 1 1937 (Ashton and Stockwell 1996, 260).
[299] CO 822/83/11, no 1 1937 (Ashton and Stockwell 1996, 260).

grounds.[300] This would indicate that these discourses were powerful and that they informed visions of progress. Instead, education was justified as necessary to development conceived in scientific and technical terms. The colonies needed to produce their own experts to solve the problems of labour, land, health, and hunger on their own. This would help them reach "maturity" as peoples enjoying scientific and technological modernity.

The entry of new social objects changed the cosmological basis of colonial governance, making possible a new vision of development not as natural sociocultural evolution, but as scientific and technological progress. But this basic idea had yet to be translated into a coherent alternative policy.

Detribalisation Necessitates Development, 1938–1945

A new doctrine of interventionist state-led development was consolidated in British colonial policy between 1938 and 1945. This doctrine combined new anthropological thinking with the object-based ontology. As Malinowski and his students developed and positioned functionalist anthropology over the 1930s, they abandoned most of the presuppositions of evolutionism. Instead, they embedded the "changing native" in a complex set of social and economic ties that were to be adjusted and readjusted in a complex "transformation" of a society.[301] There was still a presumption that societies would develop, but this was no longer an automatic or predictable process. Instead, it was a set of changes to be guided by expertise in the service of the colonial state.

In his 1929 essay on practical anthropology, Malinowski called for a new "anthropology of the changing native."[302] In doing so, Malinowski expressed concern that this would identify "detribalized" communities that would be "extremely difficult to manage."[303] There, he advocated for indirect rule because it was "the only way of developing economic life, the administration of justice by Native to Natives, the raising of morals and education on indigenous lines, and the development of truly African art, culture, and religion."[304] Malinowski ridiculed the "magical" assumption that "you can transform Africans into semi-civilized pseudo-European citizens within a few years."[305] Rather, "all social development

---

[300] For example, Mill justifies education as necessary for human freedom, flourishing, and improvement. See Mantena 2007a, 302.
[301] Mair 1934; Richards 1935; Fortes 1936; Malinowski 2014 [1938]; Malinowski 1943.
[302] Malinowski 1929, 36.
[303] Malinowski 1929, 35. See Stocking 1995, 415–417.
[304] Malinowski 1929, 24.
[305] Malinowski 1929, 23.

is very slow, and that it is infinitely preferable to achieve it by a slow and gradual change coming from within."[306]

Over the course of the 1930s, Malinowski's top students developed the anthropology of the changing native. In doing so, they challenged the presuppositions of indirect rule and development along native lines, necessitating a new strategy to deal with African problems. The most influential of these students was Lucy P. Mair, who received her PhD at LSE under Malinowski in 1932. In the same year, she was appointed as the LSE's first lecturer in "Colonial Administration."[307] Like Malinowski, Mair rejected the use of a "hypothetical process of evolution" to understand and evaluate colonial policy.[308] Instead, anthropologists needed to study the actual process of change over time. Starting here, Mair argued that the normal process of development had been disrupted:

> Most native societies are now undergoing a process of rapid and forcible transformation comparable only to the violent changes of a revolution, and entirely distinct from the gradual, almost imperceptible, process of adaptation in which the normal evolution of human cultures consists.[309]

For Mair, "the crucial problems ... arise just where the traditional system has been forcibly wrenched away."[310] Mair, as a good functionalist, saw society as a "mechanism" that was normally in "good working order" but as a result of cultural contact, African societies were in a "pathological" condition.[311] Thus, societies could not simply resume the gradual process of change.

In study after study, this new generation of anthropologists demonstrated that rapid cultural change was undermining traditional African societies.[312] These new studies were published in *Africa*, which Malinowski had set up as the house journal of the African Institute. It was through the Institute that the findings were disseminated throughout the academic and government circles concerned with colonial policy. Taken together, the analyses suggested that African societies were being "detribalized," undermining the basis for indirect rule. This opened the door to a new theory of development to be led not by the natural course

---

[306] Malinowski 1929, 23.
[307] After the war, the lectureship was upgraded to a readership. Mair had worked at the Royal Institute for International Affairs during the war, but returned to LSE in 1947. See Dimier 2006.
[308] Mair 1934, 417.
[309] Mair 1934, 417.
[310] Mair 1934, 418.
[311] Mair 1934, 422, 418.
[312] Richards 1935 provides a good review. See, amongst others, Culwick and Culwick 1935; Fortes 1936; Hunter 1934; Wagner 1936.

The Emergence of State-Led Development

of sociocultural evolution, but by the colonial state. The result was a new purpose of state-led development that emerged in the late 1930s and early 1940s. The new discourse combined the more complex understanding of the development process put forward by Malinowski, Mair, and others, with a modernist faith in economic, labour, education, and health expertise.[313]

Lord Hailey's *An African Survey*, published in 1938, was the first major text to combine the object-centred perspective and the new anthropological thinking into a compelling defence of state-led development.[314] Hailey influentially argued that whereas British colonies have been governed using laissez-faire principles, the government should now "take an active part in developing the resources of the territory or in the organization of measures for improving the standard of living."[315] In Hailey's *Survey*, development was not a simple evolutionary process:

> The development of Africa is not a process of gradual evolution; it assumes ... something of the character of a transformation, and its achievement inevitably demands more than a routine application of existing knowledge.[316]

Here, Hailey adopted the arguments and vocabulary of the functionalists.[317] If development was no longer automatic, it would require specialized knowledge from a variety of fields. Just as Malinowski had a decade earlier, Hailey concluded by calling for an integrative programme of research into "African problems." Hailey's report is divided into chapters centred on epistemic objects: "native administration," "agriculture," "forests," "water," "health," and "education." These represented the central problems of African development and the core areas in which improved knowledge would benefit colonial policy. In concluding, Hailey argued for the expansion of colonial research in each of these fields.

Hailey made the case for a programme of state-led development that would "involve the Colonial Office in an effort to maintain common standards of progress and development in the Colonial Empire."[318]

---

[313] Malinowski himself had put these elements together in a 1943 paper: "Man lives in his culture, for his culture, and by his culture. To transform this traditional heritage, to make a branch of humanity jump across centuries of development, is a process in which only a highly skilled and scientifically founded achievement of cultural engineering can reach positive results" (Malinowski 1943, 650).

[314] Cosmo Parkinson, by now permanent under-secretary at the Colonial Office, made the text required reading for his staff (Sanger 1995, 148).

[315] Hailey 1979 [1940–1942], 3.

[316] Hailey 1938, 1612.

[317] The report was written by a team of authors, including two Malinowski students, Lucy Mair and Audrey Richards (Mills 2002, 165). So, it is not surprising that the report adopts functionalist positions and the section on anthropology favours the functionalist school. See Hailey 1938, Ch. 2.

[318] Hailey 1979 [1940–1942], 4.

Hailey could now posit a "common standard" as the goal of development because "the native" had been constructed as a universal actor shaped by social forces and deserving of a European standard of living. As one official in 1938 argued, "our duty to the people themselves is to promote their social and economic welfare, to stimulate the desire for ... a higher standard of living."[319] Another argued that "it is an imperative duty to do all that is practically possible to raise the standard of living of such people, even during the war period, alike for humanitarian, political, economic and administrative reasons."[320]

If Hailey laid out the ends, colonial officer Sydney Caine laid out the means. The state was to be the locus of control: "the state ... must develop machinery ... needing specialized qualities different from those of ordinary administration and needing continuous thought."[321] This machinery could be created along the lines of the Russian planning commission or the American Tennessee Valley Authority. But that necessitated a "revolution in Colonial Research": "we need to be able to send freely and promptly experts of every kind to particular Colonies to report on particular possibilities of developments."[322] Continuous feedback would establish a "habit of investigation."[323]

Thus, in 1939 colonial officials began pushing for a successor to the 1929 Colonial Development Act that would provide monies disbursed by a "strong advisory committee to examine schemes."[324] One official remarked, "unless the fund for financing is administered by a specially appointed committee, responsible only to the Secretary of State and the Treasury, we shall not get the services of the best scientific advisers."[325] New realities necessitated "expert personnel required for conducting preliminary surveys ... on approved schemes."[326] A report on colonial research concluded the "basic data [is] required in all fields," and this should be collected by surveys on the "incentives" and "standard of living" in peasant communities.[327] On this basis, experts must work on the "extension and application of social and economic theory and methods developed in the older industrial countries to the particular conditions of the colonies."[328] These experts would enable some control over domains

---

[319] CO 852/214/13, no 1 1939 (Ashton and Stockwell 1996, 78).
[320] CO 852/482/6, no 11 1941 (Ashton and Stockwell 1996, 128).
[321] CO 852/588/1, nos 1 & 2 1943 (Ashton and Stockwell 1996, 168).
[322] CO 852/588/1, nos 1 & 2 1943 (Ashton and Stockwell 1996, 170).
[323] CO 852/588/1, nos 1 & 2 1943 (Ashton and Stockwell 1996, 170).
[324] CO 859/19/18, nos 1 & 2 1939 (Ashton and Stockwell 1996, 105).
[325] CO 859/19/18, nos 1 & 2 1939 (Ashton and Stockwell 1996, 106).
[326] CO 852/482/6, no 11 1941 (Ashton and Stockwell 1996, 132).
[327] CO 859/79/11, no 3 1943 (Ashton and Stockwell 1996, 355, 360).
[328] CO 859/79/11, no 3 1943 (Ashton and Stockwell 1996, 360).

previously ruled by supernatural forces: "There has been no staff to look ahead and to direct; prosperity and depression have been gifts from God or Satan ... economic development of the colonies deserves to be carefully planned and as carefully controlled."[329]

In October 1939, amidst the commotion of the new war, the Labour government's Secretary of State for the Colonies, Malcolm MacDonald, called a meeting at the Carlton Hotel that was attended by officials from the Colonial Office and the intellectual giants of colonial administration, Lord Lugard, Lord Hailey, and Margery Perham.[330] MacDonald was eager to imbue the British colonial mission with newfound legitimacy.[331] He felt the enterprise would come under threat during the war and wanted to articulate a humane and progressive "forward policy." The policy issues raised in the Carlton Hotel meeting, such as when African societies might be ready for self-government, were left unresolved. However, the attendees reached a consensus on the need to obtain more funds for development projects and research on African problems. The meeting inspired MacDonald to craft a new development act. Such an act would take up the project proposed by Hailey: a reorientation of colonial policy towards development informed by a massive new programme of colonial research. In effect, Hailey's ideas gave structure and content to MacDonald's desire to relegitimate colonial policy. The purportedly scientific backing provided by Western natural and social scientists would shore up the legitimacy of colonial governance by promising to increase human well-being throughout the colonial world.

In 1940 Parliament passed the Colonial Development and Welfare Act. The bill made available £5,000,000 per annum to back loans to colonial governments. In addition it provided £500,000 for research per annum for the period 1940–1951. This marked the end of the policy of self-sufficiency and extended the principles of the welfare state to noncitizens. The Act included provisions for a Colonial Social Science Research Council that funded and guided research in the postwar era.[332] After the war, the 1945 Colonial Development and Welfare Bill raised the development fund to £12,000,000 per annum for ten years.

---

[329] CO 852/369/3, no 50 1942 (Ashton and Stockwell 1996, 152).
[330] Sanger 1995, 148–149; Mills 2002, 163–164.
[331] Hyam 2010, 232.
[332] Hailey recommended Raymond Firth, an economic anthropologist who was a student of Malinowski, for a spot on the council. The influence of functionalist anthropology in colonial policy was maintained by Firth, Mair, Richards, and others. See Mills 2002; Dimier 2006. Perham and Mair remained particularly influential in the training of colonial officers. They worked to found the "science of colonial administration" that would both provide knowledge for development and ensure colonial staff had the appropriate expertise to oversee development. See Dimier 2006.

This provided both a technical model and a moral foundation for postwar development lending to the newly independent countries of the global south.

By 1945, Darwinian vestiges had dropped out of British colonial discourse. The primary documents themselves give an account of the transformation. One report argues that in the 1920s British colonial policy was shaped by the doctrine of indirect rule on the underlying assumption that "indigenous societies adapt themselves instinctively of their own volition, to changing conditions – and therefore the task of government is merely to guide, not create development."[333] However, by the 1940s, as one official put it, it was "no longer reasonable to think of colonial policy in terms of the gradual adaptation of traditional societies carefully preserved against radical change."[334]

The hope had been that societies, suitably protected by British trusteeship, would be shepherded into modernity on their own terms: "indigenous institutions should be preserved intact from the brunt of Western influences so that they may evolve spontaneously."[335] But with life in "tribal society" deteriorating rapidly, there was no option but to press ahead with "positive action" by the government to spur economic and political development. Colonial officials concluded that political and economic development would build a new society where old tribal life had been. The changes that had wrought natives from their traditional contexts now pressed them to pursue development: "Detribalisation goes hand in hand with economic development."[336]

On the new view, development was a technical exercise of expanding economic output. The complex array of objects that structured the landscape of 1930s development policy could now be redefined and reconceptualized as economic problems, amenable to statistical knowledge and productivist solutions. The work in the 1930s to reduce food to calories and "protein content" made this reconfiguration possible.[337] This had broad consequences. If food was just calories abstracted from local and cultural contexts, then it did not matter where those calories came from. Thus, any barriers to making food policy on the basis of agricultural science and the economic doctrine of comparative advantage could be removed. The Labour Party argued in 1943 that agricultural policy in African countries should focus on developing "crops for which they are best suited, irrespective of whether these are foodstuffs for local

---

[333] CO 847/35/9 1947 (Hyam Pt. 1, 149).
[334] CO 847/38/3 1947 (Hyam Pt. 1, 155).
[335] CO 847/38/3 1947 (Hyam Pt. 1, 151).
[336] CO 847/38/3 1947 (Hyam Pt. 1, 155).
[337] CAB 129/9 1946; see CAB 24/250 1934.

consumption, or produce for export."[338] On this view, nutrition is purely abstract:

> [T]he securing of an optimum diet is an object not so much of agricultural policy ... as of general social and economic policy, and is to be attained not by trying to produce as much as possible at home but by disposing the productive resources of the territory as to secure its inhabitants optimum nutrition and ... well being.[339]

Cash crops provide an abstract quantity of calories or nutrition that, without cultural context, is just as good as any other form of food.

Abstract thinking about the universality of well-being legitimized ambitious, expert-led development schemes. In the postwar era, development projects grew in size and scope, aiming to transform landscapes and societies.[340] The projects represented a fusion of expertise and state-led intervention that treated societies as objects of development, abstracted from the social and cultural particularities. The British conducted massive resettlement schemes that displaced thousands of Africans in order to improve public health, reform land tenure and land use, mechanize agriculture, and improve welfare.[341] For example, in East Africa, the groundnut scheme aimed to clear three million acres of bush for groundnut cultivation. However, the experts provided facts and figures that were so ridiculous, so removed from the actual agricultural, climatic, topographical, and technological conditions, that they were unusable. In the end, the plan caused widespread erosion and never produced the promised increases in agricultural productivity.[342] Such projects were seen as justified, even necessary, in an era of detribalization.

Scott explains resettlement schemes and large agricultural projects as emerging from a "high modernist" ideology.[343] However, the account here shows that high modernism itself was a contingent configuration that drew on new cosmological elements circulating in twentieth-century political discourses. The construction and spread of high modernism was made possible and desirable by a process of cosmological conflict that transformed the epistemes, ontologies, and purposes of colonial and postcolonial states alike.

The emergence of interventionist social knowledge displaced the vision of automatic progress that had supported both laissez-faire liberalism

---

[338] CO 323/1858/9 1943 (Ashton and Stockwell 1996, 159).
[339] CO 323/1858/9 1943 (Ashton and Stockwell 1996, 160).
[340] Scott 1998; Bonneuil 2001; Mitchell 2002.
[341] Scott 1998, 225–228; Bonneuil 2001, 261–264.
[342] Scott 1998, 228–229.
[343] Scott 1998.

and evolutionism. A new discourse of state purpose oriented politics towards state-led development by combining the new concept of time, a modernist episteme, and the new object-based ontology. Development was no longer an inevitable process of achieving moral and commercial civilization, but rather a process of harnessing science and technology to achieve calculable, statistical standards of welfare. Development, at home and in the colonies, was to be driven by state interventions guided by expertise.

Conclusion

At first, the rise of biological and Darwinian cosmological elements only reconfigured the liberal vision of human progress as proceeding through a series of stages. The idea that progress was an automatic process lived on into the twentieth century. However, colonized peoples resisted colonial rule and the rebellions of the 1920s forced a re-examination of this idea. In response, British colonial officials articulated a new vision of state-led development. The colonial state would marshal the power of knowledge to educate and advance primitive peoples towards scientific and technological modernity. This change in state purpose was made possible and desirable by cosmological shifts originating in new forms of social knowledge. These new forms of social knowledge were premised upon an epistemic modernism in which humans could harness science and technology to resolve the complex problems of modern politics.

In a process led by the British hegemon, these ideas were embedded in international order in the League of Nations. The League sanctioned the notion that the legitimate goal of all states was the peaceful pursuit of civilizational development. This state purpose did not displace balance of power politics or prevent the rise of fascism that destroyed the League. But henceforth, international order has been built on the illegitimacy of war, shifting the centre of discursive gravity from security concerns towards economic development. The League does not mark the height of epistemic modernism, but it does mark the emergence of development as a primary purpose of the state. That is, in the nineteenth century, the goal of improvement was embedded in balance of power discourses. However, by the early twentieth century, the goal of advancing the welfare of individuals had emerged as an end in its own right. In the League of Nations, security concerns were still paramount, but they existed alongside the ends of civilizational development in a new discourse of state purpose. Thereafter, states could no longer legitimate themselves only in terms of security. They also had to advance the nation morally and materially by harnessing knowledge to the development of human welfare.

The realist and liberal alternative explanation is that a doctrine of development was necessitated either by the demands of international power politics or a constellation of domestic interests. Indeed, the Depression and the exigencies of the Second World War made a compelling case for revising the colonial doctrine of self-sufficiency and increased demand for expert knowledge that could be applied to the problems of government. But the functionalist logic that underlies this alternative explanation cannot explain the outcomes here. States cannot have interests in developing forms of knowledge that do not exist or that officials do not know exist. Instead, representations of reality constitute those interests and structure the response to the problems of governance. So what appears to be in the natural interests of states seems that way because the world has already been arranged conceptually to privilege those ends.

The history of colonial development had a lasting effect on international order. As Grovogui and Anghie argue, modern international law emerged from the negotiation of problems in colonial order.[344] On this view, international law is less a body of universal principles extended to non-Western spheres by colonialism than an improvised set of distinctions and idioms created in response to the needs and ideals of the colonial powers.[345] Similarly, development expertise grew out of the "science of colonial administration."[346] For Anghie, the Mandate System created a new norm of international governance: it was legitimate for international organizations to collect data for the purposes of governing states.[347] That is, the very notion that international governance structures can and should intervene in the lives of states emerged from the colonial era. Of course, this norm was a centrepiece of the post-Second World War order constructed by the American hegemon. Thus, the basic norm of international expert intervention has been taken for granted for decades, hiding its colonial origins. But the colonial history here helps explain why such a principle, which violates the basic premises of cultural and legal sovereignty that are supposed to underwrite international orders, became so widespread and natural.

Colonial governance practices also had a direct effect on the postwar era. Some experts and colonial officials moved from working in the British

---

[344] Grovogui 1996; Anghie 2004. See also Keene 2002.
[345] Anghie 2004, 6–7.
[346] Anghie 2004, 156–192; Dimier 2006; Staples 2006.
[347] Anghie 2004, 183–192, 264. See Pedersen (2007, 1104) for some important correctives to Anghie's account.

colonial context to positions in postwar global governance.[348] John Boyd Orr, for example, was a British public health scholar who supervised studies in colonial medicine in the 1920s. His ideas were influential in the British Colonial Office and shaped the emerging discourse of state-led development. Orr went on to serve as the first director-general of the Food and Agriculture Organization.[349] In addition, the postwar financial institutions inherited the technologies of management from colonial administration and the League Mandate System.[350] However, whereas the League mandate was restricted to specific countries, the World Bank and the International Monetary Fund universalized the economic management of states in the global south.[351] So the postwar American order directly incorporated personnel, discourses, and techniques from the late colonial and interwar orders.

More generally, between 1815 and 1945, states and international organizations shifted from promoting a tenuous balance of competition amongst the great powers to encouraging the goal of development conceptualized inchoately in terms of scientific and technological progress. The new discourse of late colonial policy oriented the goals and values of international order under British and later American leadership to moral and material development. As we shall see in the next chapter, this revolution established the ideological grounds on which the Cold War was fought. The economic stakes of the Cold War in turn laid the foundation for the triumph of Western capitalist modernity after the fall of the Soviet Union.

The colonial origins of development ideas would be of only historical interest if not for the fact that the development discourse today is premised upon the same implicit standards of scientific and technological superiority that legitimated colonial trusteeship. Adas' study of colonial discourses argues that in the fifteenth and sixteenth centuries Europeans defined themselves in relation to peoples from Africa and Asia in predominantly religious terms.[352] However, after the Enlightenment and the industrial revolution, the terms of comparison shifted to highlight the superiority of Western science and technology. The fruits of "civilization" were the fruits of scientific discovery and proof of Western superiority.[353] Moreover, scientific and technological achievement became

---

[348] Mazower 2012, 143.
[349] See Worboys 1988; Brantley 1997; Staples 2006.
[350] Anghie 2004, 191.
[351] Anghie 2004, 191–192.
[352] Adas 1989, 31.
[353] Adas 1989, 134–142.

a measurable standard with which to classify and rank the progress of other civilizations.[354] Adas argues that modernization theory is merely the postwar version of the colonial civilizing mission because it assumes that eliminating peasant society and fostering industrial development are necessary. The basic logic was the same in 2015: capitalist, scientific, and technological society is superior and should be the goal of all peoples via a transfer of knowledge and expertise from north to south. The constitutive mechanisms of knowledge transfer now operate through the profession of economics and the international financial institutions, but the colonial assumption that all people should adopt modern Western civilization is unchanged. Moreover, the homogenizing impulse of development ideology remains unchanged. By incorporating diverse societies, against their will, into the expansion of Western modernity, colonialism displaced alternative modes of organizing life. The expansion of global capitalism, while producing new syncretic forms of capitalism mixed with elements of local institutions, spreads a technologically oriented and consumerist lifestyle that takes on similar characteristics the world over.

Thus, colonialism and neo-colonialism unfolded as processes of cosmological contestation that repressed and eradicated indigenous knowledge throughout the world. This eradication proceeded both through the systematic destruction of indigenous cultures and through the dissemination of Western discourses. As Kalpagam argues, colonial administration served as a conduit for cosmological elements:

> The dense administrative discourses of colonial governance were not merely representations of modern power enabling certain kinds of interventions but served as carriers of Western categories of space, time, measure, reason, and causality that constitute modern sciences and that were not hitherto part of the epistemological fabric of those societies.[355]

Similarly, Chatterjee argues "the implantation" of colonial ideas in other contexts changes power relations and frameworks of thought in the dominated society.[356] In the case of British colonialism in India, Chatterjee argues that the British doctrine of "improvement" became the basis of collaboration between the colonial government and educated Indians to sideline revolutionaries and their rejection of capitalist social order.[357] Thus, British concepts of time and purpose were incorporated into

---

[354] Adas 1989, 144.
[355] Kalpagam 2000.
[356] Chatterjee 1986, 27.
[357] Chatterjee 1986, 26.

Indian nationalism. Such nationalist movements surely reconfigured Western discourses and cosmologies, but they did not compete with indigeneous traditions on an even ground. Backed by power, the doctrines and practices of colonial development accelerated the transformation of colonized societies, introducing cosmological elements that in turn set off further cultural changes, reconstituting the basis of all future international orders.

5 Neoclassical Economics and the Rise of Growth in the World Bank and Postwar International Order, 1945–2015

> We must embark on a bold new program for making the benefits of our scientific advances and industrial progress available for the improvement and growth of under-developed areas ... The old imperialism – exploitation for foreign profit – has no place in our plans. What we envisage is a program of development based on concepts of democratic fair dealing ... Greater production is the key to prosperity and peace. And the key to greater production is a wider and more vigorous application of modern scientific and technical knowledge.
> – Harry Truman, 1949[1]

Introduction

On a freezing cold Chicago day in December 1942, Enrico Fermi brought years of work on atomic physics to their culmination by performing the first human-made nuclear chain reaction:

At first you could hear the sound of the neutron counter, clickety-clack, clickety-clack. Then the clicks came more and more rapidly, and after a while they began to merge into a roar; the counter couldn't follow anymore. That was the moment to switch to the chart recorder. But when the switch was made, everyone watched in the sudden silence the mounting deflection of the recorder's pen. It was an awesome silence. Everyone realized the significance of that switch; we were in the high intensity regime and the counters were unable to cope with the situation anymore. Again and again, the scale of the recorder had to be changed to accommodate the neutron intensity which was increasing more and more rapidly. Suddenly Fermi raised his hand. "The pile has reached critical," he announced. No one present had any doubt about it.[2]

This moment, as much as the more famous explosions at Los Alamos, Hiroshima, and Nagasaki, captures the cosmological significance of the Manhattan Project. On that day, Fermi "had controlled the release of energy from the atomic nucleus."[3] This was a crucial step in the process

---
[1] From the 1949 Inaugural Address. Quoted in McCarthy 2007, 4.
[2] Rhodes 1986, 440. The eyewitness report is from Herbert Andersen.
[3] Rhodes 1986, 440.

208 Neoclassical Economics and the Rise of Growth

Figure 5.1 Scientists observing the world's first self-sustaining nuclear chain reaction in the Chicago Pile No. 1, 2 December 1942, Gary Sheahan, 1957

of constructing a nuclear weapon, but it was also cosmologically meaningful in that it demonstrated the power and reach of human control over the forces of nature.

The image of an object in a field of forces under precise control dominated intellectual developments in postwar physics, engineering, and the social sciences. In a more general form, it underwrote the rise of cybernetic-systems thinking. The cybernetic-systems view represented the world as a series of objects or systems that could be modelled, predicted, and manipulated.[4] As we saw in the previous chapter, the social sciences had already started to adopt this mode in the 1920s and 1930s, but it was bolstered after the war. Social scientists had worked alongside physicists and engineers in expert committees during the war and so they had been exposed to the latest systems thinking. Following the war, social scientists, especially economists, developed techniques for modelling, projecting, and forecasting objects. From the social sciences, these new modes of thought entered international political discourses and transformed ideas about state purpose.

Cybernetic-systems thinking was cosmologically significant because it was used to reconfigure ideas about human agency and time. First, object-centred thinking contributed to the sense that society, the economy, the climate, and humans themselves could be precisely controlled.[5]

[4] Lilienfeld 1978; Gerovitch 2002; Heyck 2015; Kline 2015; Rindzevičiūtė 2016.
[5] This is Scott's (1998) "high modernism."

Introduction

Few, surely, thought that the social sciences could control objects as precisely as Fermi had controlled the uranium pile on that cold Chicago day. Nevertheless, cybernetic-systems thinking contributed to the modernist faith in the power of human knowledge to solve problems.

In addition, these techniques altered conceptions of political time by bringing the future into the political present.[6] Scientific and technological progress ceased to be, as it had been in the nineteenth century, conceived of as an automatic or autonomous process. Instead, the dream of unending linear progress could be enacted and controlled from the present. But more deeply, human nature came to be conceived in terms of scientific and technological progress. Whereas in Condorcet, the light of reason served human progress, in neoclassical economics, human progress came to be equated with the unleashing of scientific and technological progress itself.

This cosmological shift made possible and desirable the construction and dissemination of a new state purpose: economic growth. The purpose of growth was recursively institutionalized in international order throughout the postwar era under the auspices of American hegemony. First, American economists and policy-makers embedded the goal of expanding production in the Bretton Woods institutions. Then, the transnational rise of neoclassical economics backed by American hegemony drove further associational changes, spreading and reconfiguring the discourse of growth.[7] While it was first developed to measure productivity during the war, economic growth came to be equated with human progress itself.

There are a number of possible explanations for the rise of growth as a purpose of states.[8] Realists might argue that states value economic growth as a proxy for military power, which is necessary for their survival.[9] In particular, the goal of increasing production might have been necessary to undergird the expansion of the military-industrial complex in the context of a global geopolitical competition with the Soviet Union.[10] Marxists might argue that the dynamics of capital accumulation necessitate economic growth.[11] Rationalists might argue that the growth imperative is driven by domestic political dynamics.[12] That is, in Western

---

[6] Mitchell 2014.
[7] On economists generally, see Fourcade 2006, 2009. On economic growth, see Maier 1989; Collins 2000; Coyle 2014; Yarrow 2010; Lepenies 2016.
[8] For a good overview of these issues, see Purdey 2010, 30–56.
[9] Mearsheimer 2001. See also Waltz 1979; Gilpin 1981; Grieco 1988.
[10] Collins 2000, 24. The role of NSC-68 was important here. On this view, economic growth was just the latest concept to codify the goal of pursuing wealth and military power at the same time. See Kirshner 1999; Baldwin 1985; Earle 1986.
[11] Foster 1992, 2015. For a discussion, see Kallis 2017, 84–86.
[12] Frieden 1991; Simmons 1994; Moravcsik 1997. For a good discussion, see Kahler 2002, 41–43.

democratic societies politicians found that growth-based policies delivered increases in personal consumption that were electorally beneficial.[13]

While each of these can help explain why economic growth is a powerful force in modern politics, they leave important aspects of the rise of growth unexplained. First, the concept of economic growth is relatively recent. Timothy Mitchell has shown that the concept of "the economy" emerged only in the 1930s and the idea that it could grow was not well-formed until the 1940s.[14] Moreover, states have pursued a wide variety of economic and military goals since the sixteenth century. As late as the 1930s, the central end of international economic policy was not growth, but balance and stability. In the United States, this was underwritten by a stagnationist school in economics that advocated organizing and redistributing wealth rather than expanding production. To explain why the pursuit of wealth and power came to take the form of economic growth in the latter half of the twentieth century, we have to investigate the cosmological shift and associational changes that gave it form and content. As Ruggie points out, "in order to say anything sensible about the *content* of international economic orders ... it is necessary to look at how power and legitimate social purpose become fused."[15]

A discursive analysis of growth as a purpose of international politics reveals that it depends on a powerful configuration of concepts and ideas: an ontology that specifies objects like "the economy"; mechanical, systems thinking to theorize economic transactions and relationships; statistics, tables, and models to map the transactions in summary; an absolute, open plane of time upon which to chart progress; the notion that an object can grow and develop over time; and a modernist ethos underwriting a set of economic and policy levers to manipulate the economy. These elements were brought together under a highly specific set of cosmological and geopolitical conditions in the middle of the twentieth century. In this sense, the postwar settlement consolidated and extended shifts in the discursive basis of international order that had been ongoing since the sixteenth century.[16]

To trace the construction and operations of the growth configuration at the meso-level, this chapter examines the role of growth in the World Bank discourse.[17] The World Bank offers an important case study

---

[13] Block 1977, 33–36; Maier 1987, 179; Ruggie 1982, 388–390. See also Ikenberry 1992.
[14] Mitchell 2002, 2005, 2014.
[15] Ruggie 1982, 382.
[16] In tracing this gradual change, my account goes beyond the punctuated equilibrium logic of other accounts of international change. See Gilpin 1981; Ikenberry 2001; Osiander 1994; Reus-Smit 1999.
[17] At the associational level, the evidence presented in this chapter combines a discourse analysis of primary Bank documents with process-tracing techniques. The text sample for the discourse analysis contained fifty official reports (one per year), fifty country reports

Introduction

of associational change in the post-Second World War era for three reasons. First, the Bank played a central role in the promotion of economic growth and its attendant cosmology. As an IO, the Bank does not provide direct evidence of state goals. But as a central provider of policy advice in the US-led order, the Bank played a key role in defining the legitimate ends of state action. In this role, the Bank consistently reflected American economic and intellectual hegemony, while operating relatively autonomously from the US government itself.[18] Second, the internal dynamics of change in the Bank demonstrate the effects of economists in creating and constituting the international discourse of state purpose. Third, the Bank case reveals how the purpose of growth was defended in the face of challenges from the basic human needs approach. As such, it helps uncover the cosmological elements used to justify and naturalize growth in international order.

In 1968 Robert S. McNamara left the US Department of Defense to become President of the World Bank. Upon taking office, he set out to reorient the Bank away from economic growth for its own sake towards projects that would directly alleviate poverty. However, McNamara's administrative reforms ended up promoting neoclassical models, along with their cosmological implications, that reoriented the Bank to growth understood as scientific and technological progress. While the Bank agenda broadened after 1990, Bank policy remained focused on bolstering economic growth. Why, despite McNamara's reformist zeal and organizational power, did the World Bank fail to institutionalize poverty alleviation policies in the 1970s and 1980s? Why did it reverse course and promote the goal of unending economic growth in all countries?

I argue that coercion and learning mechanisms alone cannot explain the pattern of associational change in the Bank. Instead, my story highlights the role of a constitutive mechanism – the rise of neoclassical economics through personnel changes and bureaucratic reforms. As we shall see, the ends of growth were not imposed from the outside, but emerged unintentionally from a combination of McNamara's reforms. McNamara strategically deployed scientific ideas as part of reforms intended to reorient Bank purposes. But by hiring neoclassical economists and imposing

---

(twenty-five from 1950 to 1975 and twenty-five from 1975 to 2000), and 100 project reports (fifty from 1950 to 1975 and fifty from 1975 to 2000, including fifteen structural adjustment loan reports). I used secondary accounts of Bank history, privileging works that had internal access to the everyday life of the organization, to reconstruct the processes of associational change. For more information see the Methodological Appendix.

[18] On American hegemony and the World Bank, see Wade 2002. On epistemic authority and IO autonomy, see Barnett and Finnemore 2004.

212 Neoclassical Economics and the Rise of Growth

quantitative standards he created a configuration of interlocking elements that privileged statistics and mathematical modelling. This configuration made it easier to measure and value growth measured in terms of gross national product (GNP) or gross domestic product (GDP) than to measure and value poverty alleviation. Moreover, neoclassical economists directly advocated for trade-led growth and worked to marginalize direct poverty alleviation. These factors explain the success of strategic deployment in the Bank, but they do not account for the discursive reconfigurations that generate new purposes. To explain how growth emerges in the postwar era not just as an end but as a purpose, I show how growth was naturalized via cosmological narratives of scientific and technological progress. Tracing the history of growth in the Bank helps us to understand how the broader discourse of state purpose changed between 1945 and 2015.

The Cosmological Shift in Physics, Engineering, and Economics, 1930–1960

The successes of physics and engineering during the Second World War meant that the atomic era, for all its anxieties, was a time of scientific and technological optimism in which experts had great authority.[19] In both the United States and the Soviet Union, the valorization of engineering began in the 1920s and 1930s when high profile public projects demonstrated its potential.[20] But the epistemic authority of physics and engineering peaked when they were credited with winning the Second World War. In the United States, wartime scientific research was led by Vannevar Bush. In 1940, Bush pressed Roosevelt to create a committee that would coordinate scientific research for the coming war. Bush was appointed as the head of the resulting National Defense Research Committee. The committee created national laboratories and sponsored basic research on armour, bombs, communication, and radar. In 1941, the Office of Scientific Research and Development (OSRD) superseded the committee. It was the OSRD that made science synonymous with the central technological achievements of the war effort: radar, synthetic rubber, and the nuclear bomb.[21]

After the war, many scientists took jobs in the government and served the American security establishment.[22] However, some scientists rejected

---

[19] Gilman 2003; Fourcade 2009; Backhouse 2010.
[20] Rindzevičiūtė 2016, 16.
[21] Gilpin 1964, 5.
[22] Leslie 1993; Evangelista 1999; Friedberg 2000.

such a role, claiming that the destructive potential of the atomic bomb was too dangerous to be left in the hands of the politicians. Niels Bohr argued that an international organization of scientists should be created to facilitate cooperation amongst the nuclear powers: "only by creating an 'open world,' an international order based on the kinds of cooperation that existed in the scientific profession, could an arms race and a nuclear war be circumvented."[23] Members of the Manhattan Project concluded that there should be an international agency of scientists with "police powers."[24] Vannevar Bush and James Conant, both of whom served as advisors to President Roosevelt during the war, argued for a similar system:

Sovereignty would be replaced by an association in which an international community of scientists would play a major role; such an arrangement was thinkable in 1944–45 because the colossal benefits and enormous dangers conferred on civilization by science seemed so apparent.[25]

While these utopian visions of scientific hegemony may strike us now as naively ignorant of power politics and the tendency of political leaders to jealously guard national sovereignty, they nonetheless speak to the authority of science in the postwar era.

The authority of physics and engineering meant that the image of the scientific object as a self-contained system in a field of forces was adopted by other natural and social sciences. This inspired a broad intellectual movement: the rise of systems theory. General systems theory grew out of Ludwig von Bertalanffy's (1901–1972) research in the 1930s.[26] He defined a system as a "complex of interacting components" that could be either closed or open to exchange with its external environment.[27] Bertalanffy operationalized systems as a series of interlinked differential equations such that any variable in the system was a function of all the other variables.[28] He argued that macro-level systems exhibited characteristics not visible by looking only at individual elements. Systems were governed by a series of general laws to be uncovered by general systems theory. In a similar development, Norbert Wiener's (1894–1964) early research on computers inspired him to create the field of cybernetics.[29] Cybernetics modelled phenomena as input–output machines that could achieve homeostasis through feedback loops.[30] The simple images of

---

[23] Manzione 2000, 29.
[24] Manzione 2000, 29.
[25] Manzione 2000, 30.
[26] Lilienfeld 1978, 17. See also Bertalanffy 1951.
[27] Lilienfeld 1978, 23.
[28] Lilienfeld 1978, 24.
[29] Conway and Siegelman 2005, 171–194.
[30] Lilienfeld 1978, 63–65.

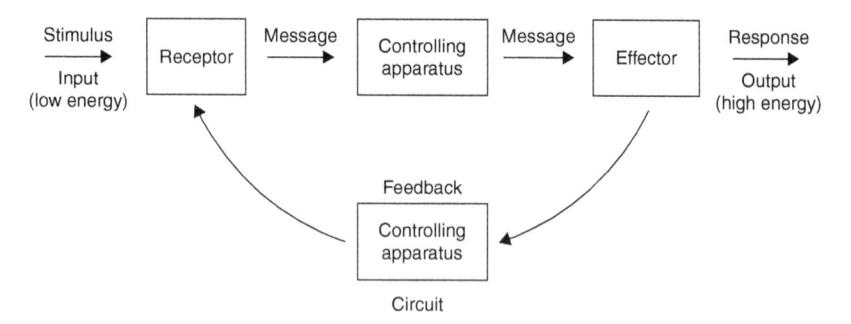

Figure 5.2 A simple feedback arrangement, Ludwig Bertalanffy (Bertalanffy 1951, 349)

cybernetic systems (Figure 5.2) represented objects using simple control diagrams borrowed from engineering. Systems theory and cybernetics were combined and recombined with emerging discourses and techniques from information theory, game theory, and linear programming, all of which encouraged a macro-level perspective.[31]

The confluence of physics, systems theory, cybernetics, and game theory in the development of "operations research" also had a lasting influence on the postwar world. During the war, American economists and social scientists were deployed alongside physicists and engineers in operations research groups that served the military.[32] To serve the war effort, British and American governments created new advisory committees. In these committees, civilian scientists and experts would provide advice and practical problem-solving to military officials.[33] In Britain, one such committee was named the operations research group because it applied the scientific method to military operations.[34] Initially the group was composed of physicists who focused on problems such as where to place radar sets, how to maximize the efficiency of flights, and so on. But the British military saw value in the method and operations research groups were soon working on anti-aircraft targeting, logistics problems, personnel assignment, and so on.

From Britain, operations research was exported to the United States via academic exchanges organized by the American National Defense Research Committee. The first American operations research groups were formed in 1942. In the United States, operations research brought

[31] Lilienfeld 1978, 1.
[32] Backhouse 2010, 38; Mirowski 1999, 2002; Galison 1994, 248–251.
[33] Fortun and Schweber 1993, 599.
[34] Fortun and Schweber 1993, 600–601.

together physics and new techniques of mathematical economics such as game theory.[35] Game theory was applied to maximize the efficiency of bombing and reconstruction efforts by taking into consideration the actions of the adversary.[36] After the war, the operations research approach was generalized into "systems analysis." Systems analysis, broadly construed, was a science concerned with the "diagnosis, design, and management of complex configurations of men, machines, and organisations."[37] Systems analysis then travelled through the corporate world as a general, Taylorist tool for the rationalization of organizations.

John von Neumann (1903–1957) and Norbert Wiener were at the centre of all these intellectual developments. Von Neumann was instrumental in every great scientific research programme from 1920 through 1950. In the 1920s, he did foundational work on the mathematics of set theory, geometry, lattice theory, and operator theory. He made seminal contributions to mathematical economics, linear programming, and invented game theory almost in his spare time. In the early 1930s, he produced a rigorous mathematical basis for quantum mechanics. He then became a leading expert in the mathematics of explosions and played an important role in the Manhattan Project. There, von Neumann modified early computers so they could perform explosion simulations. In the 1940s, von Neumann also worked with the operations research group attached to the US Navy. There, he adapted his theory of games to efficiency problems and made further contributions to mathematical economics. After the war, von Neumann built a new computer, MANIAC, and used it to produce the first weather models. These models formed the basis of the first climate models.

Before the war, Wiener was an ambitious mathematician working on probability theory, prediction, and computing. During the war, Wiener worked on operations research problems under the auspices of the National Defense Research Committee.[38] Wiener's project was to design an "anti-aircraft predictor" which would calibrate anti-aircraft fire based on predictions of the position of enemy aircraft.[39] Wiener's innovation was to mathematize the strategic action of enemy pilots and Allied gunners as "proprioceptive and electrophysiological feedback systems."[40] In short, humans and machines were both modelled as "servomechanisms" or systems governed by positive and negative feedback loops.

[35] Mirowski 1999, 695.
[36] Fortun and Schweber 1993, 604.
[37] Lilienfeld 1978, 111.
[38] Conway and Siegelman 2005, 108–109.
[39] Galison 1994; Conway and Siegelman 2005, 110–115.
[40] Galison 1994, 229; Conway and Siegelman 2005, 135.

Wiener knew that the anti-aircraft predictor was more than a mathematical model; he saw it as inaugurating a whole new discipline of "communications engineering."[41] After the war, Wiener generalized this vision of humans and machines as systems governed by feedback into cybernetic theory.[42]

Implicit in von Neumann and Wiener's work was a vision of the natural and social world as calculable and predictable.[43] Von Neumann advanced the mathematical and technological infrastructure necessary to imagine the world as a series of interlocking systems. His computer models vividly depicted the system behaviour of interacting parts. His theory of games made it possible to imagine the human world in the same terms, wherein the behaviour of any individual is made dependent on the behaviour of any other. His work in linear programming gave the sense that these complex problems of human and natural interaction could yield unique solutions. Wiener's cybernetics generalized systems theory so that humans, machines, and social systems could be modelled as servomechanisms. As such, all entities could be understood and predicted with mathematical models. Von Neumann and Wiener's world was complex, but knowable and controllable with mathematical tools. Together, linear programming, game theory, and computing "facilitated the metaphorical understanding of world politics as a sort of system subject to technological management."[44]

The rise of systems thinking, cybernetic ideas, and computational modelling had lasting effects on the social sciences in general and on American economics in particular. In general, systems thinking reconfigured the study of social objects into the study of input–output systems comprised of rules, rational actors, feedback loops, and so on. These entities were articulated as quantified systems, amenable to prediction, intervention, and control. On one hand, cybernetics borrowed ontological and epistemic elements from biology. On the other hand, it stripped out the historicity and specificity of biological knowledge. In cybernetics and systems ontology, everything is an interchangeable, functional machine. In this worldview, the human is not a specific species. Rather, it is a natural entity, but a natural entity redefined in the language of physical machines. Humanity itself is reduced to functional processes like reproduction, calorie-conversion, and other elements that can be represented in causal chains and measured in statistical terms.

[41] Conway and Siegelman 2005, 116.
[42] Galison 1994; Gerovitch 2002.
[43] On some of the differences between Wiener and von Neumann, see Gerovitch 2002.
[44] Edwards 1996, 7.

It was in this context that the object "the economy" was forged.[45] John Maynard Keynes (1883–1946), based on his work in the India Office, provided the first representation of the "colonial economy." His 1913 study of problems in India conceptualized "the new totality not as an aggregation of markets in different commodities, but as the circulation of money ... the sum of all the moments at which money changed hands."[46] The designation of the economy as a separate sphere was then naturalized via the rise of statistics. The first major instance of national accounting was the Dawes Commission's attempt to estimate Germany's ability to pay reparations to the Allies.[47] The committee initially set out to measure Germany's wealth in materialist terms, as in mercantilist-substantialist conceptions. The approach ran into difficulties, however, because it was not clear how to define economic activity in quantitative terms. After Keynes' breakthrough, the task became much easier: count income, not wealth. Income provided a better analog to "money changing hands" and so could be used to create aggregate statistics to measure the emerging object.[48]

In the 1930s, the Great Depression and war planning renewed the quest for a method of producing complete, up-to-date national income statistics. Simon Kuznets, a University of Pennsylvania economist, led the first systematic efforts at the National Bureau of Economic Research.[49] Kuznets' estimates gave policy-makers a sense of how much damage the 1929 crash had caused and, later, how successful the recovery efforts were. By the late 1930s, "national income" had entered American public discourse.[50] Initially, the US Department of Commerce adopted Kuznets' method for calculating national income, but during the Second World War Kuznets' method was dropped. Kuznets' method defined income primarily in terms of consumption measured by market prices. During wartime, this created a problem because government spending on nonconsumable goods like tanks and bombs reduced the national income. To resolve this problem, economists in the Department of Commerce drew on Keynesian logic to include state spending in the national income so that spending on armaments increased rather than reduced the national income.[51] The war and the requirements of state

---

[45] Mitchell 2002, 91–99.
[46] Mitchell 2005, 135.
[47] Mitchell 2005, 135.
[48] Mitchell 2005, 135.
[49] Wesley Mitchell worked at NBER to measure national income in the 1920s. He handed his work over to Simon Kuznets in the run up to the Second World War. See Breslau 2003; Carson 1975; Collins 2000, 32–35; Yarrow 2010, 33.
[50] Lepenies 2016, 63–74.
[51] Lepenies 2016, 76–77.

power shaped the precise form that growth took when it emerged in the 1940s and 1950s.

The increasing availability of aggregate national statistics bolstered the view that the economy was a self-contained object that could grow or contract under the influence of government policy. The conception of the economy that emerged after the Second World War combined the statistical description of the national accounts with new cybernetic, systemic metaphors that entered economics via wartime collaborations with physicists and engineers. Before the Second World War, American economics was largely historical, contextual, informal, and institutional.[52] During the war American economists were deployed alongside physicists and engineers conducting operations research for the government.[53] In interdisciplinary environments where economists and scientists worked together, as they did at RAND, economists came to see their task as a form of "social engineering."[54] Moreover, the general authority of physics and engineering meant that when economists sought to bolster their epistemic authority, they turned to ideas and methodologies from the natural and mathematical sciences.[55] Economists' wartime experiences alongside physicists, engineers, and mathematicians facilitated the introduction of modelling, computational tools, and cybernetic concepts into economics.[56] The metaphor of the cybernetic system operationalized Keynes' intuition that the economy was best represented as a series of monetary exchanges. On the new view the economy was a series of mechanical flows between individuals and businesses, governed by feedback loops.

The synthesis of statistics and cybernetics contributed to the creation of rigorous macroeconomic models that dominated both abstract academic debate and the development of policy tools. In the 1950s, the Cowles Commission, which received support from private business, the Rockefeller Foundation, and RAND, grew into an interdisciplinary forum for the mathematization of economics.[57] Cowles channelled axiomatization, general equilibrium analysis, and linear programming from mathematics into economic theory.[58] The proliferation of techniques soon converged in the "postwar neoclassical orthodoxy" built upon the

---

[52] Fourcade 2009; Ross 1991. On early mathematical approaches in economics, see Morgan 2003; Breslau 2003.
[53] Backhouse 2010, 38; Mirowski 1999, 2002; Galison 1994, 248–251.
[54] Backhouse 2010, 41–42; Morgan 2003, 289.
[55] This, of course, was a long tradition. Mirowski 1989, 2002; Gilman 2003.
[56] Mirowski 2002.
[57] Fourcade 2009, 88; Düppe and Weintraub 2014, 460–465.
[58] Düppe and Weintraub 2014, 460, 473.

rigorous neoclassical-Keynesian synthesis of Paul Samuelson (1915–2009).[59] Samuelson's macroeconomic General Equilibrium Analysis underwrote the Keynesian consensus that dominated postwar theory and policy. At the Massachusetts Institute of Technology (MIT), Samuelson oversaw the creation of a neoclassical research programme oriented to growth.[60] The new object "economy" was to be represented in deductive models that would be tested against historical data by econometric techniques. Within these models, individuals were depicted as optimizing, rational actors whose actions aggregated into macroeconomic phenomena. As Mirowski puts it, "neoclassical economics became more formal, more abstract, more mathematical, more fascinated with issues of algorithmic rationality and statistical inference, and less concerned with the fine points of collective action or institutional specificity."[61]

The same cosmological elements arose in Soviet thought. Cybernetic thinking emerged in the Soviet Union under Andrey Kolmogorov (1903–1987). Kolmogorov, like Wiener, was a mathematical prodigy at a young age. Throughout the 1930s, Wiener and Kolmogorov's work on prediction theory ran in parallel as the two developed new theorems almost simultaneously.[62] But it was Wiener who first articulated cybernetic theory during the war. After the war Kolomogorov and others worked to import the American science of cybernetics into the Soviet Union. This was challenging because American science was often refuted as "bourgeois nonsense."[63] Nonetheless, under various guises, Soviet mathematicians introduced the core concepts of cybernetic and systems modelling into military and economic planning.[64] In so doing, they conceptualized "the Soviet economy in cybernetic terms as a giant control system."[65] Indeed, the Soviets strategically deployed cybernetic ideas in governance precisely because they "appeared to promise *more* control" than alternatives.[66]

So the economy emerged within a broader, transnational epistemic and ontological movement to redescribe the natural and social world within the mathematical language of cybernetic systems and models. The change had cosmological implications because it redefined the place of humanity within this universe of systems. Humans were, on the cybernetic

---

[59] Backhouse 2010, 49–58.
[60] Boianovsky and Hoover 2014.
[61] Mirowski 1999, 686.
[62] Gerovitch 2002, 56–58.
[63] Gerovitch 2002, 16.
[64] Rindzevičiūtė 2016.
[65] Gerovitch 2002, 270.
[66] Rindzevičiūtė 2016, 13.

view, just like other machines and so they were knowable, representable, and manipulable in mathematical terms. Economics on both sides of the Cold War naturalized certain understandings of human behaviour and excluded others.[67] By privileging the rational and calculable features of the world, this way of thinking introduced representational constraints that defined humans and their societies in aggregate statistical terms. This discourse displaced the more varied and messy conceptual world of pre-war economics, redefining political goals in terms of increases in aggregate statistical indicators.

The shift to a political discourse centred on objects and models was also cosmologically significant because it reconfigured notions of time and reoriented politics towards the future. Statistics and models made it possible to develop economic projections that could form the basis of political and bureaucratic action. In 1947 the US Council of Economic Advisers started producing statistical projections.[68] At first, such projections were little more than extrapolations from current data. But in the 1950s, computational advances and improved statistical models made it possible to calculate future dynamics. These projections and models brought the future into the present and reoriented politics to prediction and control of the future.[69] Moreover, the future could be used to adjudicate disputes between labour and business in the present: "[e]xposed to the changing metrics of the economy, businessman would be pressured to consider the impact of excessive price increases, and workers would be forced to moderate wage demands."[70] For Mitchell, this was an important shift in modernity itself because it helped to represent the future as stable, knowable, predictable, and thus controllable.[71] Thus, economists were able to argue that they could design policies that would increase production and affluence: "armed with their macroeconomic models, economists now claimed to be able to deliver economic growth and full employment."[72]

The cosmological shift introduced by cybernetic-systems thinking made possible the rise of the economy and with it the reorientation of state purpose towards productivist economic policies. Before the war, economic policy was focused on a small number of limited tasks: "governments generally were taken to be responsible for trade policy, for keeping their own spending within budget, and for monetary and exchange

[67] Amadae 2003, 296.
[68] Collins 2000, 35.
[69] Mitchell 2014.
[70] Mitchell 2014, 492.
[71] Mitchell 2014, 497.
[72] Fourcade 2009, 85.

rate policy."[73] Through the 1920s and into the 1930s, the principal factor in the health of national and international economic life was considered to be the effective management of the currency.[74] Thus, the high politics of international economic policy in the early twentieth century was centred on maintaining currency stability, not advancing growth.[75]

In Anglo-American domestic politics, this was supported by a school of thought centred on the idea that economic development in the West had run its course and that industrial states needed to manage "secular stagnation."[76] Drawing on this, in 1932, Roosevelt argued that the central problem in American economic policy was not the need to grow or further exploit natural resources, but "adapting existing economic organizations to the service of the people."[77] Restrictions designed to dampen and balance the recovery were a central feature of New Deal policy-making.[78] There had always been opposition to these efforts from "expansionists" and "maximalists" who wanted to increase production. But stagnationism was also taken seriously and there was a live debate about which goals could or should be pursued. After the war, this debate disappeared and growth became the natural and taken-for-granted end of economic policy.

My argument is that the growth imperative emerged from a constellation of cosmological ideas and political forces that came together in the period 1940 to 1948. First, the rise of social policy, the invention of the economy, Keynes' *General Theory*, and the development of new mathematical techniques made the concept of growth possible and desirable. Over the course of the twentieth century, a series of social welfare initiatives for poverty alleviation, education, and health care become important political issues in the United States and Europe.[79] The growing middle class demanded increased prosperity and welfare protections from the state. These demands intensified after the First World War as powerful labour movements pressed for state-led industrial policy.[80] Then, Keynesian theory shifted the emphasis of economic thought from managing the business cycle to increasing household spending, which in turn, Keynes argued, could be influenced by government spending.[81] After the Second World War, new policy tools such as the Cowles

---

[73] Morgan 2003, 288.
[74] Morgan 2003, 288. See also Eichengreen 1996.
[75] Eichengreen 1996.
[76] Collins 2000, 6.
[77] Quoted in Collins 2000, 5.
[78] Collins 2000, 7.
[79] de Swaan 1998; Hicks 1999; Huber and Stephens 2001.
[80] Gallarotti 2000.
[81] Backhouse 2010, 52.

Commission planning models and Samuelson's Keynesian synthesis made it possible for governments to claim they could control the economy. In this "high period of the economist as engineer," the range of government tools expanded to include dampening economic cycles, stabilizing growth, lowering unemployment, reducing inflation, and maintaining the balance of payments.[82] Increasingly these tools were applied not to maintain balance but to create economic growth.

Second, a series of political and economic forces locked-in growth as a central end of the United States and European countries. First, the Second World War jolted American economic policy, necessitating a policy centred on maximum production.[83] Wartime manufacturing realigned production networks and business interests into a system centred on producing as many goods as possible. This pushed the United States decisively away from stagnationism towards a growth-oriented policy. Then, through the period 1945–1948, the United States embedded productivism in the core sites of the postwar order and pressed productivism in Europe through bilateral ties and, ultimately, the Marshall Plan.

In a democratic age, Western governments adopted productivism and growth-oriented policies as an electoral and political tool to meet the demands of both labour and business.[84] In short, expansion allowed Western governments to promise social and economic goods to workers, while pleasing business interests. In Europe and the United States, explosive ideological conflict between labour and business was replaced with a broad social consensus on policies to expand production. Politics was transformed: "As Western leaders looked more and more to economic growth, increasingly presupposed, first as automatic, and second, as the major index of a society's welfare, the stakes of politics narrowed."[85] Growth-oriented policies displaced broader debates about stagnation, redistribution, and expansion. In the United States, scientific progress and economic growth was seen as essential to national security in the early Cold War. So, the central political question was whether to pursue growth within the bounds of stability and inflation (the Republican position) or to pursue all-out growth (the Democratic response).[86] In Europe, radical parties on the left and right were marginalized and centrist Christian and Social Democrats "alternated officeholding in a consensual politics that debated only whether the anticipated dividends of

[82] Morgan 2003, 294.
[83] Collins 2000, 10–15.
[84] Maier 1987, 161; Ikenberry 1992, 299.
[85] Maier 1987, 179.
[86] Collins 2000, 40–61.

economic growth should be devoted to social-welfare consumption or ploughed back into private investment."[87]

This constellation of cosmological and political elements was institutionalized in the post-Second World War settlement. Under the auspices of the new American hegemon, British and American policy-makers created an open economic order designed to maximize production in Western economies. First, the Anglo-American designers of the Bretton Woods institutions ensured that the new order would be premised upon increasing production and economic growth.[88] The question was how to maintain currency stability, the old challenge, while accommodating the rise of the welfare state and the goal of full employment. The lesson of the interwar years was that international economic order needed to prevent tariff wars while allowing states to protect citizens from the domestic dislocations and instabilities usually associated with free trade. But, full employment policies were in tension with the deflationary requirements of the gold standard. If states wanted to spend on welfare, they would need an alternative scheme for achieving currency stability and balance of payments equilibrium.[89]

So in the talks leading up to Bretton Woods, Harry White and John Maynard Keynes set out to combine trade openness with a full employment system.[90] In the end, it was the goal of expanding production that provided the lynchpin of the design.[91] White pressed the American position that a productivist order would both fuel international trade and allow states to provide welfare goods at home. Keynes was sceptical at first because he thought that expansionary policies would either cause chaos or require something like a gold standard, which he felt would be unworkable.[92] But, under pressure from White, Keynes came to believe that a regulated system of international capital could be compatible with a productivist order. In the end, White and Keynes agreed that their aims could be achieved through trade openness tethered to a system capital controls and supported by an international stabilization fund.[93]

The Bretton Woods negotiators drew on a variety of advisory committees and bureaucracies that channelled discursive configurations from cybernetic-systems thinking and associated cosmological elements into

---

[87] Maier 1987, 179.
[88] On the economic basis of the postwar settlement, see Best 2005; Block 1977; Bordo and Eichengreen 2007; Eichengreen 1996; Gardner 1956; Helleiner 1994, 2006; Ikenberry 1992, 2001; Maier 1987; Ruggie 1982.
[89] Ruggie 1982, 390–393; Ikenberry 1992, 299.
[90] Best 2005, 33–35; Ruggie 1982, 390–393.
[91] Maier 1987.
[92] Best 2005, 38–40.
[93] Ikenberry 1992, 298; Helleiner 1994, 33; Block 1977, 46–51.

international order. This marked a shift in the epistemic basis of international discourses. In the eighteenth and nineteenth centuries, scientific cosmologies were carried into political discourse through informal channels between aristocratic and governing elites. Starting in the twentieth century, new ideas entered through expert brokers working in state agencies. Thus, as Britain handed off hegemonic leadership to the United States, the mode of expertise underlying international order shifted.[94] The classically and anthropologically trained elites of Cambridge, Oxford, and the London School of Economics were replaced by a rising transnational class of experts trained in the formal mathematical tools of economics and the positivist social sciences.[95]

In this context, the ethos of order building after the Second World War was influenced by American policy-makers' experiences with New Deal planning and their affinity for social science.[96] The State Department under the leadership of Leo Pasvolsky and Cordell Hull set up a Division of Special Research as early as 1941 to begin planning for the postwar order.[97] Drawing on work by the Brookings Institution and the Council on Foreign Relations, the Division's proposals formed the basis of the United Nations. The future world organization was to improve upon the League by embedding collective security in the balance of power and expanding the League's bureaucratic work to fight poverty, hunger, and humanitarian crises.[98] While Cold War realities placed strict limits on the role of expertise in postwar international politics, the vision of the United Nations was "an organisation that combined the scientific technocracy of the New Deal with the flexibility and power-political reach of the nineteenth-century European alliance system."[99]

While it is tempting to treat the postwar order as a new creation following a huge shock to the system, the new growth-based order was built on the epistemic and cosmological foundations laid by the League. The League had spread modernist presuppositions among states and laid the groundwork for the collective management of problems. So while after the war the secondary rules of international order were revised to reinstitutionalize great power management, bolster economic cooperation, and make the world safe for democracy, the underlying discourse of state purpose and cosmological elements exhibited considerable continuity from the nineteenth into the twentieth century.

[94] Mazower 2012, 199.
[95] Mazower 2012, 199.
[96] Helleiner 2006; Mazower 2012.
[97] Mazower 2012, 198.
[98] Mazower 2012, 199, 204.
[99] Mazower 2012, 213.

The same basic cosmological shift also shaped Soviet policy discourses. The Cold War then played out as an ideological struggle between two productivist visions of international order.[100] Both superpowers offered a vision of Western modernity as a universal end for all states. Capitalism and communism were both premised on the modernist claim that control over nature, social engineering, and scientific and technological progress would generate growth and prosperity.[101] Superpower competition in the Cold War was intensified by the superpowers' efforts to defend and export their respective models throughout the world. In the United States, Truman expressed the cosmological basis of American foreign policy in the global south when he declared that the United States would offer the "benefits of our scientific advances" for the "improvement" of underdeveloped states.[102] In the Soviet Union, planning was obsessed, almost naively, with the ability of science and technology to solve all problems.[103] Soviet industrialization plans aiming to "overtake and surpass" the United States served as an appealing model for economists and policy-makers throughout the world.[104] Despite their differences, the American and Soviet models were, as Nehru put it, "branches of the same tree."[105]

Despite both superpowers' anti-imperialist pretensions, their drive to spread productivist, modernist purposes offered a continuation of the late colonial push towards the "control and improvement" of the global south.[106] The postwar imposition of productivist discourses carried Western cosmological elements into the global south where they displaced and transformed indigenous cosmologies and political purposes. This should be no surprise, given that both the United States and the Soviet Union were premised on the eradication or suppression of alternative cosmologies within their borders.[107] The spread of productivism was justified and characterized as necessary because development and modernization were seen as universal processes that all societies undergo.[108] However, experts and policy-makers no longer believed, as they had in the nineteenth century, that the process of development was automatic. Instead, development would have to be guided, not by colonial rule, but by natural and social scientists who analysed the obstacles and

---

[100] Engerman 2004, 24; Westad 2007, 4.
[101] Scott 1998; Westad 2000; Westad 2007.
[102] From the 1949 Inaugural Address. Quoted in McCarthy 2007, 4.
[103] Rindzevičiūtė 2016, 14–15; Graham 1993.
[104] Gerovitch 2002, 17; Engerman 2004, 41.
[105] Quoted in Chatterjee 1986, 158.
[106] Westad 2007, 5, 69–75.
[107] Westad 2007, 12–13, 50–51.
[108] Inayatullah and Blaney 2004, 97; Chatterjee 1986, 158; Gilman 2003, 272.

problems countries faced and recommended policies that would accelerate the process.

But it was the American version of the productivist vision and its attendant cosmology that was recursively institutionalized in the Western and then post-Cold War international order. American hegemony drove associational changes through a variety of multilateral and bilateral policies over the course of the post-Second World War period. In doing so, the United States drew on a variety of forms of power. It used military power to overthrow regimes that resisted its economic plans.[109] It constructed bilateral and regional development plans like the Marshall Plan in Europe and the Alliance for Progress in Latin America to support economic ties.[110] It drew on cultural capital, such as the appeal of jazz and rock, to attract the masses and their leaders.[111]

In addition, and most importantly for our purposes, the United States served as the centre of the postwar natural and social sciences. In particular, the transnationalization of economics and policy ideas emerged from American economic and cultural power.[112] Throughout the post-war era, economics fuelled the recursive institutionalization of international order, inscribing and naturalizing economic growth as a central end of states. That is, once embedded in the core sites of international order, economists channelled shifting ideas about production and growth into states, IOs, and other associations, transforming international discourses and spreading the growth imperative. Some of this work happened through bilateral ties, but international financial institutions like the World Bank played a leading role. So organizations like the Bank participated in recursive institutionalization in two ways. First, they were outcomes of recursive institutionalization, undergoing associational changes that were driven by American interests and changes in economic ideas. Second, they in turn spread the purpose of growth to other associations.

In sum, new ideas from physics, engineering, and economics set off a cosmological shift. The authority of physics and engineering inspired a new ontology of quantitatively defined systems and objects that reoriented politics to a calculable, controllable economic future. This was an intensification of the epistemic modernist discourse that we saw emerge in the last chapter, the origins of which can be traced back to Enlightenment ideas about progress and improvement. So the concept

---

[109] Grandin 2006.
[110] Gilman 2003; Engerman 2004; Latham 2000; Maier 1989.
[111] Poiger 2000.
[112] Burgin 2012; Fourcade 2006, 2009; Gilman 2003; Latham 2000.

of growth as it emerged in the second half of the twentieth century represented the coming together of a series of cosmological changes that had been ongoing since the sixteenth century. Old ideas were reconfigured in a powerful new political discourse. Mechanistic representations were updated in the form of cybernetic servomechanisms. Newtonian absolute time and biological ideas about growth as a law of nature structured thinking about economic progress. The rise of statistics made possible the depiction of the economy in tables and figures. Together, these elements rendered the ideal of unending, infinite economic production as a natural part of the universe.

Cosmology and Development Discourse, 1950–1970

The importance of the cybernetic-systems cosmological shift for reconstituting state purpose can be clearly seen in international development discourses. Over the course of 1950s and 1960s, converging academic, organizational, and cultural processes produced a configuration of discursive elements, modelling devices, and institutional imperatives that marginalized poverty alleviation efforts and bolstered growth-centric policies in the World Bank. This configuration cannot be explained as the outcome of coercive pressures or learning processes alone. Rather, the strategic deployment of economic ideas set off discursive changes that embedded and strengthened the growth imperative in the Bank.

The Bank was a key part of the productivist order constructed at Bretton Woods and Dumbarton Oaks. The International Monetary Fund would help maintain balance of payments equilibrium, while the Bank would provide funds for reconstruction and development.[113] The articles of agreement signed at Bretton Woods specified that the Bank was to be oriented to "the expansion, by appropriate international and domestic measures, of production, employment, and the exchange and consumption of goods which are the material foundations of the liberty and welfare of all peoples."[114] Here, the goals of the Bank (relief and reconstruction and expanding production) were still naturalized in terms of liberal ends: freedom and welfare.

The Bank spread productivism through loans to countries for reconstruction and development projects. The bankers, lawyers, and engineers

---

[113] Helleiner 2006, 958. Helleiner challenges the received wisdom that the Bank was primarily a tool for lending to Europe that was later adapted to lending for development. See Alacevich 2009.

[114] Quoted in World Bank 1957, 3. See World Bank 1947, 5; World Bank 1948a, 10.

who ran the early Bank were not ignorant of the problem of poverty, but they nonetheless oriented Bank discourse and lending to expanding production.[115] Further, they implemented personnel and organizational changes that privileged a Bank culture oriented not to social programmes but to "creditworthy" projects that increased productivity.[116] Addressing poverty directly was thought to entail value-laden decisions that would violate the principle of Bank neutrality.[117] Projects were to "contribute directly to a productive capacity."[118] Moreover, the Bank "normally does not finance community projects of a primarily social character."[119] The implicit theory of growth was that predominantly agricultural societies must build infrastructure to spur industrialization.[120] So Bank loans throughout the 1950s and 1960s focused on construction projects that would remove the obstacles to expanding production and economic development.[121]

These productivist ends were supported by a linear conception of time along which progressive development unfolded. Development was equated with the achievement of urban, technological modernity. For example, a report on Brazil from 1948 mapped the rural–urban divide onto a distinction between backward and industrial forms of life: "Brazil is a country with a low standard of living, with striking contrasts between rural and urban life, and between backward inefficient methods of production and up-to-date industrial techniques."[122] The same report encouraged the exploitation of natural resources for "development" on the "fascist, Nazi, and Soviet example."[123] The implicit *telos* of modernity is to be realized by the injection of capital and technocratic management on the authoritarian model. This would allow Brazil to conquer nature and thus more effectively exploit its natural resources: "Large expanses of apparently fertile land remain to be cleared, her forest resources are tremendous, she is rich in many minerals and in water power and her people are quick to learn."[124]

In the 1950s, the underlying episteme and ontology of the Bank combined informal modes of decision-making, narrative analyses, and

---

[115] Konkel 2014, 280; Alacevich 2009, 40–41.
[116] Chwieroth 2008, 495–500.
[117] Konkel 2014, 280–281.
[118] World Bank 1957, 43.
[119] World Bank 1957, 43.
[120] World Bank 1953, 2–3; World Bank 1963, 10–13.
[121] World Bank 1951; World Bank 1960; World Bank 1965a; World Bank 1965b.
[122] World Bank 1948b, §II.1.
[123] World Bank 1948b, §I.1a–b.
[124] World Bank 1948b, §I.1. The gendered language here points to the themes raised by Carolyn Merchant (1980) in her seminal feminist history of the environment.

economic modelling techniques. Loans were to be made on a "sound business basis" rooted in "objective economic appraisal."[125] However, this appraisal was to be determined, not by rigorous science, but by "an exercise in judgment."[126] This is because creditworthiness is in part determined by an "intangible factor of the country's attitude ... willingness to maintain debt service at the expense, if necessary, of sacrifices in consumption standards."[127] Thus, "the situation in every country must be considered on its own merits."[128]

Country reports were centred on a narrative summary of the country's natural resources, productive capacity, and government effectiveness.[129] The reports often began with an account of the geography, topography and climate, population, history, politics, and debt levels. This empiricist, chorographic mode of analysis first emerged in European discourses in the seventeenth century when it was used in travel reports and colonial dispatches.[130] So as late as 1950, the ontology of the World Bank was not yet dominated by quantitative representations of the economy, population, and so on.

The concern with production was slowly transformed into a concern with economic growth over the course of the 1950s as cybernetic-systems thinking entered the Bank. Consistent with Mitchell, the concept "economic growth" did not emerge in Bank discourse until the latter half of the 1950s.[131] When it did, it entered alongside concepts drawn from cybernetic-systems thinking. A 1955 report on "world growth" notes that countries now have useful "instruments of control over the functioning of the economy." These controls can "correct deviations from internal and external equilibrium."[132] The language of "control" and "function" from cybernetic modelling here shows the deployment of modernist cosmological ideas in the Bank.

The report also orients Bank policy towards the future. As Mitchell leads us to expect, it discusses the prospects for growth in order to arbitrate disputes in the present. In effect, it uses the "outlook" for growth to persuade countries to allow and encourage foreign investment. However, it does so in a humble way, noting that it offers an account only of "what is *possible*," not an account of "what will happen."[133] While this report

---

[125] World Bank 1957, 38.
[126] World Bank 1957, 38.
[127] World Bank 1957, 39.
[128] World Bank 1957, 42–43.
[129] E.g. World Bank 1948b.
[130] Brown 2016.
[131] The first instance in my sample is in World Bank 1955. See Mitchell 2002, 2014.
[132] World Bank 1955, 4.
[133] World Bank 1955, 13. Emphasis original. See Mitchell 2014 on how the future is used to govern the present.

clearly maintains that expanding production is a central end of states, the rate of output growth is the last item mentioned. Instead, agricultural and industrial production, unemployment, the balance of payments, and a review of monetary policy all appear first. So at this time economic growth was one element in a broader, multidimensional understanding of development.

Throughout this period, Bank lending was guided by the 1946 Harrod-Domar economic model. The Harrod-Domar "two-gap" model predicted that the rate of production growth is proportional to capital investment (domestic savings plus foreign investment).[134] Thus, it was consistent with Paul Rosenstein-Rodan's argument that a big push of foreign investment could produce the economies of scale necessary to break poor countries out of poverty traps.[135] Later, these claims were supported by Arthur Lewis' argument that developing countries possessed surplus labour in the agriculture sector that would flow into the industrial sector if the government invested there. In the 1960s, Hollis Chenery, who would later serve as Chief Economist of the World Bank, adapted the Harrod-Domar model into a tool for determining lending levels.[136] Chenery's model calculated the "financing gap" necessary to achieve a target growth rate. Chenery's equations were later incorporated into the Revised Minimum Standard Model which was used in the Bank to set lending levels in country analyses for decades.[137] These models served as an "anchoring device" that oriented the actual practices of Bank lending to growth rates, despite broader changes in the political and economic justifications for Bank actions.[138]

The episteme and ontology of the Bank grew more technical and formal over the course of the 1960s. Country reports after 1960 begin with a standard menu of statistical "Basic Data" including area, population, GNP, GNP per capita, exports, imports, government expenditures, government revenues, and debt that structured reports in the 1960s and 1970s.[139] These statistical indicators emerged alongside the concept of the economy and new models intended to predict and control growth.[140] Meanwhile, Bank project lending became increasingly focused on "growth rates" and "gross domestic product."[141] So, over the course of

---

[134] Easterly 1999, 427.
[135] Rosenstein-Rodan 1943. While Rosenstein-Rodan himself was marginalized in the 1950s, his theory of development remained important. See Chwieroth 2008, 496–500.
[136] Stern and Ferreira 1997, 530; Easterly 1999, 424.
[137] Easterly 1999, 424.
[138] See Swidler 2001.
[139] World Bank 1963.
[140] See World Bank 1968a, 1.
[141] World Bank 1965c, 3; World Bank 1968b, iv.

the 1960s Bank discourse shifted from an informal emphasis on production and expansion to the more precise ends of development conceptualized as GNP growth.

Changes in Neoclassical Economics, 1950–1980

The discourse of the early Bank reflects the fact that development economics in the 1950s and 1960s was a diverse discipline featuring a variety of historical and sociological approaches.[142] As Ascher puts it, "neoclassicists were conspicuously absent" from development economics until 1970.[143] However, neoclassical economists had nonetheless been developing growth models with important policy implications for development economics. When these models entered the World Bank and other international associations, they introduced new cosmological elements into the discourses underlying international order. In short, neoclassical economics naturalized growth as the inherent, spontaneous product of scientific and technological progress. This claim reconceptualized humanity as the authors of progress, inscribing growth into the human condition. This claim built on the cosmological shifts initiated by early modern natural philosophy and the nineteenth-century historical sciences. It relied on the mechanism, materialism, and modernism of those earlier shifts, while reproducing quantitative ontologies, calculative epistemes, and progressive understandings of time and human nature.

In the 1950s, Robert Solow (1924–) created a neoclassical model that aimed to trace the effects of full employment on growth.[144] In doing so, he assumed a "single universally available technology" that drove economic growth.[145] In a 1957 application of the theory, Solow explained seven-eighths of American growth from 1900–1950 with technological change.[146] The latter paper demonstrated how to use neoclassical growth theory to explain or account for economic growth by quantifying the contributions of various factors. Solow's model of economic growth was the most intellectually and politically dynamic expression of Samuelson's general equilibrium analysis. At MIT, Samuelson and Solow oversaw the development of growth theory into an academic cottage industry.[147] Although Solow did not directly apply his model to developing countries, Solow's papers had an indirect but important influence on economics

---

[142] Ascher 1996, 315–320; Crafts 2001, 302–313.
[143] Ascher 1996, 320.
[144] Solow 1956. On the influence of this piece, see Boianovsky and Hoover 2009.
[145] Toye 2009.
[146] Solow 1957.
[147] Boianovsky and Hoover 2014.

in the Bank. Solow's model clearly predicted that the incomes of rich and poor countries would converge in the long-run. Since all countries have access to the universal library of technology, rates of return in poor countries will be high relative to rich countries as poor countries adopt technology and increase their productivity.

Solow's model had strong resonances with Walt Rostow's modernization theory. Rostow joined the Center for International Studies at MIT in the 1950s and built it into an important player in the American policy-making process. While Rostow's work was rooted in a historical rather than a technical approach, it was influential in the Kennedy administration.[148] Like Solow's model, Rostow's seminal *Stages of Growth* closely connected growth to scientific and technological development. Rostow defined a "traditional society" as one "whose structure is developed within limited production functions, based on pre-Newtonian science and technology, and on pre-Newtonian attitudes toward the physical world."[149] To reach take-off, societies must "exploit the fruits of modern science."[150] A country could reach the highest stage of development, "maturity," when it "demonstrates that it has the technological and entrepreneurial skills to produce not everything, but anything that it chooses to produce."[151] The implicit end of both neoclassical and modernization theory was that all states and therefore all people should join Western scientific and technological modernity.

In cosmological terms, modernization theory updated and modified Adam Smith's stage theory by grafting it onto an image of linear time as unfolding between a backward past and a modern future. But it was Solow's work that explicitly opened up a plane of infinite progress. In neoclassical growth theory, science does not reveal God's will or uncover human and natural history, but provides the means to deliver unending prosperity. It links an image of progressive time to the epistemic modernist ideas we saw emerge in the British Colonial Office. For Solow, it is the power of scientific and technological discovery that moves humanity across the infinite plane of progressive time. This places humanity in time and defines the purpose of humanity in scientific and technological terms.

After 1970, the demand for rigour in American economics grew. New literatures on "rational expectation" and "information" built on mathematical microfoundations had a broad effect on the discipline's epistemic norms. Heterodox economists that criticized growth and

[148] Gilman 2003, 198–212. See Latham 2000.
[149] Rostow 1990 [1960], 4.
[150] Rostow 1990 [1960], 6.
[151] Rostow 1990 [1960], 10.

free markets from social and environmental perspectives were neutralized on technical grounds.[152] Moreover, neoclassical Keynesian policy seemed unable to cope with the economic conditions of the 1970s. Keynesian models could offer no theory to explain or policy to address rising inflation.[153]

These disciplinary and political conditions produced a window of opportunity for neoliberal theories that promoted free markets and advocated for a reduction in the role of government in economic life.[154] While Samuelson and Solow built neoclassical, Keynesian growth theory at MIT, economists at the University of Chicago pursued neoclassical theory in its purest form.[155] Whereas traditional neoclassical theory portrayed perfect markets as an unattainable ideal but useful regulative principle for the design of policy interventions, the Chicago view was that most real-world markets were an embodiment of perfect competition except to the degree that states interfered in their operation.[156] Chicago economists understood that the neoclassical theory and mathematical techniques of Samuelson and the Cowles Commission had an obvious affinity with state planning, although this was downplayed by the scientific rhetoric of the discipline.[157]

For neoclassical economists in the Chicago School, the implication of Solow's technological assumptions was that tariffs and other trade protections prevented the free flow of technology and harmed economic development. They were easily represented as barriers to the perfect market competition that would drive growth. In the 1970s, this theory was studied at the World Bank under Bela Balassa and at the National Bureau of Economic Research under Jagdish Bhagwati and Anne Krueger.[158] Balassa, Bhagwati, and Krueger were the most influential economists at the Bank between 1970 and 1990.[159] Their central insight was a corollary of the technological optimism embedded in Solow's growth theory: "the case for trade raising the long-run growth rate has to suppose the existence of a technology gap that trade can subsequently close."[160] Neoliberalism in the Bank was a product of neoclassical growth theory, not the other way around.

---

[152] Backhouse 2010, 58.
[153] Backhouse 2010, 58–59; Morgan 2003, 294. See also Blyth 2002; Best 2008.
[154] Burgin 2012; Mirowski and Plehwe 2009; Stedman Jones 2012.
[155] Fourcade 2009, 93–96.
[156] Fourcade 2009, 96.
[157] Düppe and Weintraub 2014.
[158] Toye 2009, 228.
[159] Stern and Ferreira 1997, 547, 601; Kapur et al. 1997, 483–484.
[160] Toye 2009, 228.

Strategic Deployment: McNamara's Reforms

The Bank's growth-centric policies were challenged in the 1960s as a diverse network of academics and philanthropists inside and outside the organization pressed the Bank to take up direct poverty alleviation.[161] There was a wide-ranging debate about the merits of trickle-down economics in which an influential group of structuralist economics persuasively argued that "economic growth had failed to bring benefits to the majority."[162] If the organization had learned anything from the experiences of the 1950s and 1960s it was that expanding production and increasing growth was good, but that it was insufficient if the goal of Bank policy was to increase standards of living and alleviate poverty.

Robert McNamara was appointed World Bank President in 1968. McNamara was trained as an economist and loved cost–benefit analysis. He had been an early believer in systems analysis and scientific management techniques, which he implemented as the President of Ford Motor Company and Secretary of Defense.[163] He is symbolic of an age in which the strategic deployment of science and expertise was considered a requirement of rational organization. However, his reforms unintentionally contributed to the formation of an interlocking configuration of discursive and institutional elements that privileged growth.

At first, McNamara valued economic growth as the rightful goal of Bank policy.[164] His early years at the Bank were focused on the population problem. He was deeply concerned that population growth would wipe out income gains produced by economic growth.[165] However, in the early 1970s, McNamara shifted from thinking that poverty was a symptom of underdevelopment, to recognizing it as a distinctive problem that had to be confronted directly.[166] Moreover, whereas in the 1960s McNamara defined poverty as "per capita income," by the early 1970s he expressed a complex, biological view of poverty.[167] In his famous 1973 Nairobi speech McNamara made a distinction between relative and absolute poverty. Relative poverty was defined as international and domestic income inequality. Absolute poverty, which

[161] Shapley 1993, 505–507.
[162] Hürni 1980, 26; Konkel 2014, 287. See Chenery et al. 1974, xiii.
[163] Finnemore 1997, 212; Weaver 2008, 84.
[164] McNamara 1981 [1968], 4.
[165] McNamara 1981 [1968], 4. See also World Bank 1978, iii.
[166] Shapley 1993, 504–505. Shapley states that Barbara Ward, a Harvard academic and personal friend of McNamara's, persuaded him to take up the cause of poverty alleviation (1993, 506–507).
[167] McNamara 1981, 137, 152, 194.

McNamara wanted to focus on, was "a condition of life so degraded by disease, illiteracy, malnutrition, and squalor as to deny its victims basic human necessities."[168] He argued that "growth alone" could not solve the problems of absolute poverty.[169] Moreover, GNP statistics alone could not "measure the achievement of multiple development objectives."[170]

So McNamara came to agree with critics who argued that the Bank was too focused on growth and sought to mainstream poverty alleviation. Nonetheless, growth was to remain an important part of development policy. For McNamara, economic growth and direct poverty alleviation were the "twin objectives of development."[171] So, McNamara set out to rebalance growth and poverty alleviation while expanding Bank lending.[172] To accomplish these goals, McNamara led a series of organizational reforms and strategic deployments of expert ideas. First, he centralized control by changing the organizational structure and introducing scientific management techniques. As Kapur, Lewis, and Webb put it: "[b]y reducing multiple, complex, individual decisions and transactions to summary statistics, McNamara sought to make the institution more transparent and subject to control."[173] Just as he had at Ford and the Department of Defense, McNamara introduced statistical controls to help him monitor and guide the organization.[174]

In addition, McNamara introduced systems analysis and demanded precise, quantitative indicators throughout Bank work.[175] In short, he instructed his staff "to quantify project costs and benefits."[176] Both traditional and new social projects were to be quantified:

Reluctant staff were not allowed to apply a double-standard – using "precise" figures when they referred to expected output effects and vague estimates when claiming "poverty reduction." Despite much grumbling, statements about poverty impact were required to be backed up by statistics, and staff had to scurry to find or create data on income distribution and on the number and living standards of project beneficiaries.[177]

Bank staffers would need to express their core arguments and claims in quantitative terms.

[168] McNamara 1981 [1973], 131.
[169] McNamara 1981 [1970], 131
[170] McNamara 1981 [1973], 243.
[171] World Bank 1978, iii.
[172] Sharma 2013.
[173] Kapur *et al.* 1997, 245.
[174] Finnemore 1997, 213.
[175] Shapley 1993, 499.
[176] World Bank 1974, 48.
[177] Kapur *et al.* 1997, 219.

Finally, McNamara hired large numbers of economists.[178] By 1986, there were 682 economists in the Bank, out of a total of 2,808 staff.[179] Non-economists complained that their ideas had to be translated into quantifiable, cost–benefit terms in order to be taken seriously.[180] This influx of economists created a channel for neoclassical academic economics to enter the Bank and transform its episteme, ontology, cosmology, and purposes. Furthermore, McNamara consolidated economic expertise into a central Research Department that displaced the expertise that had previously been housed in the Central Projects Department. This empowered economists at the expense of project managers working close to the ground.

To demonstrate the rise and effects of neoclassical economists in the Bank, I conducted an analysis of research in the World Bank between 1965 and 1985. I made a list of economists performing research in or for the Bank during this period and coded whether or not they received a neoclassical training.[181] Following Chwieroth, I coded all economists trained in the top fifteen American economics departments as "neoclassical economists."[182] This coding scheme is conservative because a number of lower-ranked American schools and non-American departments, which Chwieroth does not count as neoclassical, train their students in neoclassical theories and methods. Even on this conservative accounting, between 1965 and 1985, 54.5 per cent of the economists in my sample were trained in neoclassical economics departments. The rest were trained in other American economics departments (24.5 per cent), foreign economics departments (18.2 per cent), and non-economic fields like education or sociology (2.7 per cent).

As we shall see, these strategic deployments channelled important cosmological elements into the Bank. First, they brought the formal,

---

[178] Stern and Ferreira 1997.
[179] Baldwin 1986. Baldwin's survey seems complete. Stern and Ferreira's 1991 survey of a single department (Policy, Research and External Affairs) revealed that 252 of 465 staff members received their highest degree in economics (1997, 586). There do not seem to be good surveys that track the number of economists over time.
[180] Kardam 1993, 1777; Bebbington et al. 2004, 44–47; Weaver 2008, 77–78.
[181] I began by making a database of all first authors with two or more publications in the World Bank online archives (documents.worldbank.org). I then recorded where they obtained their PhD, using the Worldcat and Proquest dissertation databases or by locating their CV online. I am grateful to Ian Gustafson for research assistance.
[182] Chwieroth 2010, 89. Chwieroth bases his conclusions both on the secondary literature and a study of economic publications. Chwieroth identifies the top fifteen departments between 1963 and 1980 by their rate of publication in the top economics journal, *American Economic Review*. Chwieroth's rationale is that "during this period the *AER* was primarily publishing articles that employ neoclassical assumptions and models." Thus, scholars who published in the *AER* most likely came from neoclassical schools.

mathematical version of the cybernetic-systems view into the Bank. It was, principally, neoclassical economics that carried and reproduced this ontology and episteme in the Bank. The institutional imperative to create a quantitative culture and expand lending further empowered the incoming cohort of formally, mathematically trained neoclassical economists because they too demanded mathematical rigour. These ontological and epistemic changes constrained and shaped Bank goals through the period 1970–2000. Second, neoclassical economics imported cosmological themes about time and scientific and technological progress that transformed these ontological and epistemic changes into new purposes backed by cosmological ideas about time and the place of humanity in the universe.

Discursive Reconfigurations, 1970–2000

In one sense, McNamara was a successful leader. He used his position to reform the organization by controlling policy, realigning personnel, and marshalling resources.[183] Bank lending did double and the Bank's role in international economic policy expanded considerably. However, McNamara's strategic deployment of scientific management techniques and neoclassical economics did not have the intended effect. The Bank struggled to operationalize poverty alleviation and displace growth-centric policies. Instead, McNamara's reforms initiated a set of unintended discursive reconfigurations.

In this section, I trace the constitutive mechanisms that altered World Bank discourse. First, personnel changes strengthened the power and authority of neoclassical economists who then worked to delegitimate and discredit poverty-centric work. Second, the tools of neoclassical economics were adopted by structuralists and other economists. But these tools made it difficult to operationalize the concept of absolute poverty. That is, the influx of economists and quantitative standards created a set of representational constraints that made it difficult to represent and value poverty. Moreover, the formal mathematical structure of neoclassical modelling allowed poverty, basic human needs, and other issues to be incorporated into the growth-centric framework, albeit in a subordinate role. The new discursive configuration did not reorient policy towards the alleviation of what McNamara called absolute poverty. Instead, it privileged policies that promoted growth measured in terms of GNP. Finally, neoclassical theory was used to legitimate and naturalize GNP growth not as a means to some broader end, but as an end in

---

[183] Finnemore 1996b; Nielson *et al.* 2006; Chwieroth 2008, 2010.

its own right, expressing the human purpose of unleashing scientific and technological progress.

Marginalizing Direct Poverty Alleviation, 1970–1977

First, neoclassical economic professionals in the Bank drew on methodological standards of rigour to marginalize approaches that privileged basic human needs over growth. The basic needs approach seized on the radical implications of McNamara's biological concept of poverty to advocate for direct poverty alleviation. Mahbub ul Haq, an early proponent of this view in the Bank, questioned the value of growth because, in his view, growth in Pakistan had failed to improve the lives of the poor. At first, McNamara berated ul Haq for questioning growth at a time when McNamara believed the core mission of the Bank was to advance growth. However, McNamara was eventually persuaded by key aspects of ul Haq's arguments and made him a trusted advisor.[184] Ul Haq argued that poverty alleviation could not wait and that the basic needs of food, clothing, shelter, and education needed to be provided directly to the poor. Paul Streeten and Norman Hicks worked to make this argument tractable for economic analysis and Bank policy.[185]

However, despite some success, the basic needs approach was never mainstreamed.[186] Why? One possibility is that the basic needs approach was incompatible with McNamara's goal of doubling Bank lending.[187] It is true that the institutional imperative to increase disbursements benefited growth-centric policies promoted by other groups. However, this cannot explain why these programmes were continually marginalized after 1973 when McNamara had already achieved his goal of doubling Bank lending.[188] A more significant factor in the marginalization of the basic needs approach was the withering critique offered by neoclassical economists.

In the late 1970s T.N. Srinivasan, an influential neoclassical voice in the Bank, mounted an attack on basic needs approaches.[189] He first argued that poverty alleviation policies might hurt growth "if the redistribution cuts too much into savings for growth."[190] Redistribution towards poor people would contribute to consumption, but not saving.

---

[184] Konkel 2014, 290; Shapley 1993.
[185] E.g. Streeten *et al.* 1980.
[186] Stern and Ferreira 1997, 550; Konkel 2014, 296–297.
[187] Finnemore 1996b, 105; Sharma 2013.
[188] Kapur *et al.* 1997, 216, 971.
[189] Srinivasan 1977a, 1977b.
[190] Srinivasan 1977a, 20–21. For a response from the basic needs camp, see Hicks 1979.

Since saving was considered necessary for capital accumulation and growth, redistribution could hurt growth. For Srinivasan, this was an empirical question.[191] But Srinivasan's deeper objection was with the serious "conceptual and measurement problems in quantifying basic needs."[192] Given the potential costs of poverty alleviation policies, demonstrating their efficacy would be essential. But, the "extreme fuzziness" of the basic needs approach made it difficult to assess its impact empirically.[193] Moreover, because the concept was so vague, efforts to "define universal standards," develop "global modelling," or produce cost estimates were "futile."[194]

Indeed, the new configuration of personnel and institutional rules privileged scientific rigour over moral and ethical arguments for poverty alleviation. Nicholas Stern reports that during his time at the Bank economists who wanted to promote poverty concerns were labelled "bleeding hearts" or "social planners."[195] William Ascher's interviews from the time provide evidence that economists' professional commitment to rigour led them to reject the poverty alleviation approaches:

> [T]hose staff members most skeptical about the validity of the objectives of the "basic needs approach" (i.e., attending to the nutritional, educational, and housing needs of the population even if the economic payoff is only the delayed result of human resource improvement) were most likely to question the technical feasibility of basic-needs-oriented projects.[196]

Scientific standards provided a reason for economists to delay and subvert poverty alleviation initiatives.

In the course of these battles, some structuralists working in and around the Bank adopted more rigorous modes of analysis to demonstrate their arguments. However, not all structuralist concerns and insights were compatible with the mathematical standards. Take for example Hollis Chenery's efforts to operationalize the concept of absolute poverty. Chenery was the head economist at the Bank from 1970 until 1982. Although Chenery was a structuralist, he supported McNamara's push for quantification and was happy to hire neoclassical economists. Under Chenery, research became more data-driven as he pushed to bring structuralism in line with the empirical standards of mainstream neoclassical economics.[197]

---

[191] Srinivasan 1977a, 21.
[192] Srinivasan 1977a, 25.
[193] Srinivasan 1977a, 21.
[194] Srinivasan 1977a, 20.
[195] Stern and Ferreira 1997, 535.
[196] Ascher 1983, 425.
[197] Konkel 2014, 291; Chenery 1975, 312.

At the same time, Chenery sought to operationalize poverty alleviation initiatives within this new framework. So, Chenery led a research project to quantify poverty that culminated in the 1974 publication, *Redistribution with Growth*.[198] The report argued that while growth in developing countries had been impressive, its benefits had not been evenly distributed.[199] The report purported to address absolute poverty by describing living standards in terms of calorie intake, nutrition levels, clothing, sanitation, health, and education.[200] However, it argued that "[t]o be socially meaningful, minimum levels cannot be defined according to some absolute biological standards but must necessarily vary with the general level of economic, social, and political development."[201] Thus, some relative economic indicators were necessary.

Thus, *Redistribution with Growth* operationalized the condition of absolute poverty as living below a "poverty line" of US$50 or US$75 a year. Although these poverty levels were supposed to be linked to biological needs, in practice they were not.[202] Thus, Konkel concludes, "poverty alleviation became a matter of raising a country's per capita GDP ... [which] is exactly what the Bank's understanding of the development process had been all along."[203] Thus, in the 1970s and 1980s, alleviating absolute poverty and food insecurity was made to be consistent with "efforts to accelerate growth, through adjustment assistance, policy reform, and productive investment."[204]

This was part of a more general phenomenon. As the 1974 policies and operations manual argued, "to be able to help countries in all phases of their development, the Bank has to know and understand the economies of its member countries ... [it must] obtain a comprehensive picture."[205] The manual laid out new standards for country analysis. Economic data on any country had to include "national accounts data; socioeconomic data indicators; commodity (production, export, import, and price) information; and general capital flows and debt statistics."[206] The idea was that this common template "permits intercountry comparisons, and provides a 'global framework' of projections."[207]

---

[198] Chenery *et al.* 1974. For background on the report, see Ayres 1983, 74.
[199] Chenery *et al.* 1974, xii–xvi.
[200] Chenery *et al.* 1974, 10.
[201] Chenery *et al.* 1974, 11.
[202] Konkel 2014, 293–294.
[203] Konkel 2014, 294.
[204] World Bank 1986b, v.
[205] World Bank 1974, 42.
[206] World Bank 1974, 70.
[207] World Bank 1974, 70.

These reforms made it difficult for the Bank to account for "the benefits of projects ... that did not involve direct production."[208] In response to McNamara's initial shift in emphasis from growth to poverty, there were proposals to expand "basic needs" lending for population control, health, and nutrition, and agricultural sector projects that would directly target the living conditions of the poor.[209] However, the proposals for health and nutrition projects failed in part because they were difficult to represent in a discourse dominated by growth models. Consider, for example, a 1976 special expert report, "Measurement of the Health Benefits of Investments in Water Supply." The report argued it has been "unequivocally demonstrated" that clean water reduces the incidence of disease and is necessary for public health.[210] But the report investigated "whether the impact of water supply investments on health could be reliably predicted and quantified."[211] The report concluded that the benefits of health investments "have not been quantified to permit the derivation of reliable formulas."[212] Even though the "difficulty lies in measurement rather than in qualitative trends" the report could not recommend clean water projects.[213] In short, even though the benefits of direct poverty alleviation projects for human well-being were clear, such projects were not to be mainstreamed because the benefits were difficult to quantify within the representational constraints of Bank discourse.

An important debate about changing the criteria for project evaluation provided a second potential avenue for poverty alleviation to be taken seriously: a proposal to replace financial cost–benefit analysis with a "social" cost–benefit approach.[214] Social cost–benefit analysis includes a system of weights and shadow prices that would allow project evaluations to consider the income-levels of the project's beneficiaries.[215] In standard cost–benefit accounting one dollar of consumption is worth the same no matter who consumes it. However, if the goal is to improve the lives of the poor, it makes sense to value a dollar for the poor more than a dollar for the rich.[216] Throughout the 1970s and early 1980s economists within the Bank debated the merits of social cost–benefit analysis. Ascher's interviews reveal that economists opposed the new standards "on the grounds that the data were insufficiently precise."[217] According to the professional

[208] Kapur *et al.* 1997, 39.
[209] Finnemore 1996b.
[210] World Bank 1976, 2.
[211] World Bank 1976, i.
[212] World Bank 1976, 1.
[213] World Bank 1976, 2.
[214] Hürni 1980, 68–79; Kapur *et al.* 1997, 38–39.
[215] Little and Mirrlees 1974, 52–60; Little and Mirrlees 1994, 207.
[216] Hürni 1980, 69.
[217] Ascher 1983, 425. See also Lal 1974.

and discursive norms of the economists, differential weights were seen as "less rigorous" and therefore "arbitrary and judgmental."[218] In the end, the approach was deemed "too complicated" and the effort to mainstream social cost–benefit analysis was abandoned.[219] The methods were "hardly ever used except in an experimental manner."[220]

In the end, basic needs proposals were shelved and agricultural sector lending became the locus of Bank support for the poor.[221] But this lending looked a lot like old agricultural sector lending, which was easy to quantify and plug into growth models. Quantitative standards of representation imposed by economists' demands for rigour made it more difficult to represent and consider the needs of the poor. All of this combined with the institutional imperative to increase lending, the necessity of quantitatively demonstrating benefits, and the prevalence and authority of neoclassical economics to bolster the Bank's focus on growth.

In a cosmological sense, the rise of quantitative culture in the Bank reduced humanity from a biological being with basic human needs to an individual possessing an income. In the previous chapter, we saw how Rowntree and others first represented humans in reductive, statistical terms such as "man-values" of calories needed for labouring. That conception retained some link to the idea that humans possess biological needs. But throughout the 1970s and 1980s, thinkers like Ward, ul Haq, Streeten, and Hicks struggled to articulate this view of humanity within the strictures of neoclassical, mathematical models. However, because these alternative features of humanity were not easily legible to statistical tables and models, they fell by the wayside. In the end, the Bank traded robust concepts of human needs for statistically reliable indicators.

The Rise of Growth, 1978–1986

In addition to intensifying demands for scientific rigour, neoclassical tools and concepts introduced a form of general equilibrium analysis tied to a theory of trade-led growth. The theory of trade-led growth emerged from neoclassical studies inside and outside the Bank. Inside the Bank, Bela Balassa led a growing group that linked traditional neoclassical concerns with microeconomic efficiency and macroeconomic openness in a clear, unified argument: eliminating protections and liberalizing trade

---

[218] Ascher 1983, 428.
[219] Hürni 1980, 74. See Kapur *et al.* 1997, 39.
[220] Little and Mirrlees 1994, 207.
[221] McNamara 1981 [1974]; Kapur *et al.* 1997, 412–415. See, e.g., World Bank 1969.

encourages growth by eliminating price distortions and disincentives to competition.[222]

While neoliberals later took up this argument, it is important to emphasize that most neoclassical economists, statist or otherwise, accepted the core of this argument.[223] For example, the work of Bank economist Helen Hughes made the case for trade-led growth without promoting a strict trade liberalization doctrine.[224] On one hand, she conceded that protection for the purpose of encouraging industrialization makes sense in some circumstances.[225] On the other, she argued that the "neomercantilist" policies of the industrialized, developed world constrained and harmed developing countries.[226] Nonetheless, for a neoclassical economist like Hughes, the benefits of trade were unquestioned. Protectionism creates "distortions" and establishes "[i]nefficient productive capacity."[227] That is, within the cybernetic-systems mode of equilibrium analysis import restrictions and pricing schemes are domestic distortions by definition. The ontological and epistemic presuppositions of economic analysis privileged free trade arguments.

These internal Bank studies were supported by new research taking place under Anne Krueger and Jagdish Bhagwati at the National Bureau of Economic Research.[228] Subsequently, both Krueger and Bhagwati entered the Bank and worked hard to dismantle the arguments for infant industry protection and press the case for trade liberalization as a path to growth.[229] Their core claim was that liberalizing trade would force the domestic reallocation of resources in line with macroeconomic equilibrium and microeconomic efficiency. Neoclassical economists understood that this would generate "adjustment costs," but these were necessary for growth.[230] These neoclassical arguments provided the intellectual foundation of the structural adjustment approach.

The effects of this work on Bank policy and purposes are striking. McNamara's 1977 speech to the Board of Governors appropriated this research directly. He argued that trade-led growth was a necessary condition for poverty alleviation.[231] McNamara used the arguments of Balassa, Krueger, and Bhagwati to argue that free trade forces a beneficial

---

[222] Stern and Ferreira 1997, 601. See, e.g., Balassa 1971a, 1971b.
[223] See World Bank (1993b) for a version compatible with state intervention.
[224] Hughes 1973.
[225] Hughes 1973, 9.
[226] Hughes 1973, iii.
[227] Hughes 1973, 14.
[228] Toye 2009, 228. See Krueger 1978; Bhagwati 1978; Bhagwati and Srinivasan 1978.
[229] Krueger and Tuncer 1982; Bhagwati 1986.
[230] Bhagwati and Srinivasan 1978, 23–24.
[231] World Bank 1977, 18–19.

reallocation of labour and capital. In promoting "greater efficiency," trade liberalization would help developing countries.[232] However, the reallocation of resources created "adjustment costs" that require foreign assistance to ameliorate.[233] In 1979, McNamara laid out the Bank campaign for initiating "structural adjustment" for the first time. He argued that the unfavourable conditions for growth in the late 1970s necessitated "rapid adjustment of each country's pattern of production to its evolving comparative advantage."[234] Exports had to increase by "11 to 13% a year" to finance "reasonable GDP growth rates."[235] To increase exports at that rate, developing countries would have to accept "difficult structural adjustment internally."[236] But it would also require large program loans from the Bank, which suited McNamara's organizational goal of increasing Bank lending.[237] From this vantage point, the neoliberal turn in the Bank grew out of the neoclassical tools and concepts that formed an integral part of the new organizational configuration in the Bank.

As Sharma has shown, the Bank's shift to a macroeconomic perspective centred on structural adjustment "reflected the revival of neoclassical approaches."[238] As laid out in Chapter 2, associational change depended on an alignment between a problem and a solution. In effect, the rise of neoclassical economists led to discursive reconfigurations that privileged certain macroeconomic, growth-centric definitions of the problem and its solutions. The 1980 *World Development Report* (*WDR*) provides an illustrative example. At first, the report seems to be oriented to poverty alleviation while making the case that developing countries must "adjust" to new global economic realities. In his preface, McNamara notes that adjustment should not come at the expense of the living standards of the poor. But after the opening pages the concern for poverty disappears. The body of the report offers an analysis of how countries can raise GNP via structural adjustment policies.[239] GNP is favoured because it can be modelled and statistically represented in "illustrative projections."[240] After all, "[i]t is difficult to measure the extent of poverty" because the "data are inadequate" due to incomplete household surveys.[241] This demonstrates

[232] World Bank 1977, 19.
[233] World Bank 1977, 19.
[234] World Bank 1979, 38.
[235] World Bank 1979, 38.
[236] World Bank 1979, 38.
[237] Sharma 2013, 2017.
[238] Sharma 2017, 113. For neoclassical arguments in the 1960s, see World Bank 1965d.
[239] World Bank 1980a, 3–6.
[240] World Bank 1980a, 33.
[241] World Bank 1980a, 33.

the constraints of quantitative discourse in action: even though actors want to promote poverty alleviation, the model in-and-of-itself imposes a focus on growth because precise tools to further poverty alleviation do not exist and cannot be easily ported into neoclassical models.[242]

In 1981, the landmark "Berg report" reoriented Bank policy in Sub-Saharan Africa.[243] The Berg report argued against state-led development strategies and made the case for trade-led growth. States were advised to improve public sector performance (make government more efficient) and adopt policies designed to raise "agricultural and export growth rates."[244] The report did not emerge from a comprehensive process of internal learning in the Bank. Rather, it reflected the rise of neoclassical economists who sought to challenge the statist policies that had been proposed by African leaders in the *Lagos Plan of Action*.[245] Indeed, it is not clear how such a learning process could have taken place in the absence of adequate household-level data. Comprehensive time-series data on household consumption is necessary to test the effects of growth on poverty, but such data, as the 1980 WDR pointed out, was patchy.[246] Under these conditions, the shift towards growth-based policies in the 1970s and 1980s could not have emerged from a rational, empirical learning process.[247]

The preface of the Berg report states that the goal of its prescriptions is to "raise the living standards ... reverse the stagnation and possible decline of per capita incomes which are projected for the 1980s."[248] But the body of the report "focuses on how growth can be accelerated and how the resources to achieve the longer-term objectives set by Africa governments can be generated."[249] Here, "social and development goals" are relegated to the margins as longer-term objectives the countries can attend to on their own.[250] The main argument of the report is that "[a] reordering of postindependence priorities is essential if economic growth is to accelerate." To achieve growth, African governments must rely on the private sector and reduce "the widespread administrative

---

[242] The 1985 *WDR* exhibits the same themes.
[243] World Bank 1981a. For the context, see Taylor 2010.
[244] World Bank 1981a, 30.
[245] Taylor 2010, 116.
[246] Deaton 2001, 2010; Reddy and Pogge 2010. In general, advocates of multidimensional measures have had a difficult time "proving" the value of household data in development economics. See Clegg 2010, 486.
[247] Nonetheless, my account here is compatible with a weak or constructivist version of the learning argument on which neoclassical solutions were constructed as the answer to problems in poverty alleviation and economic development.
[248] World Bank 1981a, v.
[249] World Bank 1981a, 1.
[250] World Bank 1981a, v.

overcommitment of the public sector."[251] This will help countries use resources more efficiently: "Economic growth implies using a country's scarce resources – labour, capital, natural resources, administrative and managerial capacity – more efficiently."[252]

So the growth imperative and the macroeconomic perspective of neoclassical economics were firmly entrenched in the Bank by 1981. This was before neoliberal economists took control of Bank policy. Anne Krueger was appointed as Chief Economist in 1982 at the behest of the Reagan administration, and so American influence over the Bank shaped Bank policy throughout the 1980s.[253] Krueger hired like-minded neoclassical economists and pushed trade-led growth. Both public and project documents reveal the shift. The *WDR*s in 1983, 1985, 1986, and 1987 all promoted price, trade, and tax reforms to increase efficiency and economic growth.[254] Bank projects through the 1980s also focused on growth. While one project in my random sample explicitly oriented itself to "social achievements in relation to per capita income" and another to "more equitably distributed growth," most projects promote the purpose of growth in various guises: "economic growth," "rapid growth," "resumption of growth," "restoring economic growth," or "stimulate growth."[255]

Legitimating and Naturalizing Growth as Scientific and Technological Progress, 1986–2000

Starting in the late 1980s, actors inside and outside the Bank contested the focus on structural adjustment and growth.[256] As a result, gender, environmental degradation, bottom-up participation, good governance, and poverty alleviation all moved up the Bank's agenda. While the public rhetoric of the Bank changed once again to accommodate these challenges, the fundamental discourses and models used to guide policy remained anchored in the trade-led theory of growth.[257]

The concepts and models of neoclassical economics proved adaptable to the political and organizational pressures introduced by the new movements. At the centre of the Bank's ideological response throughout

---

[251] World Bank 1981a, 4; v.
[252] World Bank 1981a, 24.
[253] See Babb 2009. For an alternative perspective on which the US Congress is the external force pushing the Bank *towards* poverty alleviation and human rights conditionality, see Lavelle 2011, 119.
[254] Stern and Ferreira 1997, 539.
[255] World Bank 1981b, 2; World Bank 1983, 3; World Bank 1985b, i; World Bank 1985c, i; Romer 1989, 1, 5; World Bank 1990, 2.
[256] Kapur *et al.* 1997, 357–360.
[257] See World Bank 1990.

the 1980s and 1990s was Solow's insight that technological development drives growth.[258] This claim was used to legitimate and naturalize growth via links to cosmological ideas about scientific and technological progress. The central argument was that trade leads to economic growth because it facilitates scientific and technological progress in poor countries. By liberalizing trade, a country enters a market with scientifically and technologically advanced countries. Competition and investment then encourage technology transfers that raise productivity and increase growth. For example, the 1986 *WDR* argued that foreign direct investment is a necessary part of economic development in part because it facilitates "transfers of technology."[259] Government subsidies distort this process because they "encourage the wrong mix of inputs and misdirect technological change."[260] This is the core of neoclassical growth theory: exposing an economy to competition in the global marketplace naturally spurs scientific and technologically-led growth.

The 1987 *WDR* begins by arguing that "technological advance" depends "on scientific progress," which is best achieved by societies integrated into the global economy.[261] It asserts that sound "macroeconomic policy" is needed to "promote smooth adjustment to [global] imbalances and to lay the groundwork for other reforms that would raise productivity."[262] In this perspective, though there are multiple paths to industrialization, "the same basic policy prescriptions apply to every country."[263] All of this "requires little action by the government, relying largely on self-regulating market mechanisms to prevent disequilibria."[264] In this report, neoclassical equilibrium models drive the conclusion that growth is a natural corollary to universal policies that aim to achieve scientific and technological progress. Growth is depicted as emerging effortlessly and automatically from the human propensity to gather scientific and technological knowledge. Thus, these *WDR*s connected neoclassical growth theory to wider cosmological ideas about scientific and technological progress. These connections legitimated growth by depicting it as a natural, inevitable part of the universe.

The influential *East Asian Miracle* report also drew on ideas about scientific and technological progress.[265] The report valorizes Indonesia, Malaysia, Singapore, South Korea, Taiwan, and Thailand for achieving

[258] Solow 1956.
[259] World Bank 1986, 53.
[260] World Bank 1986, 53.
[261] World Bank 1987, 6.
[262] World Bank 1987, iii.
[263] World Bank 1985a, iii.
[264] World Bank 1980b, i.
[265] See Wade 1996.

Western-style development through trade-led growth. The report argues that scientific and technological developments were central to the success of these countries. South Korea, for example, is touted for "forward-looking" projects in which "only current technology was imported, and US-trained, Korean scientists and engineers were recruited."[266] Firms like Samsung were successful because they drew upon scientific and technological expertise from the West. Similarly, Taiwan's success in high-tech industries is credited to "close coordination of industrial, financial, science and technology, and human resources policies" and revising "university curricula to strengthen science, mathematics, engineering, and computer education."[267] Furthermore, export-oriented strategies eventually exposed these industries to "the forces of international competition and technological change."[268] In short, human capital investments like education and technological advances spurred growth in East Asia.

The report marks a shift towards the view that there is a role for the state in development because state intervention can, like trade, produce the necessary human capital. But the broader effect of the report is to reinforce the neoclassical legitimatory background with its roots in modernist cosmology. The report, like Bank documents throughout the period, draws on Solow's fundamental insight that scientific and technological progress fuels growth. This claim relies on modernist ideas about the power of knowledge that can be traced back to the rationalism of Petty and Temple in the seventeenth century and the evolutionist doctrines espoused by Lugard and others. That is, for over three hundred years, ideas about scientific and technological progress had bolstered the incipient idea that the purpose of the state is to improve, progress, and grow. Here, the argument also draws on the nineteenth-century idea that development and growth are natural elements of the universe. Neoclassical theory drains the biological elements out, replacing them with the cybernetic-systems ontology of objects and models. Nonetheless, neoclassical ideas place humanity in linear time as progressing towards the future, just as nineteenth-century geology and biology had.

The ascendance of growth as scientific and technological progress was not unchallenged. Poverty and a multidimensional view of development reemerged on the Bank agenda in the 1990s. First, the 1990 *WDR* resurrected the poverty line first introduced in 1975 and defined absolute poverty as living on less than one US dollar a day.[269] This represented yet another attempt to mainstream absolute poverty, but as in the 1970s,

---

[266] World Bank 1993b, 129.
[267] World Bank 1993b, 133.
[268] World Bank 1993b, 132.
[269] World Bank 1990, 5, 41–42.

it reduced poverty to income per capita. Second, as in the *Miracle* report, poverty appeared as an element of human capital that would spur neoclassical growth.[270] This line of thinking emerged from developments within neoclassical thought. In a 1989 World Bank working paper Paul Romer offered an influential reinterpretation of Solow's work, showing that the total stock of human capital explains growth better than investment.[271] Following Romer, education, women's empowerment, and other pro-poor initiatives could be conceptualized as investments in "human capital," which became just another input or factor in the neoclassical growth model. This legitimated state interventions designed to promote education etc., but did not fundamentally alter the foundations of development discourse. Whereas McNamara set out to ensure that growth served poverty alleviation, in the 1990s neoclassical discourse, poverty alleviation came to serve growth.

In any event, poverty did not return within a social, basic needs framework. The potentially radical edge of McNamara's conception of absolute poverty had been blunted by institutional imperatives and neoclassical ideas. Instead, the purpose of the state articulated in Bank discourse is to spur growth by unleashing scientific and technological progress. On the neoclassical view this can be done either by opening up the country to flows of scientific knowledge and technology from abroad or by investing in scientific and technological education at home. Both the neoliberal 1987 *WDR* and the 1993 *Miracle* report marginalized development approaches that aimed to improve the lives of the poor without modernizing or altering their way of life. That is, both neoliberal and Keynesian economic approaches are embedded within a larger modernist discourse in which progress is defined in scientific and technological terms. In this frame, the necessity of and desire for growth is naturalized and often taken for granted.

Thus, despite the appearance of a more multidimensional concept of development, most project documents in the 1990s emphasize growth and macroeconomics.[272] Eastern European countries in particular were invited by the Bank to join economic modernity. Poland, the Bank suggested, must turn towards Europe to continue its transformation: "Projected GDP growth of 4.5% for 1994 looks realistic and could even be bettered. Policies to sustain recovery and allow Poland to complete her transformation into a modern, market-oriented, democratic society must capitalize on the strengths of recent performance, and

---

[270] On the former, see World Bank 1990.
[271] Romer 1989.
[272] World Bank 1996, i; World Bank 1998, 1, 6.

remedy emerging weaknesses."[273] In projects, the dominant goals were macroeconomic health and "economic growth," though some projects did explicitly emphasize poverty alleviation as a central target.[274] Moreover, despite the considerable broadening of the Bank agenda in the 1990s, social scientists and anthropologists continued to report that they were marginalized and had to make their work amenable to quantitative representation in order to gain credibility.[275] Engel's review of the "post-Washington Consensus" in the Bank shows that alternative approaches to development were tolerated only to the extent that they did not challenge the growth imperative.[276]

In sum, McNamara's reforms created institutional imperatives that combined with neoclassical ideas and personnel to constrain Bank ideas about state purpose. Poverty alleviation was marginalized and growth was privileged. Growth was then naturalized through links to a modernist cosmological backdrop: human control over the forces of science and technology underwrites the pursuit of endless quantitative progress.

Scientific Cosmology and International Order, 2015

Over the course of the postwar era, cybernetic-systems thinking spread throughout the international system. The case of the World Bank provides a window onto the processes of recursive institutionalization that reconfigured the discourses underlying international order. The World Bank, alongside the International Monetary Fund, the General Agreement on Tariffs and Trade, and countless multinational corporations and consulting firms, strategically deployed neoclassical economics during the Cold War. The fact that economists had become embedded in these organizations meant that as academic economics changed – adopting new cosmologies, means, and ends – so too did states and international organizations.[277]

Moreover, these organizations worked to export their purposes and analytical tools to other associations. By producing knowledge, imposing conditions on loans, playing favourites with certain countries,

---

[273] World Bank 1994, i.
[274] World Bank 1996, I; World Bank 1998, 6; World Bank 2000; World Bank 2003, 2.
[275] Broad 2006; Kardam 1993, 1777; Bebbington et al. 2004, 44–47; Weaver 2008, 77–78. Best (2014) argues that expert authority in the Bank was eroded after the Asian financial crisis, leading the Bank to adopt a "provisional" mode of governance. This raises the possibility that after 2000, when my discourse analysis ends, the Bank underwent an epistemic shift in a less modernist direction.
[276] Engel 2012, 74.
[277] Fourcade 2006; Suddaby and Viale 2011.

and so on, these organizations spread the idea that all states should pursue unending economic growth.[278] These changes gradually transformed the discourse of state purpose over the period 1945–2015. Even in the absence of an order-building moment, the processes of recursive institutionalization reconfigured the purposes embedded in international order. As a result, international order went from being premised upon a general commitment to productivism to being based on the idea of economic growth as scientific and technological progress.

This process was fuelled by a number of political-economic pressures and crises that necessitated changes in the Bretton Woods system and its embedded liberal foundations. In the 1960s, the rise of multinational manufacturing and financial firms created a constituency that demanded capital mobility.[279] In the late 1960s, speculative flows overwhelmed the system of capital controls.[280] In addition, American domestic spending drove inflation and the dollar became overvalued. These forces were used to create a crisis in the Bretton Woods financial system and in August 1971, the United States decided to abandon it. After 1971, the switch to floating exchange rates and the collapse of capital controls marked the death of Bretton Woods.[281] In the absence of capital controls, domestic spending created inflationary pressures that were harder to control. Moreover, international finance lobbied hard for cuts to public spending in order to balance budgets and curb inflation. Finally, the debt crisis in the global south made countries beholden to the Bank and the IMF, which imposed structural adjustment policies in exchange for emergency financing.

For Best, each of these problems was driven by the fact that policymakers had forgotten Keynes' central insight: that economic problems have an intersubjective, discursive dimension.[282] The rise of neoclassical economics not only introduced new ontological categories and purposes; it constituted a new, limited, and technical approach to economic problems. In effect, the neoclassical synthesis drained out Keynes' more radical argument that the economy was governed by expectations and the irrationality of investment. In effect, it reduced Keynes' *General Theory* to a set of technical interventions: "the right combination of fiscal and monetary policy could alter demand sufficiently to eliminate inflationary

---

[278] Nelson 2014, 2017; St. Clair 2006a, 2006b; Wade 2002; Woods 2006. On the spread of science, see Finnemore 1993.
[279] Verdier 2002; Fourcade-Gourinchas and Babb 2002, 538.
[280] Helleiner 1994, 103.
[281] Ruggie 1982, 405–406; Helleiner 1994, 103; Fourcade-Gourinchas and Babb 2002, 538.
[282] Best 2008. See Blyth 2002.

or deflationary gaps between current and full-employment."[283] In effect, this reduction led policy-makers to address economic problems with a series of "technical fixes" rather than confront their structural and psychological bases.[284] Nonetheless, neoclassical ideas in the hands of neoliberals, monetarists, and supply-siders provided apparent solutions to problems in developed and developing countries alike. Keynesianism was unable to explain stagflation in the developed north or help states address cascading social and economic crises in the global south.[285] This weakened its ideological appeal and perceived effectiveness.

So alterations in transaction flows, political-economic conditions, and policy discourses combined to create an opportunity for the spread of neoliberal policies. The global spread of neoliberalism was conditioned by domestic factors such as whether reigning economic discourses were Keynesian or liberal, the severity of inflation, and the relative political strength of business and labour. For example, Britain and Chile acceded to neoliberalism quickly because they had suffered badly under inflation and the labour movements there were weaker.[286] Mexico, by contrast, had long been led by a successful socialist government that rejected liberalization.[287] Moreover, Mexico had not experienced debilitating inflation. So it did not adopt strict liberalizing policies until it was crippled by the debt crisis in 1982. But in each of these cases, economists trained predominantly in American schools facilitated domestic policy change. In Mexico, for example, 60 per cent of Cabinet members were banking or planning bureaucrats, 44 per cent of which had training in economics, mostly from American schools.[288] So neoliberalism spread along transnational networks of economists.

Despite the broad changes in political-economic factors and economic thought that drove neoliberalism, the underlying purpose of the postwar order remained constant.[289] Productivism in its various guises was the dominant purpose in the postwar order from 1945 through 2015. But there was also variation in the spread of productivist discourse. The goal of growth and its attendant valorization of scientific and technological modernity entered societies and organizations in a variety of ways. For example, the British colonizers tried to impose a specific version of Western scientific modernity on Indian society.[290] After independence,

[283] Best 2008, 55.
[284] Best 2008, 51.
[285] Centeno and Cohen 2012, 323; Backhouse 2010, 58–59.
[286] Fourcade-Gourinchas and Babb 2002.
[287] Woods 2006, 100–102.
[288] Woods 2006, 91.
[289] Ruggie 1982.
[290] Chatterjee 1986; Kalpagam 2000.

both modernization theorists like Rostow from the United States and economic planners like M.I. Rubinshtein from the Soviet Union worked with Indian Prime Minister Nehru's government on development plans.[291] Developing countries drew some elements of modernist ideas into domestic configurations, while leaving others aside. In Brazil, as in India, the quest for scientific and technological autonomy was linked to a nationalist, statist project of development.[292] But in Argentina, elites debated about what progress and modernization meant, and the ideal of scientific and technological progress was not consistently adopted.[293]

So throughout the period 1945–2015, many associations persistently, if unevenly, took up modernist, productivist, and neoclassical ideas. Processes of associational change did not unfold in the same way everywhere. Nonetheless, a transnational network of economists, dominated by students and professors from the top American programmes, served to channel new cosmological elements into states, IOs, and NGOs throughout the postwar world. Thus, by 2015, the practices and institutions of international order drew on epistemic and ontological elements from the systems-cybernetic and neoclassical social sciences.

In this section, I provide a window onto international order in 2015 by examining the discourses of the core IOs.[294] This is admittedly a partial view, but one likely to reveal the discourse of state purpose as it existed. In 2015, the United Nations (UN) remained the political and cultural centre of international order. The world's most powerful states, the United States and China, sat permanently on the Security Council alongside two strong European voices, the United Kingdom and France, and a declining revisionist power, Russia. Germany and Japan were excluded from the Security Council, but were important members of other multilateral fora such as the Group of 7 (G7). The North Atlantic Treaty Organization (NATO) remained a core element of the Western hegemonic alliance. The core postwar financial institutions – the IMF, World Bank, and World Trade Organization (WTO) – still played a central role in organizing economic policy throughout the world, extending the influence of US-led Western hegemony into the twenty-first century. The Asian Development Bank (ADB) had emerged as an important regional lending arm. Finally, 2015 witnessed the emergence of the

---

[291] Engerman 2004, 34. Nehru himself creatively recombined post-Enlightenment rationalism with indigenous Indian thought. See Chatterjee 1986, 138–141.
[292] Adler 1987, 8–9.
[293] Adler 1987, 8.
[294] See the Methodological Appendix for more details on text selection.

Asian Infrastructure Investment Bank, which alongside the Shanghai Cooperation Organisation served as a non-Western vehicle for Chinese influence.

Discourse of State Purpose

There are two dominant discursive configurations in the 2015 discourse of state purpose. First, there is a liberal security configuration oriented to "international peace and stability." Second, there is an economic configuration oriented to growth. The security discourse is more prominent in the UN Security Council and NATO, but in policy neutral fora like the G7 where any issue can be placed on the agenda, the centrality of economic growth in the discourse of state purpose is clear. These dominant themes reflect the continued hegemony of the United States and Western states over the institutional and discursive bases of international order.

In 2015, security goals are embedded within the liberal discursive formation. Security is portrayed as a defensive goal aligned with peace and stability.[295] The NATO report is at pains to emphasize that NATO's military capabilities "reinforce defence."[296] But NATO does not merely defend its member states. NATO exists to defend "the safety, stability, and well-being of people around the world" by defending international order.[297] Within the liberal discourse, military aggression and conquest are delegitimized.[298] Russian aggression in Ukraine is not just illegal, but threatens the stability of the international order.[299] However, the use of military force, defence spending, and military training are still important and legitimate means.[300]

But military means are secondary to legal norms and principles which constitute the "rule-based" international order.[301] The liberal discourse maintains that legitimate international action must move through diplomatic channels, legal frameworks, and multilateral forums.[302] The NATO report argues that emerging "threats challenge the international order we have built since the fall of the Berlin Wall – an order that embodies our democratic values and is vital for our way of life."[303] NATO countries are "open societies" threatened by violent extremism and Russian

[295] G7 2015, 8.
[296] NATO 2014, 5.
[297] NATO 2014, 5.
[298] NATO 2014, 4.
[299] NATO 2014, 4; G7 2015, 6.
[300] NATO 2014, 3.
[301] NATO 2014, 18; OECD 2015, 7.
[302] G7 2015, 6.
[303] NATO 2014, 3.

aggression.[304] Russia's aggressive actions in Ukraine "disregard international law and violate security arrangements," thereby undermining the basis of international order.[305]

The dominance of Western liberal hegemony is evident in the UN Security Council, NATO, and the G7. Here, liberal and humanitarian norms promoting democratic governance, human rights, protection of civilians and marginalized groups, freedom of movement, and the well-being of people predominate. The G7 declaration recognizes "the importance of freedom, peace and territorial integrity, as well as respect for international law and respect for human rights."[306] Terrorism is a threat because it "denies tolerance, the enjoyment of universal human rights and fundamental freedoms."[307]

In 2015, the goal of wealth is embedded within a neoclassical economic discursive formation. The central economic end of international order is economic growth, although development, poverty reduction, employment and well-being are also prevalent. In the IMF and ADB, growth is simply economic growth or GDP.[308] The IMF's *World Economic Outlook* is a straightforward analysis of world economic growth. The ADB's *Asian Development Outlook* is focused on regional economic growth in a global context. In the joint declaration of the G7 summit, the principal term is "inclusive growth" but "sustainable growth" is also mentioned.[309] The WTO mentions inclusive growth as well. The OECD argues there is a "need for a new growth narrative; one that is more sustainable, more inclusive and focused on people's well-being."[310] The Shanghai Cooperation Organisation communique promotes "sustainable socio-economic development" and the "living standards of citizens."[311]

Despite this diversity, it is economic growth operationalized as increases in GDP that dominates the discourse. The WTO report on trade and development argues "integration into the world economy" fuels market competition, technological transfer, and investment in developing countries leading to "economic lift off."[312] This argument is supported by an economic history in which international orders rooted in trade openness lead to periods of global economic growth.[313] Growth has come in three waves,

---

[304] NATO 2014, 3.
[305] NATO 2014, 4.
[306] G7 2015, 6.
[307] G7 2015, 9.
[308] WTO 2014, 5–6; ADB 2015a, 2.
[309] G7 2015, 1.
[310] OECD 2015, 6.
[311] SCO 2014, 2.
[312] WTO 2014, 40.
[313] WTO 2014, 41–51.

each of which was underpinned by free trade orders: the industrial revolution in Europe, the post-Second World War order, and the rise of the rest into the Western-led economic order.[314] The central image is of an "accelerating and widening circle of development" made possible by an "open and integrated" order.[315] For the WTO, the history of international order is the economic history of growth. To join the order is to achieve sustained high rates of GDP growth through scientific and technological innovation. The WTO report represents integration into the global economy, especially by achieving a position in global value chains (GVCs), as the central goal of all developing states.[316] The report argues that successful states such as Korea, Taiwan, India, and China have all "increased their participation in GVC trade."[317]

The discourse also includes references to improving living standards, reducing poverty, or enhancing well-being. The ABD's report *Together We Deliver* argues that the ADB has always aimed "to free the people of Asia and the Pacific from poverty."[318] The report contains stories from projects aimed not necessarily at scientific and technologically-led growth, but at improving rural livelihoods.[319] The World Bank's twin aims in 2015 are "promoting development and combating poverty."[320] The OECD is oriented to "promoting growth, development and well-being."[321]

One could interpret the discourse as suggesting that growth is a means to further well-being, prosperity, poverty reduction, health, or any number of other ends. My discourse analysis revealed one instance in which growth was mentioned as a means to other ends: "Growth is an essential precondition for employment and wealth creation."[322] But other references to growth and other economic ends like well-being or employment reverse the arrows. The G7 declaration states that the G7 has a "growth agenda" and that growth requires "the protection of our climate, the promotion of health and the equal participation of all members of society."[323] Here, as in the World Bank discourse, alternative issues like environment and health serve economic growth, rather than the other way around.

[314] WTO 2014, 44.
[315] WTO 2014, 44.
[316] WTO 2014, 15.
[317] WTO 2014, 6.
[318] ADB 2015b, xi.
[319] ADB 2015b, 20.
[320] World Bank 2015, 3.
[321] OECD 2015, 6.
[322] OECD 2015, 14.
[323] G7 2015, 2.

In the majority of instances, the discursive justification for growth as generating well-being or poverty reduction drops out. Instead, growth is reduced to the quantitative indicator of GDP, which is treated as an end in itself. For example, the WTO report states that "development goes beyond higher GDP per capita" and that other indicators of well-being include "life expectancy at birth, infant mortality, nutrition, literacy, gender inequality and employment."[324] However, in the remainder of the report, these alternative indicators are not mentioned and GDP is privileged. In the Shanghai Cooperation Organisation communique the opening paragraphs highlight "sustainable socioeconomic development" but the document later focuses on "economic growth" via policies clearly directed at improving production alone: "invigorating investment cooperation by implementing concrete projects in high-tech sectors of economy, transport logistics, information communications and other promising areas."[325] This is similar to the findings of the World Bank discourse analysis above: even when other ends like poverty reduction are mentioned, it is the quantitative indicators that drive policy formation and evaluation. Thus, states naturalize GDP as an end, even though the discourse also contains statements that recognize the limits of GDP.

Cosmological Elements

The purpose of growth as GDP is supported cosmologically by the cybernetic-systems orientation to models and projections. The IMF outlook presents page after page of charts and tables that project current trends into the future. World output is projected through 2016.[326] GDP, inflation, exchange rates, and exports are simulated through 2020.[327] These simulations are used to show relationships between variables: if currencies appreciate, GDP will decline. Together, projections and simulations produce a "forecast horizon" that orients policy and expectations to the future.[328] Growth and its underlying causes can be precisely modelled and represented in economic analyses. Economic analyses, therefore orient policy to growth rather than the more slippery ends of well-being and quality of life.

Another important ontological theme is the division of the world into classes of states. As during the Cold War, the world is divided into three groups of states: advanced economies, emerging markets, and developing

[324] WTO 2014, 6.
[325] SCO 2014, 3.
[326] IMF 2015, 2.
[327] IMF 2015, 9.
[328] IMF 2015, 11.

countries.[329] The implicit telos of the division is that developing states should strive to become emerging markets (by securing a position in GVCs) and emerging markets should become advanced economies. This recalls the League of Nations Mandate System of classification in which countries were ordered by their place along a temporal line of civilizational development. Such a division of the world into classes suggests that the purpose of states is to develop along the scale of progressive time.

The underlying demarcation of the world into countries conceived as separate, ontologically distinct systems is taken for granted. An alternative ontology would take seriously the relational view that states and their histories are bound up with one another. Rather than positing a simplistic narrative in which trade drives up growth, it might recognize the colonial origins of trade and growth in the nineteenth century and the relational factors underpinning Japanese and Chinese growth in the twentieth century. The standard story tells this as a tale of "export-led" growth but what is not mentioned is the relational fact that not all countries can be net exporters. But these historical and relational links are downplayed in an ontology that treats policy as a universal, one-size-fits-all strategy that all countries can pursue independently. Reducing the unit of analysis from historical relations to individual countries has value, but it is a partial view and one amongst many possible alternatives.

The epistemic foundations of the discourse are modernist and resonate with neoclassical growth theory. Thus, as in the World Bank discourse of the 1980s and 1990s, scientific and technological progress is tightly linked with growth. The central driver of the WTO's teleological history of trade-led growth is "technological diffusion" from Europe to Asia, Latin America, and Africa. While the colonial origins of Western pre-eminence are alluded to in the WTO narrative, the central message is that Western states have given technology and access to developing states, permitting them to achieve high rates of growth. The goal of these states today remains borrowing technological modernity from the West. To gain a place in GVCs, a state must "be sufficiently close to having the capacity to produce at world standard quality and efficiency levels" and thus "technology and knowledge transfers" are necessary.[330] Research and development is also necessary to advance agricultural productivity and create spillover effects.[331] The OECD justifies itself as provider of expertise to the Western states: better knowledge leads to better policy

---

[329] See IMF 2015, xv, 12. This echoes the findings of Ferguson 1990.
[330] WTO 2014, 7.
[331] WTO 2014, 9–10.

which leads to growth, development, and well-being.[332] Perhaps this emphasis on its own epistemic authority predisposes it to faith in science and technology. For the OECD, productivity is essential to growth and science and technology are central to productivity.[333] Science and technology also advance health and environmental governance. The G7 calls for more research and development on tropical diseases.[334] To meet climate change goals "low-carbon technologies" are needed.[335] The Shanghai Cooperation Organisation calls for "fundamental and applied scientific research on organic food production technology" and "high-tech innovation projects."[336]

Science and technology are cosmologically significant as the means for creating progress. For example, the G7 declaration states, "[t]o ensure that G7 countries operate at the technological frontier in the years ahead, we will foster growth by promoting education and innovation, protecting intellectual property rights" and so on.[337] Here, education and innovation are means to growth, and growth is a means to operating "at the technological frontier." The WTO's narrative that the implicit telos of developing states is to gain technological modernity from rich, Western states via trade places scientific and technological modernity as a crucial goal in development. A place in world order cannot be secured without achieving a certain standard of scientific and technological development.

The importance of science in international order is also demonstrated by the rise of climate change onto the international agenda. The G7 declaration argues that "[u]rgent and concrete action is needed to address climate change, as set out in the IPCC's Fifth Assessment Report."[338] The authority of science was necessary to put climate on the agenda of states. However, while environmental aims are mentioned in the discourse, environmental ends are in the second-tier, alongside health and gender issues. They are more likely to be mentioned as components of economic growth than they are as ends in themselves. Indeed, sometimes the texts depict environmental protection as means to economic growth. For example, in the Shanghai Cooperation Organisation communique, environmental cooperation is directed at "preserving the nature and resource potential of the SCO member states."[339]

---

[332] OECD 2015, 6.
[333] OECD 2015, 17.
[334] G7 2015, 11.
[335] G7 2015, 12.
[336] SCO 2014, 4.
[337] G7 2015, 1.
[338] G7 2015, 12.
[339] SCO 2014, 5.

While it by no means forms the basis of a competing discursive configuration, the language of biology has reentered international discourses. This is not really evident in the references to the environment, which tend to be cloaked in the language of economic policy. Rather, the language of biology appears in the many references to "resilience."[340] This could be dismissed as merely the incorporation of a popular term into political discourse. However, there are deeper resonances. For example, the WTO report reveals that deep ontological metaphors have travelled with the language of resilience. It argues that "[e]conomic openness has, in turn, depended on the underlying strength and resilience of the international system – its ability to absorb rising giants, to withstand shocks, and to promote cooperation and coherence."[341] Perhaps this indicates the rise of an object-centred ontology that draws on ecological conceptions of systems, instead of the continued dominance of models and projections that emerged from cybernetic-systems thinking in the 1950s and 1960s.

Conclusion

The emergence of growth in the post-Second World War order was made possible by five centuries of cosmological shifts arising from the natural and social sciences. In the introductory chapter I outlined five features of cosmological configurations: ontology, episteme, time, the origins of the universe, and the place of humanity in the cosmos. Economic growth depends on a configuration of all these elements. First, it draws on a mechanistic, object-based ontology that emerged over the course of the nineteenth and twentieth centuries. Second, epistemically, economic growth depends on the modernist idea that objects like the economy can be represented, known, and controlled through statistical tables and mathematical models. Third, it relies on linear ideas about time and progress borrowed from Enlightenment histories. Fourth, the vision of progress inherent in the idea of growth implicitly relies on the nineteenth-century idea that growth is a natural part of the history of the universe. This is bound up with the final cosmological element: the idea that humans unleash and benefit from scientific and technological progress, which in turn furthers human well-being. In this cosmological narrative, humans drive inexorable progress by harnessing the tools and knowledges of science to their purposes. That is, faith in scientific and technological progress is central to growth discourse because only

[340] G7 2015, 12–13; OECD 2015, 15.
[341] WTO 2014, 51.

unending scientific and technological progress can explain how growth itself can continue indefinitely.

The rise of growth made it possible to redefine state purposes in terms of GNP and later GDP. Purely quantitative indicators of growth displaced moral, civilizational, and biological standards of well-being. As late as the 1950s, the goal of development was embedded in a multifaceted conception of development that included employment, agricultural productivity, and liberal values. But the imposition of quantitative standards and the rise of neoclassical economists altered the discourse to privilege a reductive goal of quantitative increase. Thus, one set of economic policies becomes embedded in institutional routines and ends rather than others. As Clegg puts it:

> Quantification plays a central role in the concretization of intersubjective understandings about aspects of economic life. In a process that is both subtle and drawn out over a long period of time, a transformation occurs whereby concepts evolve from an initially highly contested malleable form, to being regarded as representing a self-evident and pre-existing object, contested only at the margins. Through a process of abstraction, homogenization, and sedimentation, complex social phenomena are reduced to a small number of measurable features.[342]

Quantifications can be and often are contested, as they were in the Bank through the 1990s and 2000s. However, in the absence of reliable statistical alternatives or competing integrative models, the critical force of challenges to the status quo are likely to remain impotent.

Economic growth was transformed from a reified, sedimented quantitative goal into a true purpose through links to the broader cosmological backdrop. In short, neoclassical theory naturalized growth as scientific and technological progress. Whereas in the 1950s, Bank objectives were legitimated with reference to liberal values, in the 1990s, Bank objectives were naturalized through scientific and technological narratives. In effect, the purpose of humanity is equated with harnessing and advancing the development of science and technology. Scientific and technological development is a natural expression of human flourishing that drives the expansion of wealth. This tight link between growth and scientific and technological progress continues to be reproduced in the present international order as the IMF, WTO, and other IOs carry and articulate the same cosmological vision.

Moreover, neoclassical economics expresses a still grander narrative that redefines the purpose of states: all countries should strive for economic integration with Western modernity. The central route to modernity is open trade policies which channel scientific and technological

[342] Clegg 2010, 485.

progress into economies. By promoting a universal goal for all states, economics today reproduces the cosmological ideas that emerged with the colonial idea of development. However, development is no longer a law of nature conceived via biological analogy. Now, states are expected to use government policy, Keynesian or neoliberal, to deliver the benefits of scientific and technological modernity. Thus, the range of politics is limited to a debate about the best way to achieve growth via scientific and technological progress. Both Keynesian and neoliberal policies rest on modernist epistemic presuppositions. That is, they both place faith in the idea that expertise and knowledge can advance human progress.

Although this chapter demonstrates only part of the overall process, the theory of recursive institutionalization allows us to understand how the purpose of growth was spread. The World Bank was a constituent element of the US-led postwar order. But it was not a mere agent or expression of American hegemony. The discursive and organizational changes within the Bank were not controlled from the outside.[343] Changes in the Bank reflected its own politics and processes of associational change, which produced contingent configurations of meanings and practices. Moreover, it contributed to associational changes throughout the world by producing knowledge, offering technical assistance, and providing loans.

The reproduction of the cosmological narrative of human history as the history of scientific and technological progress in other states and IOs speaks to both the wider influence of the Bank and to the fact that neoclassical knowledge drove similar associational changes in other associations. From this vantage point, the central story of the postwar order is the construction of growth-oriented, internationalist capitalism via the transnational networks of neoclassical economists that formed in the 1960s, 1970s, and 1980s. This is a distinct story from the one that is usually related in historical and constructivist international political economy about the rise and fall of Keynesianism and neoliberalism.[344] Here, it is the rise of neoclassical growth theory that takes centre stage. The fates of Keynesianism and neoliberalism are still important, but they appear as subsidiary developments of a larger macrohistorical trend: the rise of growth under the auspices of neoclassical economics.

---

[343] On the autonomy of IOs, see Barnett and Finnemore 2004. On the autonomy of the Bank, see Chwieroth 2008; Nielson and Tierney 2005; Weaver 2008.
[344] Best 2008; Blyth 2002; Centeno and Cohen 2012; Chwieroth 2010; Fourcade-Gourinchas and Babb 2002; Fourcade 2006; Helleiner 1994; Woods 2006.

6 Conclusion: The Future of Cosmological Change

> Potentially the most widespread and globally synchronous anthropogenic signal is the fallout from nuclear weapons testing. The start of the Anthropocene may thus be defined by a Global Standard Stratigraphic Age coinciding with detonation of the Trinity atomic device at Alamogordo, New Mexico, on 16 July 1945 CE.
>
> – Colin N. Waters *et al.* 2016[1]

Introduction

Cosmological ideas arising from the Western natural and social sciences have repeatedly transformed the discourse of state purpose underlying international orders. Natural philosophy from Copernicus to Newton, the historical sciences and evolutionary thinking from Hutton to Malinowski, and cybernetic-systems modelling from Bertanlanffy to Solow reconfigured notions of matter, the nature of time, and humanity's place in the universe. These reconfigurations suggested new political purposes that altered the way the goals and ends of states were understood and legitimated. Policy-makers like Prince Metternich, Joseph Chamberlain, and Robert McNamara served as brokers that channelled new thinking into the associations that carry and reproduce international order. Bureaucratic realignments and personnel changes in state agencies and international organizations drove and deepened discursive changes in these associations. Postwar settlements, bilateral treaties, and diplomatic interactions embedded new purposes in the core sites of international order. These processes slowly reoriented the discourse of state purpose from God and glory to economic growth.

In this concluding chapter, I explore three sets of questions raised by the argument. First, the theoretical argument presents a multi-level account in which micro- and meso-level processes mechanisms

---

[1] From the report of the Anthropocene Working Group of the International Union of Geological Sciences.

are integrated into a macro-level theory of recursive institutionalization. What are the general implications and conditions of the theory of recursive institutionalization? Second, the empirical argument raises important questions about the future of international order. What are the prospects for a cosmological shift in the twenty-first century? In answering this question, I use the generalized version of the theory to analyse the future of international order in the age of the anthropocene. Finally, the argument suggests that the social sciences have played an important role in the processes of recursive institutionalization, raising the question of what role if any the social sciences will play in future cosmological shifts. I ask, with Weber, how does the view of science that emerges from this history help us rethink the task of the social sciences today?

Rethinking International Change

This study offers some general metatheoretical lessons for thinking about international change. First, accounts of international change must be built on a multilevel theory that places the micro-, meso-, and macro-levels of international order in systematic relationship to one another. The theory of recursive institutionalization embeds a micro- and meso-level theory of contextual associational change within the macro-level model. Brokers are constantly recombining ideas to produce new configurations of political imperatives and cosmological elements. As associations take up new knowledge, they transform the discursive bases of international order, making possible and desirable transformations in purposes, primary institutions, and secondary institutions. Only those configurations that are taken up by many associations will become lasting components of international order. In the argument here, the central source of those new configurations were cosmological shifts arising from the Western scientific tradition.

Second, change should be conceptualized as a cumulative, long-term process that features moments of order-building that consolidate and extend ongoing shifts. Recursive institutionalization foregrounds the dynamic, ongoing processes by which the international system is transformed as the associations that carry and reproduce international order change. These associational changes are then extended and consolidated by moments of institutionalization that embed new purposes in the primary and secondary institutions of international order. This image of international change thus draws from both gradual and punctuated equilibrium models of change. Gradual changes in associations drive the model, but moments of order-building are still important. While my account also features snapshots of order-building moments, they are designed to provide a window onto ongoing gradual changes in the discourses and rules underlying international order.

But the cumulative and gradual nature of associational changes should not lead us to think that change is a smooth or linear process. Instead, we might think about how the nonlinear view of change associated with the punctuated equilibrium image could be applied at the associational level. If change is always a reconfiguration of discourse, there is no reason to believe that such changes will always be gradual, predictable, or linear. After all, networks of meaning are already complex systems, with innumerable semantic and affective connections between concepts. Thus, changes in the network ripple outward, driving unexpected and discontinuous alterations in understanding and orientation. So abrupt and uneven changes in associations are always bubbling up into change at the international level. International change proceeds in fits and starts, accelerated and slowed by contests over resources, legitimacy, and purpose.

Third, stated generally, coherent change in the ideational and institutional basis of international order must be sustained by a reliably reproduced body of knowledge.[2] That is, new knowledge can change international politics if it is reproduced transnationally and taken up by the associations underlying international order. If there is a general methodological slogan guiding the theory I have introduced, it is *follow the knowledge*. This is the old slogan of the sociology of knowledge, which posited that some form of knowledge, whether specialized or folk, was necessary to explain social and political outcomes.[3] This principle follows from the fact that ideas do not float freely in international order. Rather, they must be carried and reproduced by associations oriented to the international. Constructivists have not been careful enough about specifying where ideas come from and as a result, too often take for granted the existence of a stable, reproduced body of knowledge. A processual, practical perspective on the constitution of the international system helps to remind us to specify the social location of ideas and not to assume a distribution of ideas that may not exist.

Here, the injunction to follow the knowledge can be taken to mean that major changes in state goals or the rules of primary and secondary institutions are likely to depend on some transnational group that produces and disseminates knowledge. Generally speaking, this might be either a caste of

---

[2] This is a generalization of the core argument of the epistemic communities literature. See Haas 1990; Haas 1992a, 1992b.
[3] Mannheim 1936; Berger and Luckmann 1966; Latour 1987, 2005; Adler-Nissen and Kropp 2015.

priests or an academic discipline in the natural sciences. But in order to explain major change, we have to go through the organizational culture of associations and the worldviews of individuals to the knowledge that constituted differently situated agents in similar ways. That is, if there are common ideas about purpose circulating in the international system, it is unlikely that associations from many cultural backgrounds adopted the same ideas in the absence of transnational knowledge. In order for many associations to take up similar ideas, there must be a reliably reproduced, stable body of knowledge to draw from.[4] An epistemic community might provide this, but any number of knowledge-based groups or networks can build transnational knowledge.[5]

In the case of scientific change, I argued that many associations take up similar ideas when a dynamic scientific movement inspires those associations to appropriate the new scientific ideas for strategic purposes. In early modern Europe, a transnational network of aristocratic natural philosophers maintained and reproduced a stable intellectual tradition. Later, the professionalization of the natural sciences and the creation of state bureaucracies formalized the transmission of expert knowledge from a transnational scientific profession into international order. In other domains such as security or human rights, where a transnational community may be weaker, the source of knowledge must be carefully specified and traced. For example, Adler argues that a transnational community of defence intellectuals had to be constructed over the course of the Cold War.[6] The knowledge they produced and reproduced helped to stabilize the Cold War security order. But the existence of such transnational communities cannot be taken for granted. Their existence needs to be empirically established and we need to allow differences between transnational networks to inform the mechanisms and processes of change in our theories.

Generalizing Recursive Institutionalization

In addition to these general lessons, the model of recursive institutionalization offers a series of micro-, meso-, and macro-level mechanisms and processes can be used to explain change in other cases. At the core of the theoretical model are three processes: cosmological shifts, associational change, and international institutionalization. In order to generalize the

---

[4] See Allan 2017a.
[5] Professions, disciplines, expert networks, and other groups may fail to meet the conditions of an epistemic community (consensus on causal and political beliefs) but nonetheless underwrite the diffusion of political ideas. See Haas 1992a, 2014.
[6] Adler 1992.

theory, it helps to think through the conditions under which we would expect these processes to operate. The theory can then be used to illuminate other cases of scientific change as well as cases where other forms of knowledge are in play. This takes us beyond a purely contextual approach, in which events and ideas are embedded in and related to historically unique developments, to a general form of historical theory. In my view, historical theory should proceed as inductively as possible, but at the end of the analysis, we should be willing to reflect on the general processes and mechanism the analysis reveals, and to think carefully about how those processes and mechanisms might be applied to other cases.

1. *Cosmological shifts*: under what conditions would we expect changes in knowledge to generate widespread associational change? After all, clearly not every scientific breakthrough generates change in international order. Take, for example, the emergence of quantum mechanics. Quantum mechanics is often cited as the most successful scientific research programme in history and it has produced radical ideas about the fundamental elements of the universe, the nature of time, and the place of humanity in the universe.[7] These are key preconditions for a cosmological shift and yet quantum mechanics does not seem to have fundamentally reconfigured political discourses.[8] Why?

First of all, it may simply be that not enough time has passed. It took over one hundred and fifty years for Copernicus' conjecture to begin to transform international orders (1543–1713) and it was a hundred years before the historical sciences reshaped the British-led order (1751–1860). It has only been a little over one hundred years since Planck's discovery of quanta, so we may yet see a quantum revolution in political discourses. The second answer is that quantum ideas have already made some contributions to cosmological discourses. The Heisenberg uncertainty principle helped erode determinism by suggesting that there was irreducible randomness in the universe.[9] The principle has been widely used to challenge determinism in the natural sciences and create space for non-Newtonian ontologies and epistemologies in other scientific disciplines. This and other developments were part of the reason why epistemic modernism was so successful in the period after 1930. Quantum

---

[7] For example, Christopher Fuchs' (2016) interpretation of quantum mechanics foregrounds the role of humans as participants in the universe. Humans are not mere bystanders or observers, but active agents constituting and contributing to the unfolding of the universe.
[8] On recent thinking about quantum in IR, see Wendt 2015 and Project Q, led by James Der Derian and Jairus Grove (http://projectqsydney.com/).
[9] See Porter 1986.

mechanics opened up the natural sciences and created philosophical space for human agency.[10] Third, and more deeply, quantum ideas have limited political power because the implications of quantum mechanics are difficult to translate into simple political ideas and purposes. While the core ideas of quantum mechanics (nonlocality, uncertainty, the wave function, etc.) are strange enough to drive major conceptual changes, it is not clear what these ideas mean for politics. However, the cosmological implications of a body of thought are not objectively given or determined exogenously. Instead, there are a number of political processes that shape the development of knowledge, steering it towards or away from the dominant cosmological views.

Consider the development of quantum thought and quantum technology in the post-Second World War period. Throughout the Cold War, the American military funded technological applications of quantum research, including nuclear weapons, masers, lasers, and quantum electronics.[11] The flood of military money into quantum research pulled the best minds and the best-equipped laboratories out of basic research, which might have spoken more directly to cosmological questions, into applied research.[12] The most advanced quantum research was conducted with an eye towards technological applications that would extend the American military's control over land, air, sea, and space in the Cold War. Instead of challenging the modernist elements in twentieth-century international discourses, the use of quantum mechanics to enhance control over reality may have bolstered those elements. To the extent that the quantum revolution was given a concrete cosmological content, that content tended to reinforce the mechanist, modernist cosmological backdrop of the twentieth century. The important point here is that in the case of quantum mechanics, state interest in and funding for science steered the cosmological implications of a scientific movement in some directions rather than others. Further research would be needed to specify how states and other associations shape knowledge production in ways that might further complicate the model of recursive institutionalization.

Finally, it is not clear what formal institutional channel would transmit quantum ideas into associations and therefore international politics. In the theory of recursive institutionalization, powerful brokers or politically embedded transnational networks have to transmit ideas from the natural sciences into political discourses. In the twentieth century, the cybernetic-systems model from physics and engineering entered via

---

[10] See Mirowski 1989, 89–90, 391–392.
[11] Forman 1987.
[12] This was generally true in the Cold War. See Allan 2017a; Leslie 1993; MacKenzie 1990.

economics. In the nineteenth century, Darwinian and geological ideas entered through social anthropology. In the eighteenth century, political economy transposed materialism, mechanism, and measure into growing state bureaucracies. But in the case of quantum mechanics, no politically embedded transnational network has taken up and pressed quantum ideas into political discourses.

In Chapter 2, I argued that cosmological shifts are likely to transform international orders when they are cosmologically powerful and diffused widely. The case of quantum mechanics can help us make these two conditions more precise. First, the cosmological shift needs to have clear political implications. However, this is not an objectively given fact. Specific actors in certain political, economic, and discursive contexts construct the political implications of new knowledge. In order to produce a cosmological shift with implications that might challenge the cosmology of the hegemon, the knowledge programme needs to emerge in communities that are autonomous from hegemonic power. These communities could be located in other states, or they could arise in specific sub-cultures of hegemonic society where deviant ideas are possible.[13]

In addition to being diffused widely, the new cosmological ideas need to be taken up by a transnational group of experts that are capable of serving as brokers with the political community. In each of the cases in this book, economic and social thinkers played central roles in this process. Economists like Petty and Solow as well as anthropologists like Malinowski translated cosmological ideas from the natural sciences into politically relevant ideas. This construction of useful, problem-solving knowledge then inspires associations to take up the new forms of knowledge, driving recursive institutionalization. In sum, in order for a cosmological shift to transform international politics, the breakthrough in knowledge must have implications for political thought, it must be transmitted through established channels between knowledge producers and associations, and it must inspire associations to take up those ideas.

2. *Associational change*: how might the theory of associational change be refined and elaborated? The theory of associational change included two processes: the strategic deployment of new knowledge and discursive reconfigurations that produce new purposes. In thinking about the wider applicability of the theory we need to think about both sets of mechanisms. First, under what conditions should we expect the widespread

---

[13] Berger and Luckmann 1966, 125.

strategic deployment of some set of ideas? Second, under what conditions would the theory expect strategic deployments to generate major discursive reconfiguration and the production of new purposes?

Clear answers to these questions would require a more systematic comparison of cases of associational change than has been conducted here. However, we can draw some lessons from the epistemic communities and historical institutionalist literatures on ideational change to help inform our expectations. One way to read the development of the literature on ideational change is that it has produced a long list of conditions that must be met in order for expert ideas to influence politics:

- the epistemic community or knowledge-based group must produce a consensus encompassing both causal and political beliefs;[14]
- the epistemic community must have access to decision-making channels that is established by authority, trust, or institutional capture;[15]
- political actors must be uncertain as to what their interests are in a given situation;[16]
- such uncertainty may have to be induced by a novel situation or by the onset of a political or economic crisis, making policy-makers amenable to new information and persuasion;[17]
- the epistemic community must offer a clear diagnosis of the situation or crisis;[18]
- the epistemic community must be able to point to a politically viable solution to resolve the crisis;[19]
- the solution to the crisis must either co-opt powerful interests or not face opposition from powerful or entrenched interests.[20]

These conditions are specific to scientific and expert knowledge. In generalizing the theory of recursive institutionalization, we would want to expand such an analysis to include more broad sources of organizational change. However, we can still draw important lessons from this literature for thinking about associational change.

First, what are the lessons for when we should expect strategic deployment? I noted above that cosmological shifts are likely to induce widespread associational change when they are translated into a useful set of

[14] Haas 1992a.
[15] Chwieroth 2008, 2010; Evangelista 1999; Finnemore 2003; Haas 1989; MacDonald 2015.
[16] Haas 1992a, 1992b; Adler 1992.
[17] Blyth 2002; Campbell and Pedersen 2001; Cross 2013; Helleiner 1994; Ikenberry 1992; Legro 2005; Welch 2005.
[18] Blyth 2002; Hall 1989, 1993; Ikenberry 1992.
[19] Blyth 2002; Hall 1989; Legro 2005.
[20] DeSombre 2000; Haas 1992b; Hymans 2012.

ideas. But the other side of the issue is that many associations must feel a need to solve similar problems in the first place. On this point the literature on ideational change suggests that widespread strategic deployment of an idea is likely to follow a crisis or the construction of a crisis. The crisis might be induced by major power war, but the rise of neoliberalism in the 1970s and 1980s was initiated by a more diffuse economic and constructed crisis. A variety of events can serve as an impetus to change in international discourses.

Second, what lessons can we draw about when we should expect strategic deployment to generate major changes in associational discourses? I argued earlier that there are some features of scientific ideas themselves that drive discursive reconfiguration and the construction of new purposes. Namely, ideas about nature and time are closely related to how we conceptualize the role of humanity in the universe and thus to thinking about what purposes we should pursue. In order for strategic deployments to have major effects on discourses that reconfigure purposes, the new ideas need to have cosmological implications. Though, as noted above, these connections are not objectively given and need to be made in context by specific actors.

That an idea has cosmological implications is a necessary condition of change, but this does not tell us much about the conditions under which we would expect a new idea to change an association. There are also institutional or organizational conditions that enable or constrain discursive reconfigurations. We would expect a change in ends or purposes when an association is captured by an epistemic community or knowledge-based group. The entry of a new group can alter the dominant discourses of an association, fundamentally changing its practices and goals. The simplest way for an association to be captured is through leadership change or personnel realignment.[21]

As we saw in the case of the World Bank, McNamara's leadership reoriented organizational imperatives. He deployed the resources of the presidency to alter recruitment patterns and introduce new organizational practices. As neoclassical economists entered the Bank, they reinforced his quantitative prescriptions and naturalized growth as scientific and technological progress. This did not have the effects that he hoped, but the case nonetheless shows the importance of leadership and personnel change in setting off discursive reconfigurations. In the British Colonial Office, personnel change was an important factor, but so was altering the advisory and decision-making channels of existing staff. As the training of colonial officers and the forms of advice they received were modified, the

---

[21] Chwieroth 2008, 2010; Evangelista 1999; Haas 1989; Hopf 2012.

ends of colonial governance changed. The creation of advisory committees reconstituted and resocialized officials, forcing them to rethink their means and ends. In both the Bank and the Colonial Office, capture and change in purpose was facilitated by leaders who believed that importing new knowledge would help the organization solve its most pressing problems. But in both cases leaders did not intend to change the goals of the organization in the way that they did. So there are limits to intentional control over goal change in complex organizations.

Associational capture is more likely to happen under a specific set of conditions laid out by the epistemic communities and historical institutionalist literatures. Under the conditions of uncertainty induced by a novel situation or the onset of a crisis, policy-makers are more likely to rely on experts and knowledge-based groups.[22] It is in these situations that associations are most likely to hire experts or establish advisory committees that establish channels with transnational knowledge networks. These channels will then tend to facilitate the redefinition of goals and purposes in line with those articulated in the transnational knowledge network. These changes are likely to be stable because they bring in actors who will transmit and reproduce specific forms of knowledge and this is likely to have substantial and lasting effects on associational discourses and purposes.

3. *Institutionalization*: do the pathways of institutionalization, hegemonic imposition, and horizontal change travel to other cases and what other processes of institutionalization must be theorized? In this study, hegemonic imposition and horizontal change provided a vocabulary to trace the pathways by which new purposes came to be embedded in international order. In the balance of power case, European states altered international order horizontally, through a variety of treaties and settlements between 1600 and 1815. A dominant state did not need to impose new purposes or cosmologies on other states because the European great powers had already taken up the new scientific discourses. In the nineteenth century, Britain led a hybrid process of translating already widespread historical, Enlightenment thinking into discursive support for a liberal-colonial order. Thus, it was the British hegemon's ideas that were institutionalized at Berlin and Versailles. In constructing the post-Second World War orders, the American hegemon played a central role in articulating the new growth-based order, spreading its tenets to other associations, and ensuring that its preferred purposes were embedded in international order.

[22] Blyth 2002; Cross 2013; Haas 1989, 1992b; Ikenberry 1992.

The pathways have important necessary conditions. In the case of horizontal change there must be, as pointed out above, a transnational knowledge network to circulate the new discourses amongst the relevant associations. Even cases where hegemons play a central role, the institutionalization of new purposes is only likely to be effective if the cosmological backing for these purposes has already been spread to some degree. In fact, when the British set out to construct a liberal-colonial order in the nineteenth century, other European countries had already encountered aspects of the geological and biological sciences. So, the British project to remap time as linear and progressive found a receptive audience. In the post-Second World War era, states already had experience with industrial modernity, so the backdrop for productivism and growth was already widespread. So I expect hegemonic imposition to be an easier and more effective project when the new purpose it is premised upon is comprehensible and desirable to other states or associations.

In summary, there are three sets of conditions under which I expect a cosmological shift to transform the purposes underlying international order. First, the cosmological shift must inspire many associations to take up the new ideas. Moreover, the breakthrough in knowledge must have implications for political thought that can be transmitted through established channels between knowledge producers and political associations. Second, these ideas must be backed by powerful associations that believe the new cosmological ideas will help them solve problems and reduce uncertainty. In the historical cases, great powers and hegemonic states drove the spread and institutionalization of new cosmologies. However, a powerful transnational social class, network of multinational firms, or other sets of associations can drive broad changes in discourses of state purpose. Third, new purposes are likely to be successfully institutionalized when many associations have taken up the new discourses and the new discourse is either taken for granted amongst a sufficiently large group of associations or a hegemon.[23] That is, if a recursive change is to be stable, there must be raw material for the formation of a cosmological backdrop to naturalize and legitimate the new purposes.

The theoretical architecture and causal conditions could be further refined in comparative cases that examine the role of cosmological ideas in other international systems. A comparative study of cosmology and international order could examine the East Asian systems during the Warring States period (500–200 BCE) or the Ming Dynasty (1300–1600 CE),

---

[23] Here we might, with Snidal (1985), think of a k-group and a hegemon as functionally equivalent. Of course, further research is needed to specify the properties and power of the k-group necessary to lead recursive institutionalization.

the Caliphate and other interpolity orders in the Islamic World (700–1500 CE), warfare among the Maya (250–1500 CE), diplomatic relations between the Iroquois and other American peoples (1400–1600 CE), Europe in the Middle Ages (800–1400 CE), the Greek and Roman states systems (400 BCE to 400 CE), and so on. My theory would expect that the different cosmologies underlying these orders constituted distinct state purposes, which in turn supported different sets of primary and secondary institutions. Moreover, the theory would posit that as cosmological shifts introduced new ideas into the associational basis of these systems, those purposes and institutions would shift.

But I also expect that the mechanisms and processes of recursive institutionalization could be used to illuminate other, non-cosmological cases of change. For example, it could be applied to changes in the security order between 1815 and 2015, the rise of liberalism from 1700 to 1945, the emergence of human rights after the Second World War, and the rise of environmental governance after 1970. In each case, the development of knowledge produced new ideas about ends and purposes with implications for primary and secondary institutions. The conditions just outlined would not apply directly, but the multilevel, middle-range theory could be used to inductively recover the specific causes and conditions in those cases.

Scientific Cosmologies and the Future of International Order

The empirical argument that scientific and technological ideas have formed the basis of international orders since the eighteenth century suggests that future developments in scientific knowledge will continue to shape and reshape international orders. What are the implications of my account for thinking about the future of international order? What cosmological shifts might transform international order in the twenty-first century? In this section I explore these questions by discussing two sets of issues. First, I assess the prospects for cooperation and conflict in a scientific international order. Second, I investigate the possibility that ecological thought in the era of the anthropocene will generate a cosmological shift.

Cooperation and Conflict in Scientific International Systems

Are scientific international orders stable? Are they peaceful or prone to cosmological conflict? Writing in the 1970s, Hedley Bull worried that the rise of non-Western states would shift the basis of international society away from European cultural heritage. This would generate crisis and instability because, for him, international society rested upon common beliefs and values from the European tradition. However, Bull suggested, Western and non-Western states might be able to find common ground in a world culture oriented to scientific and technological modernity:

[T]here is at least a diplomatic or elite culture comprising the common intellectual culture of modernity: some common languages, principally English, a common scientific understanding of the world, certain common notions and techniques that derive from the universal espousal by governments in the modern world of economic development and their universal involvement in modern technology.[24]

In the language I have introduced, Bull hypothesized that a common scientific ontology could support a shared orientation to the purpose of scientific and technological development. However, Bull was concerned that this would not be sufficient to support cooperation because the thin cultural agreement on scientific and technological modernity may not be thick enough to replace the common European outlook that had underwritten international society since the seventeenth century.

On Bull's view, the future of a scientific international order may be chaotic and unruly. By contrast, I have shown that the scientific ideas that constitute international order are rich enough to constitute and naturalize purposes and values. This follows from the fact that scientific ideas can do more than provide common representations or understandings; they can provide cosmological backing for purposive narratives. On this view, a peaceful transition from a European-dominated order to a truly global, cooperative one is possible. However, the claim that scientific and technological modernity is likely to underwrite a peaceful international order is an old and contested one. That is, the implicit claim here seems to fall back on the old functionalist argument that scientific and technological cooperation will spill-over into economic and security domains, driving supranational integration and stable peace.[25]

Gilpin also thought that the international system might come to rest on scientific and technological values, but concluded that this would not eliminate conflict. In rejecting the neo-functionalist argument, Gilpin contended that a scientific international order is the product of coercion and such an order would still be conflictual:

Even if modern science and technology have given mankind a new consciousness of shared values and common problems, this situation is no guarantee of

---

[24] Bull 1977, 305.
[25] For the functionalist and neofunctionalist argument, see Mitrany 1966; Haas 1964. For more nuanced versions, see Adler 1987; Haas 1990.

common interest or of a willingness to subordinate selfish concerns to the larger good. On the contrary, modern science and technology may intensify the conflict over the globe's scarce resources. But it is more important to inquire whether or not a unified humanity really exists. Unfortunately, it does not. The modern "unified world" has been a creation of the West, which has sought to impose its values and way of life on a recalcitrant set of diverse cultures.[26]

For Gilpin, the process of building an international system built on scientific and technological values was a contested, hierarchical process of hegemonic imposition.[27] The spread of cosmological ideas from Western scientific and technological discourses was bound up with the destruction of other "recalcitrant" cosmological traditions.

Gilpin's understanding of the role of scientific and technological ideas has close parallels to the argument advanced here. However, Gilpin's theory remains mired in the problem of peaceful change. In the absence of a theory that carefully links the micro-, meso-, and macro-levels into an account of how power and purpose drive change in governance arrangements, Gilpin was left without a mechanism of change beyond war.[28] So Gilpin, following E.H. Carr, argued that war was necessary to alter governance arrangements under pressure from rising powers.[29] The advent of nuclear weapons has compounded the problem Carr first identified. Under the threat of nuclear war, hegemonic war is either extremely dangerous or unthinkable. In either case, war may be unable to serve its historical function as a mechanism of change. Gilpin did not see a solution except to appeal to "wise and prudent statesmanship" to work it out.[30]

Ikenberry offers one possible solution when he argues that liberal international order is open and flexible enough to accommodate the rise of China and other emerging powers.[31] However, Ikenberry ignores the fact that the neoliberal and democratic norms of the liberal international order delegitimate and thereby exclude China from full participation in the order.[32] Thus, China is more likely to work around the existing order, creating alternative institutions to increase its influence in international order.

However, if international order could be reconstructed on the basis of a shared scientific cosmology, then common purposes and institutions acceptable to both the United States and China could be produced.

---

[26] Gilpin 1981, 225. Gilpin had long been interested in the role of science in foreign policy. See Gilpin 1964, 1968.
[27] Not unlike Ashley's (1989) critical account.
[28] See Ruggie (1982) on this point.
[29] Gilpin 1981, 206–209.
[30] Gilpin 1981, 209.
[31] Ikenberry 2011.
[32] Allan *et al.* 2018.

This is one pathway to peaceful change: the system can remain stable if the American hegemon can use scientific cosmology as a basis for order maintenance. The question then becomes, can the United States harness cosmology or help push a new cosmological shift to maintain and reinvigorate the primary and secondary institutions underlying international order?

If the United States and China are able to forge a peaceful revision of the existing international order, it is likely that cooperation will begin with economic, environmental, and other technical problems in which the issues are structured by a common understanding of the world underwritten by transnational scientific discourses. This is not because, as the functionalists supposed, it is easier to cooperate on technical as opposed to political issues. This presumption is based on the idea that scientific and technological modes of knowledge and practice are not already political and that there is some extra-political domain in which disputes can be settled.[33] Instead, the construction and spread of scientific knowledge and technological innovations is a political process of building alliances, diffusing ideas through institution building and maintenance, and other hard-won political tasks. If the United States and China do find it easier to cooperate on issues like climate change and global health it will be because both share similar cosmological discourses that constitute compatible purposes and goals.

On the other hand, shared cosmology and common purpose may not be sufficient to avoid hegemonic conflict. However, even if China challenges and replaces the US-led hegemonic order, it is likely to build the next order on scientific cosmology.[34] In this scenario, contra Ruggie, a change from American to Chinese hegemony may not lead to a revolution in purpose.[35] After all, while specific scientific and technical beliefs are sometimes challenged (as in climate denial or contestation over economic policies), no state or powerful association questions the idea that the fundamental purpose of the state and of international order is to advance scientific and technological progress. Thus, international order under Chinese hegemony is likely to exhibit considerable continuities with the present order.

There are two possible developments that might disrupt this continuity. First, a global anti-modern, anti-scientific movement could challenge the scientific cosmology at the heart of international order today. This

---

[33] The last sixty years of research in the politics of expertise (Haas 1964), history of science (Shapin and Schaffer 1985), the sociology of scientific knowledge (Bloor 1991 [1976]), and Science and Technology Studies (Latour 1987; Jasanoff 1990) contests this view.
[34] On some ideational barriers to Chinese hegemony, see Allan *et al.* 2018.
[35] Ruggie 1982.

might emerge from dark green environmentalism or religious fundamentalism, both of which contest the basic presuppositions of scientific and technological modernity. Or, it might be carried by populist movements that challenge the epistemic basis of democratic politics. Second, developments within the scientific tradition itself could transform ideas about purpose. This change from within scientific cosmology might emerge from quantum mechanics, experiments at CERN, the biological and complexity sciences, or from recent claims that we are entering a new geological era, the anthropocene.

International Change in the Anthropocene

What are the prospects for a cosmological shift in political discourses today? In August 2016, a working group attached to the International Union of Geological Sciences declared that it would formally propose that the Earth entered a new geological era called the "anthropocene."[36] The anthropocene denotes the fact that humans have emerged as a potent geological force and evidence of human involvement in geological processes will generate lasting stratigraphic markers.[37] That is, future geologists with no knowledge of human society would have to suppose the existence of a scientifically and technologically advanced species in order to explain the geologic record.

For many thinkers, the possibility that we are living in a new geological era has cosmological significance. Latour argues that we are in the midst of a counter-Copernican revolution.[38] Alexandre Koyré famously argued that the Copernican revolution created a cosmological shift by leading a movement from the closed to the infinite universe.[39] In Copernicus' universe, the Earth was transformed from a finite, temporal epicentre into a whirling celestial body in an endless universe. Latour has suggested that the ecological sciences and the discovery of global warming are returning our attention from the infinite, endless heavens back to the finite Earth.[40] The Earth is no longer a rock like any other, but an active,

---

[36] See Waters *et al.* 2016. On the possibility of the biological, complexity, and ecological sciences becoming a dominant cultural force, see Homer-Dixon 2009.
[37] The report of the Anthropocene Working Group (Waters *et al.* 2016) proposes a start date for the anthropocene around 1950 CE. There has been a wide-ranging interdisciplinary debate about the appropriate date. See Crutzen 2002; Foley *et al.* 2013; McNeill 2012; Ruddiman 2013; Steffen *et al.* 2011; Zalasiewicz *et al.* 2009.
[38] Latour 2013, 2014.
[39] Koyré 1957.
[40] "[E]xcept this time," he remarks, "there is no order, no God, no hierarchy, no authority, and thus literally no 'cosmos' ... Let's give this new situation its Greek name, *kakosmos*" (Latour 2014, 4). This raises the point that, in the West at least, cosmology has always

dynamic planet with "tipping points" and "planetary boundaries."[41] In this sense, the anthropocene is of cosmological significance because it redefines humanity's relationship to the planet, time, and the universe.

First, the ecological sciences introduce new ontological and epistemic elements by reconceptualizing physical processes. The newly intimate Earth is not governed by the classical laws of linear mechanism, but by the nonlinear dynamism of the complexity sciences. Since these processes are only partially knowable and predictable, the anthropocene challenges rationalist and modernist epistemic ideas. If the Earth is not fully knowable then it is not fully controllable.

Second, the anthropocene forces a fundamental reconceptualization of time.[42] By embedding humans in geologic time, the anthropocene narrative broadens time horizons. Situating humanity within long time horizons in turn alters our understanding of politics by parochializing the present. In geologic time, we are called to see the present industrial order as a moment in a long history of human–energy–agricultural–political configurations.[43] The anthropocene might also reconfigure our perceptions of how temporal processes unfold. The Enlightenment historical sciences were linear. By contrast, ecology and complexity theory depict a world of active, nonlinear, and chaotic processes.[44] The dominant temporality in Earth systems would not then be the linear development of the Darwinian cosmology, but a dynamic interplay of organic and inorganic matters. Indeed, the anthropocene may require us to grapple with multiple, overlapping temporalities that collide to produce unexpected outcomes.[45]

The arrival of the anthropocene also disrupts our understandings of what it means to be human.[46] Humanity is no longer, as in Darwinian naturalism, a species like any other, subject to the laws of nature. Instead, the idea of humanity as a geological force presents complex questions about the role of humans vis-à-vis the Earth. For some, the central meaning of the anthropocene is that humans are overwhelming and dominating the Earth.[47] This way of framing things shows the anthropocene story

---

been bound up with the idea of a "handsome and *well-composed* arrangement" and to the extent that the Earth is no longer seen as orderly, this would mean the end of traditional cosmology in the West.

[41] Latour 2014, 4.
[42] Foley *et al.* 2013; McNeill 2012.
[43] Chakrabarty 2015.
[44] Connolly 2013; Latour 2014, 4.
[45] Connolly 2013, 2017.
[46] Braje 2015; Chakrabarty 2009, 2015; Connolly 2013; Latour 2014; Steffen *et al.* 2011.
[47] McNeill and Engelke 2015; Rockström *et al.* 2009; Steffen *et al.* 2011; Zalasiewicz *et al.* 2009.

is bound up with rationalist and modernist beliefs about how humans have exerted control over nature. To be sure, the anthropocene concept is usually invoked in order to induce humility, but it still contributes to the sense that humans are a dominant force apart from and over nature.

On the other hand, for some, the ongoing ecological crisis presents a direct challenge to the modernist idea that humans ever controlled or dominated nature.[48] Instead, the anthropocene highlights the insignificance of human life in the long-run of the planet and the universe, the uncontrollable, unpredictable dynamic flows of Earth systems, or the impotence of humans to confront the very problems they induced. On this view, climate change, precipitous declines in biodiversity, global patterns of soil degradation, and other ecological problems call into question the long-term viability of global industrial civilization premised upon the domination of land, air, sea, and space.

Thus, the anthropocene contains two distinct cosmological narratives about scientific and technological progress that were also prominent in the Cold War. In the Cold War, the awesome, destructive power of nuclear weapons highlighted both human mastery and the fragility of humanity at the hands of its own creation. In the anthropocene, the destructive power of industrial civilization itself is at issue. This continuity is not a surprise given that the global environmental movement emerged in the 1970s, in part, as a rejection of the rationalist hubris of the nuclear age. Environmentalists then worked to highlight the dangers of a scientific and technological modernity that had reached its apotheosis in a seemingly unending arms race. It is appropriate then that the Anthropocene Working Group has identified the first nuclear blasts as the best candidate for a stratigraphic marker to start the anthropocene.[49]

The rise of anthropocene thinking in political discourses is likely not just because of its clear and important cosmological implications. States, international organizations, multinational corporations, epistemic communities, and other international associations will have to engage with and deploy ecological thinking. As cascading environmental crises create security problems, intensify food shortages, threaten existing infrastructure, and so on, political associations will increasingly take up ecological thinking in the social and natural sciences in order to deal with serious problems. The ecological sciences are now transnational, and so we can expect for these ideas to circulate horizontally into many states simultaneously. So in the theoretical vocabulary I developed above, a cosmological shift could enter the associational

---

[48] Chakrabarty 2015; Connolly 2013, 2017; Latour 1993, 2013.
[49] Waters *et al.* 2016.

basis of international order through a horizontal change. The American hegemon, as yet, has resisted taking up anthropocene ideas and imposing them on others. It is possible that after a major environmentally induced conflict, the United States, another hegemon, or a hegemonic bloc could build an imposed ecological order on other states. But to do this, ongoing associational changes will have had to generate a cosmological backdrop that could orient associations towards common ecological purposes.

In short, the anthropocene introduces a constellation of new cosmological elements that could be used to redefine state purpose: it transforms humanity's place in the universe by returning humanity to Earth; it redefines the character of the physical elements of the universe in nonlinear, dynamic, and complex terms; it places humanity within geologic time and structures nonlinear narratives of nature; it undermines representations of human mastery over nature. Taken together, these seem to offer the resources for a serious challenge to the growth imperative and the modernist ethos more broadly. Moreover, ecological crises are likely to inspire many associations to take up anthropocene thinking. However, this does not guarantee that a cosmological shift is imminent.

As we have seen, what Scott calls the high modernist imperative entered international politics in the middle of the twentieth century.[50] Thereafter, the core purpose of the state underlying international order has been to harness the power of science and technology in order to advance economic development. The anthropocene concept could be used to challenge the modernist narrative that scientific and technological development drives unending progress. But it is not clear what alternative purpose anthropocene thinking pushes us towards. What should the ends of politics in the anthropocene be? Latour orients us to "fiddling and fixing."[51] Connolly recommends a micropolitics of entangled engagement.[52] But these do not revise the fundamental purpose of the state defined as the goals for which we expect state power to be used. For Biermann, the anthropocene merely necessitates more and better global governance.[53] That it may simply require an adjustment in means rather than ends. These are not radical challenges to the purpose of the state that would require a shift away from growth or productivism.

Moreover, the meaning of the anthropocene will be contested. Recall the anthropocene contains a modernist impulse: humanity has and can control the forces of Earth. But although the Cold War contained both

[50] Scott 1998, 6.
[51] Latour 2013, 2014.
[52] Connolly 2013, 2017.
[53] Biermann *et al.* 2012.

modern and anti-modern cosmological narratives, it reinforced modernism in international order rather than challenged it. So although the anthropocene could be used to challenge the cosmological content of international discourses, that is a contingent possibility that depends on political contestation. Many states and political elites will attempt to extend modernist purposes into the anthropocene. As we saw in the case of quantum mechanics, the development of cosmological shifts is not entirely exogenous to politics. There is a danger the anthropocene could be appropriated for the purposes of extending state control over nature and society.

So, while the anthropocene idea challenges the narrative that humans, and thus the state, can exert control over reality, there is a possibility that the development of the natural sciences will be pushed in a different direction. For example, military agencies throughout the world have already taken an active interest in climate-induced conflict and proposals for geoengineering.[54] The transformation of climate change and other ecological problems into security problems could create demand for a particular kind of natural and social knowledge. The military, already an active player in the production of knowledge, could steer the development of anthropocene knowledge in a modernist direction. This would promote the view that humans must take conscious control over planetary processes. If that happens, then the radical possibilities introduced by anthropocene thinking would be muted and there would not be a cosmological shift that generated purposive change in international order. Nonetheless, as we have seen, there are other ongoing intellectual and political movements that could counter these trends and destabilize the growth-based order. The future of cosmology in international order is likely to emerge from a process of political contestation bound up with cosmological conflict and hegemonic transition.

Social Science as a Vocation

This book suggests that social scientists play a central role in the patterns of knowledge production and contestation shaping the future of cosmology and international order. As I noted above, in each empirical case social scientists operated as brokers translating knowledge from the natural sciences into political goals and purposes. This finding cuts against the Weberian distinction between facts and values in the social sciences. As such, it creates a new way to understand reflexivity and postpositivism in the social sciences.

[54] Floyd 2010.

In his vocation lectures, Weber exhorts scientists of all kinds to refrain from making evaluations or advocating for their personal opinions in their role as scientists.[55] One of Weber's arguments is that it is impossible to derive values from facts because one cannot use scientific methods to adjudicate or evaluate between ideals.[56] Thus, presenting evaluations in the guise of science is dishonest and impractical. However, Weber does allow science an important public role in "the discussion of what means to choose in order to achieve an end that has been definitely *agreed*."[57] For Weber, science can and should serve as a technical means to solve problems. This is bound up with his arguments about rationalization. Weber suggests that science drained meaning from politics because he views science primarily as technical or causal beliefs.

By contrast, I have argued that science operates as a powerful cosmological force that infuses international politics with meaning and purpose. The cases examined here show that the deployment of scientific ideas as means can change ends. When agents deploy scientific or social scientific knowledge, they also import representational constraints and cosmological elements in linked semantic networks. This has unintended effects that can produce new purposes. Thus, scientific and social scientific discourses are more than technical means that help political actors achieve their goals. They place humanity in the cosmos and constitute the ends actors pursue.

Because this is true for both natural and social knowledge, this study challenges the idea that social scientists are neutral observers of the world that they study. Rather, simply by articulating facts, social scientists participate in the construction of values in political discourses. This conclusion may strike social scientists as counterintuitive. Many social scientists are rightly sceptical of their ability to shape policy directly. Indeed, as the literature on ideational change reviewed above suggests, experts are only likely to have major short-run effects on policy under highly specific conditions. However, the argument here shows that over the long run, social and political knowledge operates as a form of productive power that structures the discursive landscape of politics.[58] This indirect route to political relevance means that social scientists have an important role in politics. But that role is less likely to arise from specific technical or causal beliefs than from the ontological, epistemic, and cosmological presuppositions that they channel into political conversations. Social scientists reproduce certain ideas that are then disseminated

---

[55] Weber 2004, 19–23.
[56] Weber 2004, 17, 21.
[57] Weber 2004, 22. Emphasis original.
[58] Haas 1990; Barnett and Duvall 2005.

through teaching, government service, and forays into the public sphere. Social scientific ideas are then picked up, recombined, and institutionalized in unintended and unpredictable ways.

From this vantage point, the deployment of scientific ideas within the social sciences takes on new significance. Appropriations of physics, biology, and mathematical modelling in IR and other social sciences cannot help but reproduce and disseminate ontological, epistemic, and cosmological elements that will have broader effects on goals and purposes. Thus, social scientists should be attentive to and critical of their appropriations of scientific ideas.[59] As we have seen, these ideas come with representational constraints and discursive slopes that tend to reproduce some ways of seeing and valuing the world.

If indeed cosmological developments are central to the politics of the future, then the articulation of new cosmologies is an important political task. As such, social scientists may have a duty not just to be critical of their own cosmological views, but to participate in the construction of new cosmologies. After all, such cosmologies will shape the future of international order in an age of ecological crisis and economic transition. In the tradition of curious cosmologists from Petty to Smith to Solow, the work of social scientists today should include the creative recombination of ideas from a variety of discourses to articulate and defend new purposes.[60]

---

[59] On recent calls for reflexivity in IR, see Hamati-Ataya 2012; Levine 2012; Madsen 2011.
[60] For work in this vein, see Connolly 2013, 2017.

Methodological Appendix

Introduction

In this brief Methodological Appendix, I explain the principles that guided the coding and the text selection for the discourse analyses and process-tracing described at the end of Chapter 2. Coding was designed to proceed in an inductive, interpretivist way. But I wanted to build in checks to my own subjective biases, so I structured my analysis around simple coding rules. This approach was intended to approximate the ethnographic ideal of overhearing conversations.[1] That is, we should aim to do as little violence to interpretive data as possible, allowing subjects to reveal their discursive-phenomenological worlds. While we cannot recover the hidden transcripts of everyday international life, we can study the traces of actions and conversations in letters, speeches, and policy documents. Since evidence of discourse can be found in "[e]verything that is said or written ... everything that is printed or talked about," my gambit was that a wide sample of documents read appropriately would reveal the purposes and cosmologies underlying associations and international orders.[2] The results of the discourse analyses were then woven with elements of process-tracing to uncover the mechanisms of recursive institutionalization.

Methods: Discourse Analysis and Process-Tracing

My argument depends first on establishing change in associations and international orders. I need to show that cosmological shifts changed associational purposes and that these new purposes were institutionalized in the discourses of international order. Evidence for this depends on the recovery of the intersubjective meanings that constitute political discourse. Discourse analysis offers an inductive

---

[1] See Hopf 2007.
[2] Angenot 2004, 200.

method that seeks to realize the ethnographic goal of capturing meanings and purposes as they were deployed. This is important for a historical analysis that seeks to recover the contexts within which actors were embedded. Discourse analysis reveals the relational structures of meaning as they are articulated in textual practices. That is, it maps the connections and hierarchies between the concepts that circulate in political discourse.

To measure discourse, I adapted Hopf's inductive method for the study of national identities and applied it to the recovery of purposes and cosmological elements embedded in discourses.[3] In his study of Soviet identity, Hopf proceeds inductively, listing all the categories of identity that appear in the texts. The result is a long list of "identity categories": modernity, class, urban–rural, ethnicity, and so on. He then moves up the ladder of abstraction to "discursive formations": clusters of concepts that order phenomena and infuse the world with meaning. Likewise, I proceeded inductively, producing long lists of purposes, epistemic concepts, ontological categories, and other cosmological themes revealed in the texts. Although much of what follows is not visible in the chapters, I followed a clear set of rules and strategies to make the analysis more reliable and transparent.

Coding Rules and Strategies

My coding aimed to uncover the basic categories of purposes, claims about the nature of the universe, representations of the world, and claims about true or desirable knowledge. I wanted to keep coding as inductive as possible, so I was guided only by five general sets of questions:

> Institutions and Instruments (means): What types of means should be used to achieve ends? What rules should govern action? What institutions guide conduct?
>
> Purposes (ends): What are the purposes of action? What goals and purposes are visible and legitimate? To what ends are state power to be used?
>
> Cosmology: What is the nature of space and time? What is humanity's place in the universe?
>
> Ontology: How is the world represented? What are the objects in the world and their relations to one another?
>
> Episteme: What is knowledge? How is reliable knowledge produced and used?

---

[3] Hopf 2002a.

Methods 287

I coded all the claims the texts in my sample made about these questions. Thus, coding produced a long list of means, ends, cosmological themes, epistemic elements, and ontological representations.

Thinking in terms of the generative structure, I sought to reconstruct the hierarchical relations between ideas so as to separate which concepts lay beneath chains of justification as foundational ends. For example, the 1957 World Bank policies and operations manual cites the 1941 *Mutual Aid Agreement between the United States and Great Britain* to state a central purpose of the organization: "the expansion, by appropriate international and domestic measures, of production, employment, and the exchange and consumption of goods which are the material foundations of the liberty and welfare of all peoples."[4] Here, a short-term goal of the Bank, expanding production, is naturalized in terms of liberal ends, freedom and peace. This illustrates the inductive nature of the coding. I allow the texts to present categories, but then seek to look beneath them, as it were, to see how those categories were constituted by hierarchies and networks of other concepts.

That is, I sought to recover the relations between categories by constructing simple diagrams of conceptual relations. This was necessary to separate means and ends, because often ends are simply means to some further ends. Constructing a hierarchy of relations between concepts allowed me to determine which concepts lay beneath chains of justification as foundational ends. Consider this example from the 1957 World Bank policies and operations:

> The assessment of repayment prospects involves an exercise of judgement after consideration of a multitude of factors. The availability of natural resources and the existing productive plant within the country are the obvious starting points, but equally important is the capacity of the country concerned to exploit its resources and operate its productive facilities effectively. The judgement required therefore involves, among other things, an evaluation of the government administration and of the business community, the availability of managerial, supervisory, and technical skills.[5]

>> Means: JUDGEMENT
>> Ends: REPAYMENT
>> Episteme: JUDGEMENT (v. scientific analysis)
>> Ontology: NATURAL RESOURCES, PRODUCTIVE CAPACITY, SKILLS

I read this as a claim about the limits of objective means for project selection in the early Bank. An assessment of repayment prospects is stated to be a judgement that analysts make after carefully balancing a number of

---

[4] World Bank 1957, 3.
[5] World Bank 1957, 38.

factors. In this instance, I created a new category under *means* and *episteme* called JUDGEMENT and would code future statements of these themes under this category. Note that I have listed under *ontology* all of the referents referred to in this discussion in order to capture Bank representations. I also coded this statement as a claim about ends: REPAYMENT. Throughout the study, but even more so in the early years, the Bank aimed to safeguard its status as a responsible lender. But this is not a foundational end for the Bank. To see foundational ends or what I think of as purposes, we must look at more rich statements. Consider the example cited above:

[The Allied powers] recognized that, if the peace were to be won, attention would have to be given not only to the immediate relief and physical reconstruction of economies disrupted by the war but also to "the expansion, by appropriate international and domestic measures, of production, employment, and the exchange and consumption of goods which are the material foundations of the liberty and welfare of all peoples."[6]

> Ends: PEACE, RELIEF AND RECONSTRUCTION, PRODUCTIVE CAPACITY, EMPLOYMENT, EXCHANGE, LIBERTY AND WELFARE
> Ontology: ECONOMIES, ALL PEOPLES, EXPANSION (v. growth)

I arranged the ends into a hierarchy: RELIEF AND RECONSTRUCTION + "MATERIAL FOUNDATIONS" (PRODUCTIVE CAPACITY, EMPLOYMENT, EXCHANGE) > LIBERTY AND WELFARE > PEACE. Here, the short-term goals of the Bank (relief and reconstruction and expanding production) are naturalized in terms of liberal ends: freedom and peace. Liberal ends are naturalized, as opposed to explicitly defended, because there is no attempt to explain the liberal basis. My coding also noted some ontological categories of interest, especially if they were not concepts I had seen in earlier periods.

I then translated these lists of categories and justificatory hierarchies into discursive configurations. Discursive configurations are constellations of means, ends, ontology, episteme, and cosmology that hang together, constituting and supporting one another. For example, the configuration supporting the end of economic growth drew on an object-based ontology, a modernist episteme which promises that knowledge can help agents control reality, and a cosmology oriented to scientific and technological progress. These elements are relationally entwined, such that they make each other possible, meaningful, desirable, and

---

[6] World Bank 1957, 3. Quoting the 1942 Mutual Aid Agreement between the United States and Great Britain.

justifiable. The discourse analysis sections of the empirical chapters begin with an overall summary of the dominant discursive configurations before unpacking the constituent epistemic, ontological, and cosmological elements.

Weaving Discourse Analysis and Process-Tracing

The discourse analyses of meso-level associations and macro-level orders provide a synchronic analysis of the relations of meaning. They establish which purposes dominate meso-level and international discourses. By taking these discursive snapshots across time, I establish that change has happened. But to explain that change, we need to conduct a diachronic analysis to show the mechanisms and processes of change in action. For these purposes, I used both primary and secondary sources to conduct process-tracing in the meso-level analyses to identify the key causal mechanisms that changed associational discourses.[7]

There are two sets of mechanisms. First, there are mechanisms that explain meso-level change in associations: strategic deployment and discursive reconfiguration. After the strategic deployment of scientific ideas, I expect that the epistemic, ontological, and cosmological discourses of states and international organizations will be reconfigured. Using primary documents, I looked for articulations that justified new ends with reference to cosmological ideas. In particular, I aimed to uncover traces of the discursive actions that drove the naturalization of new purposes. Second, moving to the macro-level of recursive institutionalization, I aimed to show that the new discourses that emerged in the associational case studies were institutionalized at the macro-level. A full demonstration of associational change and order construction is beyond the scope of the project here. However, I still want to be able to draw the inference that the associational changes played a role in the larger shift in international order. In doing so, I first want to be sure that it is the discursive configurations that arise in the associational case studies that appear in the core sites of international order. Second, I draw on the secondary literature to conclude that, indeed, the British state or the World Bank played an important role in spreading the new purposes.

Text Selection

The validity of discourse analysis depends on the sample of texts used to capture discourses. Text selection at the macro- and meso-level was

---

[7] Mahoney 2003, 360. On process-tracing, see George and Bennett 2005.

guided by two principles. First, I selected a variety of texts designed to capture how concepts are actually deployed and justified in the everyday life of meso-level associations. Second, I selected texts in a systematic way, so as not to bias the sample in favour of the theoretical argument. Where the universe of texts was small, I selected and coded all the available documents. Where the universe of texts was large, I built a database of available texts and randomly selected a small sample to analyse.

Chapter 3, 1550

I chose three kinds of texts to capture the central associational and institutional discourses of international order in 1550: international treaties, state papers and memoirs of political elites from England and France, and nonfiction texts from intellectual circles. I selected England and France, excluding the Habsburg dynasty for reasons of linguistic accessibility, because they would capture the most important associations in European politics. The challenge was in finding translated texts, and there are fewer primary documents available for the Habsburgs in this period. The purpose for selecting treaties and state papers is straightforward: they correspond directly to interstate institutions (law) and state agencies (policy-making and decision-making in state agencies). However, the intellectual background knowledge of the European elite is also an important site for the reproduction of political discourses. Reading audiences were small and printing presses were closely allied with centres of political power: dynastic states, the Church, and powerful merchants. So I also created a database of documents meant to capture this generalized intellectual discourse that surrounds and informs the political discourses underlying European political order.[8]

In selecting texts, I relied on the Early English Books Online database (http://chadwyck.eebo.com). Although ostensibly focused on English books, it contains a number of French texts that had been translated into English. To find treaties, I simply pulled all the treaties in the database by putting in the names of countries as the author name. I looked at a random sample of five of these treaties. To study elite memoirs and intellectual discourse, I set the search terms to include only English or French texts, set the date range to 1525–1575 and selected randomly 100 documents (fifty English, fifty French) from the results.

I have modernized the spelling in the quotations that appear in the text. The page numbers in the citations refer to image numbers in the database

---

[8] I am grateful to Kathryn Botto for research assistance with this.

scans, as many texts either did not have page numbers or they did not appear in the scans.

Chapter 3, 1660–1713

My intention in this section was to show the rise of interests in international politics. The goal was not to trace a specific scientific idea or effect on international order, but to provide an interlude between 1550 and 1815, to show the transitional phase. Following Bartelson and Skinner, I traced this through the French texts (Richelieu, Rohan, Louis XIV) into the Treaty of Utrecht.[9] I consider both French elite and the documents around Utrecht to be key sites of international order because they carry and codify the central political discourses of the period. Utrecht is considered by Osiander to be the moment where the role of interests and the balance were first institutionalized in international order. For an analysis of Utrecht, I read the treaty, Bolingbroke's memoirs, and Torcy's memoirs. This provides an admittedly limited snapshot of international order at this time, but this case is not central to my overall arguments.

Chapter 3, 1815

My goal here was to map text selection onto key associations and sites of international order in 1814–1815. First, it is important to look at international law, such as the Final Act of the Vienna Congress. However, it is important to add other sources, since international law captures only a partial view. Since international order from 1650 to 1850 is dominated by a transnational aristocratic elite, the memoirs and letters of that elite are central sites for the recovery of meanings in international order. For my analysis of the Congress, I read all the central memoirs and letters I could find in English translations. Since this ended up only being 1,000 or so pages of documents, I worked through it all, coding for means, ends, cosmology, episteme, and ontology. I also worked through the records of the proceedings that appear in the *British and Foreign State Papers* and d'Angeberg's collection.

Chapter 4, British Colonial Office

Evidence for the argument was based on primary documents from the Colonial Office, Foreign Office, and Cabinet Office. I have focused the analysis on the British colonial policy in Africa under the auspices

---

[9] Bartelson 1995; Skinner 1996.

of the Colonial Office to make the analysis empirically tractable. This excludes colonial policy in India, which was administered by the separate India Office, and Egypt, which was overseen by the Foreign Office. For my analysis, I studied all documents relevant to the history of African colonies in collections of primary documents edited by historians. For the period 1860–1900, the main source is correspondence between officials on the ground and the Foreign Office in the *Foreign Office Confidential Print*. I read all the documents relevant to African colonial policy between 1860 and 1885. For the period 1900–1950, the main source is correspondence between Colonial Office officials collected in the series *British Documents on the End of Empire*. I read all the documents relevant to African colonial policy in two volumes therein.

Finally, I wanted to see how ideas moved from the Colonial Office to the British executive government, so I looked at primary documents from the Cabinet. All documents from the Cabinet Papers are from the British National Archives online database. I selected ten documents manually based on citations in secondary texts and ten documents randomly from the 147 documents that met a search for "colonial development" for the period 1910–1950.

Monographs and academic articles that shaped colonial policy were selected based on what was deemed significant in the secondary literature.

Chapter 4, Conference of Berlin, 1885

I read all the conference proceedings collected in Gavin (1973). These reflect the British documentary record.

Chapter 4, League of Nations, 1920–1935

The *Official Journal of the League of Nations* contains a "procès-verbal" for all the sessions of the League. One can follow conversations between delegates naturally and easily. I randomly chose 200 pages of procès-verbal at five-year intervals from 1920 to 1935.

Chapter 5, World Bank

First, I structured my sample to capture a variety of sites of Bank discourse. The goal was to capture both the official voice of the Bank and the ideas that generated its policies on the ground. For example, if I only analysed flagship publications like the *World Development Reports* I would uncover the public voice of the Bank, but not its day-to-day operations. To ensure a balanced sample I included "country reports" that review the economy of a country and set spending priorities for project lending as well as operational reports from the actual projects of the Bank. In sum, I analysed three types of documents:

1. Official policies and publications: to capture the core policies and public voice of the Bank in the global public sphere.
2. Country reports: to capture how the Bank perceived, represented, and analysed the domestic policies of borrowers.[10]
3. Project and structural adjustment loan documents: to capture how the Bank actually spent its money on specific projects.

Second, I used randomized text selection techniques to guard against bias. Documents in the "official" category were not randomly selected. For the years before the Bank produced the *World Development Report* (until 1978) I merely selected an important document such as the annual report, Board of Governors meetings summary, official policies document, or a particularly general research report (analogous to the *World Development Report*). After 1978, I just selected the *World Development Report* unless the secondary literature identified an important report produced in that year. These reports were the Berg report (1981), the Wapenhans report (1992), and the *East Asian Miracle* report (1993). For the country reports and project documents, I created large databases of all publicly available texts. I then used a random number generator to select texts from these databases. In the case of project documents, I wanted to ensure that my sample included some structural adjustment loans, which are classified differently in Bank archives. Therefore, I made a separate database for those and randomly selected fifteen reports to read.

In total, I studied fifty official reports (one per year), fifty country reports (twenty-five from 1950 to 1975 and twenty-five from 1975 to 2000), and 100 project reports (fifty from 1950 to 1975 and fifty from 1975 to 2000, including fifteen structural adjustment loan reports). Not all of these reports are cited in the text, but they informed the conclusions.

Chapter 5, International Order 2015

The rationale for the text selection is described in the chapter. I made a list of all the important international organizations and selected a flagship publication from each institution. The analysis of the UN Security Council was performed by Michael Ki Hoon Hur. The coding results are available by request.

---

[10] These are "Economic & Sector Reports" in Bank terminology.

References

Abbott, George C. 1971. "A Reconsideration of the 1929 Colonial Development Act." *Economic History Review* Vol. 24, No. 1: 68–81.
Adas, Michael. 1989. *Machines as the Measure of Men: Science, Technology and Ideologies of Western Dominance*. Ithaca, NY: Cornell University Press.
Adler, Emanuel. 1987. *The Power of Ideology: The Quest for Technological Autonomy*. Berkeley: University of California Press.
　1992. "The Emergence of Cooperation: National Epistemic Communities and the International Evolution of the Idea of Nuclear Arms Control." *International Organization* Vol. 46, No. 1: 101–146.
Adler, Emanuel and Steven Bernstein. 2005. "Knowledge in Power: The Epistemic Construction of Global Governance." In Michael Barnett and Raymond Duvall, eds. *Power in Global Governance*. Cambridge: Cambridge University Press, pp. 294–317.
Adler, Emanuel and Vincent Pouliot 2011. *International Practices*. Cambridge: Cambridge University Press.
Adler-Nissen, Rebecca and Kristoffer Kropp. 2015. "A Sociology of Knowledge Approach to European Integration: Four Analytical Principles." *Journal of European Integration* Vol. 37, No. 2: 155–173.
Alacevich, Michele. 2009. *The Political Economy of the World Bank: The Early Years*. Stanford: Stanford University Press and the World Bank.
Allan, Bentley B. 2017a. "Producing the Climate: States, Scientists, and the Emergence of Governance Objects." *International Organization* Vol. 71, No. 1: 131–162.
　2017b. "From Subjects to Objects: Knowledge in International Relations Theory." *European Journal of International Relations* OnlineFirst DOI: 10.1177/1354066117741529.
Allan, Bentley B., Srdjan Vucetic, and Ted Hopf. 2018. "The Distribution of Identity and the Future of International Order: China's Hegemonic Prospects." *International Organization* Vol. 72, No. 3.
Amadae, S.M. 2003. *Rationalizing Capitalist Democracy*. Chicago: University of Chicago Press.
Andersen, Morten Skumrad. 2016. A Genealogy of the Balance of Power. PhD Dissertation, London School of Economics.
Anderson, M.S. 1970. "Eighteenth Century Theories of the Balance of Power." In Ragnhild Hatton and M.S. Anderson, eds. *Studies in Diplomatic*

History: Essays in Memory of David Bayne Horn. Hamden: Archon, pp. 183–198.
  1993. *The Rise of Modern Diplomacy*. New York: Longman.
  1998. *Origins of the Modern European State System, 1494–1618*. New York: Longman.
Angenot, Marc. 2004. "Social Discourse Analysis: Outlines of a Research Project." *Yale Journal of Criticism* Vol. 17, No. 2: 199–215.
Anghie, Anthony. 2004. *Imperialism, Sovereignty and the Making of International Law*. Cambridge: Cambridge University Press.
Anon. 1538. Here begynneth the pystles and gospels, of eury Sonday … Available at: http://eebo.chadwyck.com/.
  1542. Heuy newes of an horryble earth quake … Available at: http://eebo.chadwyck.com/.
  1557. A commyssion sent to the bloudy butcher byshop of London … Available at: http://eebo.chadwyck.com/.
Asad, Talal. 1973. *Anthropology and the Colonial Encounter*. New York: Humanity Books.
Ascher, William. 1983. "New Development Approaches and the Adaptability of International Agencies: The Case of the World Bank." *International Organization* Vol. 37, No. 3: 415–439.
  1996. "The Evolution of Postwar Doctrines in Development Economics." *History of Political Economy* Vol. 28, S1: 312–336.
Ashley, Richard K. 1989. "Imposing International Purpose: Notes on a Problematic of Governance." In Ernst-Otto Czempiel and James Rosenau, eds. *Global Changes and Theoretical Challenges: Approaches to World Politics for the 1990s*. Lexington, MA: Lexington Books, pp. 251–290.
Ashton, S.R. and S.E. Stockwell, eds. 1996. *Imperial Policy and Colonial Practice 1925–1945*, 2 Vols. British Documents on the End of Empire, Series A, Vol. 1. London: HMSO.
Ashworth, Lucian. 2014. *A History of International Thought: From the Origins of the Modern State to Academic International Relations*. London: Routledge.
Asian Development Bank (ADB). 2015a. *Asian Development Outlook: Supplement*. Manila: Asian Development Bank.
Asian Development Bank (ADB). 2015b. *Together We Deliver 2014: Knowledge and Partnerships to Results*. Manila: Asian Development Bank.
Atran, Scott. 2016. "The Devoted Actor: Unconditional Commitment and Intractable Conflict across Cultures." *Current Anthropology* Vol. 57, No. S13: S192–S203.
Atran, Scott and Robert Axelrod. 2008. "Reframing Sacred Values." *Negotiation Journal* Vol. 24, No. 3: 221–246.
Ayres, Robert L. 1983. *Banking on the Poor: The World Bank and World Poverty*. Cambridge, MA: MIT Press.
Babb, Sarah. 2009. *Behind the Development Banks: Washington Politics, World Poverty, and the Wealth of Nations*. Chicago: University of Chicago Press.
Backhouse, Roger E. 2010. "Economics." In Roger E. Backhouse and Philippe Fontaine, eds. *History of the Social Sciences Since 1945*. Cambridge: Cambridge University Press, pp. 38–70.

Backhouse, Roger E. and Philippe Fontaine. 2010. "Toward a History of the Social Sciences." In Roger E. Backhouse and Philippe Fontaine, eds. *History of the Social Sciences Since 1945*. Cambridge: Cambridge University Press, pp. 185–223.
Bacon, Francis. 2012 [1620]. *The Great Instauration*. New York: Start Publishing.
Bain, William. 2003. *Between Anarchy and Society: Trusteeship and the Obligations of Power*. Oxford: Oxford University Press.
Balassa, Bela. 1971a. *The Structure of Protection in Developing Countries*. Baltimore: Johns Hopkins University Press.
  1971b. "Effective Protection: A Summary Appraisal." Washington, DC: World Bank Report No. SWP101.
Baldwin, David. 1985. *Economic Statecraft*. Princeton: Princeton University Press.
Baldwin, George. 1986. "Economics and Economists in the World Bank." In A.W. Coats, ed. *Economists in International Agencies: An Exploratory Study*. New York: Praeger, pp. 67–90.
Bannister, Mark. 2000. *Condé in Context: Ideological Change in Seventeenth Century France*. Oxford: Legenda.
Barkin, Samuel and Bruce Cronin. 1994. "The State and the Nation: Changing Norms and the Rules of Sovereignty in International Relations." *International Organization* Vol. 48, No. 1: 107–130.
Barnett, Michael N. 1997. "Bringing in the New World Order: Liberalism, Legitimacy, and the United Nations." *World Politics* Vol. 49, No. 4: 526–551.
Barnett, Michael. 2011. *Empire of Humanity: A History of Humanitarianism*. Ithaca, NY: Cornell University Press.
Barnett, Michael and Raymond Duvall. 2005. *Power in Global Governance*. Cambridge: Cambridge University Press.
Barnett, Michael and Martha Finnemore. 1999. "The Politics, Power, and Pathologies of International Organizations." *International Organization* Vol. 53, No. 4: 699–732.
  2004. *Rules for the World*. Ithaca, NY: Cornell University Press.
Bartelson, Jens. 1995. *A Genealogy of Sovereignty*. Cambridge: Cambridge University Press.
  2009. *Visions of World Community*. Cambridge: Cambridge University Press.
Barth, Fredrik. 1987. *Cosmologies in the Making: A Generative Approach to Cultural Variation in Inner New Guinea*. Cambridge: Cambridge University Press.
Basu, Paul. 2015. "N.W. Thomas and Colonial Anthropology in British West Africa: Reappraising a Cautionary Tale." *Journal of the Royal Anthropological Institute* Vol. 22, No. 1: 84–107.
Bebbington, Anthony, Scott Guggenheim, Elizabeth Olson, and Michael Woolcock. 2004. "Exploring Social Capital Debates at the World Bank." *The Journal of Development Studies* Vol. 40, No. 5: 33–64.
Bell, Duncan, ed. 2007. *Victorian Visions of Global Order: Empire and International Relations in Nineteenth Century Political Thought*. Cambridge: Cambridge University Press.
Bell, Duncan. 2016. *Reordering the World: Essays on Liberalism and Empire*. Princeton: Princeton University Press.

Bell, Duncan and Caspar Sylvest. 2006. "International Society in Victorian Political Thought." In T.H. Green, Herbert Spencer, and Henry Sidgwick, eds. *Modern Intellectual History* Vol. 3, No. 2: 207–238.
Bennett, Jane. 2001. *The Enchantment of Modern Life: Attachments, Crossings, and Ethics*. Princeton: Princeton University Press.
Berger, Peter and Thomas Luckmann. 1966. *The Social Construction of Reality*. New York: Anchor Books.
Berk, Gerald and Dennis Galvan 2009. "How People Experience and Change Institutions: A Field Guide to Creative Syncretism." *Theory and Society* Vol. 38: 543–580.
Berry, Christopher J. 1997. *Social Theory of the Scottish Enlightenment*. Edinburgh: Edinburgh University Press.
Bertalanffy, Ludwig von. 1951. "Towards a Physical Theory of Organic Teleology: Feedback and Dynamics." *Human Biology* Vol. 23, No. 4: 346–361.
Best, Jacqueline. 2005. *The Limits of Transparency: Ambiguity and the History of International Finance*. Ithaca, NY: Cornell University Press.
  2008. "Hollowing Out Keynesian Norms: How the Search for a Technical Fix Undermined the Bretton Woods Regime." In John G. Ruggie, ed. *Embedding Global Markets: An Enduring Challenge*. London: Ashgate, pp. 47–69.
  2014. *Governing Failure: Provisional Expertise and the Transformation of Global Development Finance*. Cambridge: Cambridge University Press.
Bevir, Mark. 2002. "Sydney Webb: Utilitarianism, Positivism, and Social Democracy." *The Journal of Modern History* Vol. 74, No. 2: 217–252.
Beze, Theodore. 1969 [1572]. "Right of Magistrates." In *Constitutionalism and Resistance in the Sixteenth Century: Three Treatises by Hotman, Beze, and Mornay*. Trans. Julian H. Franklin. New York: Pegasus.
Bhagwati, Jagdish. 1978. *Foreign Trade Regimes and Economic Development: Anatomy and Consequences of Exchange Control Regimes*. New York: National Bureau of Economic Research.
  1986. "Export Promoting Trade Strategy: Issues and Evidence." Washington, DC: World Bank Report No. ERS7.
Bhagwati, Jagdish and T.V. Srinivasan. 1978. "Trade Policy and Development." Washington, DC: World Bank Report No. REP90
Bially Mattern, Janice. 2005. *Ordering International Politics: Identity, Crisis and Representational Force*. New York: Routledge.
Biermann, F. *et al.* 2012. "Navigating the Anthropocene: Improving Earth System Governance." *Science* Vol. 335, No. 6074: 1306–1307.
Black, Jeremy. 1983. "The Theory of the Balance of Power in the First Half of the Eighteenth Century: A Note on Sources." *Review of International Studies* Vol. 9, No. 1: 55–61.
Blanning, Tim. 2007. *The Pursuit of Glory: Europe 1648–1815*. London: Allen Lane/Penguin.
Blaser, Mario. 2013. "Ontological Conflicts and the Stories of Peoples in Spite of Europe." *Current Anthropology* Vol. 54, No. 5: 547–568.
Block, Fred L. 1977. *The Origins of International Economic Disorder*. Berkeley: University of California Press.
Bloor, David. 1991 [1976]. *Knowledge and Social Imagery*, Second Edition. Chicago: University of Chicago Press.

Blyth, Mark. 2002. *Great Transformations: Economic Ideas and Institutional Change in the 20th Century*. Cambridge: Cambridge University Press.
Boaistuau, Pierre. 1581 [1558]. Theatrum mundi. Available at: http://eebo.chadwyck.com/.
Boianovsky, Mauro and Kevin D. Hoover. 2009. "The Neoclassical Growth Model and Twentieth-Century Economics." *History of Political Economy* Vol. 41, S1: 1–23.
  2014. "In The Kingdom of Solovia: The Rise of Growth Economics at MIT, 1956–70." *History of Political Economy* Vol. 46, S1: 199–227.
Bolingbroke, Henry Saint-John. 1932 [1735–1736]. *Defence of the Treaty of Utrecht*. Ed. G.M. Trevelyan. Cambridge: Cambridge University Press.
Bonneuil, Christophe. 2001. "Development as Experiment: Science and State Building in Late Colonial and Postcolonial Africa, 1930–1970." *Osiris* Vol. 15: 258–281.
Bonney, Richard. 1991. *The European Dynastic States: 1494–1660*. Oxford: Oxford University Press.
Bonney, Richard, ed. 1999. *The Rise of the Fiscal State in Europe, c.1200–1815*. Oxford: Oxford University Press.
Boot, Max. 2006. *War Made New: Technology, Warfare, and the Course of History*. New York: Gotham Books.
Bordo, Michael D. and Barry Eichengreen. 2007. *A Retrospective on the Bretton Woods System*. Chicago: University of Chicago Press.
Borge-Holthoefer, Javier and Alex Arenas. 2010. "Semantic Networks: Structure and Dynamics." *Entropy* Vol. 12, No. 5: 1264–1302.
Bowden, Brian. 2009. *The Empire of Civilization: The Evolution of an Imperial Idea*. Chicago: Chicago University Press.
Bowler, Peter J. 1976. "Malthus, Darwin, and the Concept of Struggle." *Journal of the History of Ideas* Vol. 37, No. 4: 631–650.
Bowler, Peter J. and Iwan Rhys Morus. 2005. *Making Modern Science*. Chicago: University of Chicago Press.
Boyle, Robert. 1666. "Tryals Proposed by Mr. Boyle to Dr. Lower, to be Made by Him, for the Improvement of Transfusing Blood out of One Live Animal into Another." *Philosophical Transactions of the Royal Society* Vol. 1, No. 22: 385–388.
Braje, Todd. 2015. "Earth Systems, Human Agency, and the Anthropocene: Planet Earth in the Human Age." *Journal of Archaeological Research* Vol. 23, No. 4: 369–396.
Branch, Jordan. 2013. *The Cartographic State: Maps, Territory, and the Origins of Sovereignty*. Cambridge: Cambridge University Press.
Brandom, Robert. 1994. *Making It Explicit*. Cambridge, MA: Harvard University Press.
  2000. *Articulating Reasons*. Cambridge, MA: Harvard University Press.
Brantley, Cynthia. 1997. "Kikuyu-Maasai Nutrition and Colonial Science: The Orr and Gilks Study in Late 1920s Kenya Revisited." *The International Journal of African Historical Studies* Vol. 30, No. 1: 49–86.
Braumoeller, Bear. 2013. *The Great Powers and the International System: Systemic Theory in Empirical Perspective*. Cambridge: Cambridge University Press.

Breslau, Daniel. 2003. "Economics Invents the Economy: Mathematics, Statistics, and Models in the Work of Irving Fisher and Wesley Mitchell." *Theory and Society* Vol. 32, No. 3: 379–411.
Briggs, Robin. 1999. "Embattled Faiths: Religion and Natural Philosophy in the Seventeenth Century." In Evan Cameron, ed. *Early Modern Europe*. Oxford: Oxford University Press, pp. 171–205.
British and Foreign State Papers, 1814–15, Vol. 2. London: James Ridgway and Sons.
Broad, Robin. 2006. "Research, Knowledge, and 'the Art of Paradigm Maintenance': The World Bank's Development Economics Vice-Presidency." *Review of International Political Economy* Vol. 13, No. 3: 387–419.
Brougham, Henry. 1872 [1803]. "Balance of Power." In *Works of Henry Lord Brougham, Vol. VIII: Dissertations – Historical and Political Dissertations*. Edinburgh: Adam and Charles Black, pp. 1–50.
Brown, William. 2016. Learning to Colonize: State Knowledge, Expertise, and the Making of the First French Empire, 1661–1715. PhD Dissertation, Johns Hopkins University.
Buchwald, Jed and Mordecai Feingold. 2013. *Newton and the Origin of Civilization*. Princeton: Princeton University Press.
Bueger, Christian and Frank Gadinger. 2015. "The Play of International Practice." *International Studies Quarterly* Vol. 59, No. 3: 449–460.
Buell, Raymond. 1928. *The Native Problem in Africa*, 2 Vols. New York: Macmillan.
Bueno de Mesquita, Bruno, James D. Morrow, Randolph M. Siverson, and Alastair Smith. 2003. *The Logic of Political Survival*. Cambridge, MA: MIT Press.
Bukovansky, Mlada. 2002. *Legitimacy and Power Politics: The American and French Revolutions in International Political Culture*. Princeton: Princeton University Press.
Bull, Hedley. 1977. *The Anarchical Society: A Study of Order in World Politics*. New York: Columbia University Press.
Burgin, Angus. 2012. *The Great Persuasion: Reinventing Free Markets since the Depression*. Cambridge, MA: Harvard University Press.
Burke, Peter. 1992. *The Fabrication of Louis XIV*. New Haven: Yale University Press.
Butterfield, Herbert. 1966. "The Balance of Power." In Herbert Butterfield and Martin Wight, eds. *Diplomatic Investigations*. Cambridge, MA: Harvard University Press, pp. 132–148.
Buzan, Barry. 2004. *From International to World Society? English School Theory and the Social Structure of Globalisation*. Cambridge: Cambridge University Press.
Buzan, Barry and George Lawson. 2015. *The Global Transformation: History, Modernity and the Making of International Relations*. Cambridge: Cambridge University Press.
CAB 24/7. Food (War) Committee. "The National Food Policy." 16 March 1947.
  24/158. Secretary of State for the Colonies and the President of the Board of Trade. "Proposals for Financial Assistance to Accelerate the Development of Imperial Resources." 1923.
  24/173. Secretary of the State for the Colonies. "Report of the East Africa Commission." April 1925.

24/223. Economy Committee. "Report." 28 August 1931.
24/250. Economic Advisory Council. "Second Report of Committee on Scientific Research: The Need for Improved Nutrition of the People of Great Britain." 9 July 1934.
129/9. Chancellor of the Exchequer. "Import Programme for Mid-1946 to Mid-1947." 18 April 1946.
129/32. Chancellor of the Exchequer. "Economic Survey for 1949." 16 February 1949.
Calvin, Jean. 1561 [1536]. The Institution of the Christian Religion ... Available at: http://eebo.chadwyck.com/.
Campbell, John L. and Ove K. Pedersen. 2001. *The Rise of Neoliberalism and Institutional Analysis*. Princeton: Princeton University Press.
  2014. *The National Origins of Policy Ideas: Knowledge Regimes in the United States, France, Germany, and Denmark*. Princeton: Princeton University Press.
Cantor, Geoffrey. 1983. *Optics After Newton*. Manchester: Manchester University Press.
Carpenter, R. Charli. 2007. "Setting the Advocacy Agenda: Theorizing Issue Emergence and Nonemergence in Transnational Advocacy Networks." *International Studies Quarterly* Vol. 51, No. 1: 99–120.
Carson, Carol S. 1975. "The History of the United States National Income and Product Accounts: The Development of an Analytical Tool." *Review of Income and Wealth* Vol. 21, No. 2: 153–181.
Carvalho, Benjamin. 2016. "The Making of the Political Subject: Subjects and Territory in the Formation of the State." *Theory and Society* Vol. 45, No. 1: 57–88.
Cassirer, Ernst. 1963. *The Individual and the Cosmos in Renaissance Philosophy*. Trans. Mario Domandi. Chicago: University of Chicago Press.
Castlereagh, Viscount. 1853. *Correspondence, Despatches, and Other Papers of Viscount Castlereagh*, Vol. 10. Ed. Charles William Vane. London.
  1947 [1816]. "Circular Dispatch." *Foreign Office*, January 1, 1816. In Charles K. Webster. *The Foreign Policy of Castlereagh, 1815–1822*. London: G. Bell & Sons, pp. 509–512.
Catholic Church. 1556. This prymer of Salisbury vse is se tout along with houtonyser chyng [sic] ... Available at: http://eebo.chadwyck.com/.
Centeno, Miguel and Joseph N. Cohen. 2012. "The Arc of Neoliberalism." *Annual Review of Sociology* Vol. 38: 317–340.
Chakrabarty, Dipesh. 2000. *Provincializing Europe: Postcolonial Thought and Historical Difference*. Princeton: Princeton University Press.
  2009. "The Climate of History: Four Theses." *Critical Inquiry* Vol. 35, No. 2: 197–222.
  2015. "The Anthropocene and the Convergence of Histories." In Clive Hamilton, Christophe Bonneuil, and Francois Gemenne, eds. *The Anthropocene and the Global Environmental Crisis*. London: Routledge, pp. 44–56.
Chatterjee, Partha. 1986. *Nationalist Thought and the Colonial World: A Derivative Discourse*. Minnesota: University of Minnesota Press.
Checkel, Jeffrey. 1999. "Norms, Institutions and National Identity in Contemporary Europe." *International Studies Quarterly* Vol. 43, No. 1: 83–114.

Chedsey, Master. 1551. *A reporte of maister doctor Redmans answeres ...* Available at: http://eebo.chadwyck.com/.

Chenery, Hollis B. 1975. "The Structural Approach to Development Policy." Washington, DC: World Bank Report No. REP20.

Chenery, Hollis B., Montek S. Ahluwalia, C.L.G. Bell, John H. Duloy, and Richard Jolly. 1974. *Redistribution with Growth*. New York: Oxford University Press.

Chwieroth, Jeffrey M. 2008. "Organizational Change 'From Within': Exploring the World Bank's Early Lending Practices." *Review of International Political Economy* Vol. 15, No. 4: 481–505.

2010. *Capital Ideas: The IMF and the Rise of Financial Liberalization*. Princeton: Princeton University Press.

Clark, George. 1970. "From the Nine Years War to the War of the Spanish Succession." In J.S. Bromley, ed. *New Cambridge Modern History, Vol. 6: The Rise of Great Britain and Russia*. Cambridge: Cambridge University Press, pp. 381–409.

Clegg, Liam. 2010. "Our Dream is a World Full of Poverty Indicators: The U.S., the World Bank, and the Power of Numbers." *New Political Economy* Vol. 15, No. 4: 473–492.

Cole, Charles W. 1939. *Colbert and a Century of French Mercantilism*, 2 Vols. New York: Columbia University Press.

Collingwood, R.W. 1945. *The Idea of Nature*. Oxford: Clarendon Press.

Collins, Robert. 2000. *More: The Politics of Economic Growth in Postwar America*. Oxford: Oxford University Press.

Condorcet, Marquis de. 1795. *Outlines of an Historical View of the Human Mind*. London: J. Johnson.

Conklin, Alice. 1997. *A Mission to Civilize: The Republican Idea of Empire in France and the West*. Stanford: Stanford University Press.

Connolly, William E. 1982. *Political Theory and Modernity*. Ithaca, NY: Cornell University Press.

2011. *A World of Becoming*. Durham, NC: Duke University Press.

2013. *The Fragility of Things*. Durham, NC: Duke University Press.

2017. *Facing the Planetary*. Durham, NC: Duke University Press.

Constantine, Stephen. 1984. *The Making of British Colonial Development Policy, 1914–1940*. London: Frank Cass.

Conway, Flo and Jim Siegelman. 2005. *Dark Hero of the Information Age: In Search of Norbert Wiener, The Father of Cybernetics*. New York: Basic Books.

Cooper, Frederick. 1989. "From Free Labor to Family Allowances: Labor and African Society in Colonial Discourse." *American Ethnologist* Vol. 16, No. 4: 475–765.

1997. "Modernizing Bureaucrats, Backward Africans, and the Development Concept." In Frederick Cooper and Randall Packard, eds. *International Development and the Social Sciences: Essays on the History and Politics of Knowledge*. Berkeley: University of California Press, pp. 64–92.

Cooter, Roger. 1979. "The Power of the Body: The Early Nineteenth Century." In Barry Barnes and Steven Shapin, eds. *Natural Order: Historical Studies of Scientific Culture*. London: Sage, pp. 73–92.

Copernicus, Nicolaus. 1543. *De Revolutionus Orbium Coelestium*. Norimbergae apud Ioh: Petreium.

Corrozet, Gilles and Hans Holbein. 1549. The images of the Old Testament lately expressed ... Available at: http://eebo.chadwyck.com/.
Corry, Olaf. 2013. *Constructing a Global Polity: Theory, Discourse and Governance.* New York: Palgrave Macmillan.
Cortell, Andrew and James W. Davis Jr. 1996. "How Do International Institutions Matter? The Domestic Impact of International Rules and Norms." *International Studies Quarterly* Vol. 40, No. 4: 451–478.
Council of Trent. 1965 [1545]. "Decrees of the Council of Trent 1545." In Milton Viorst, ed. *The Great Documents of Western Civilization.* New York: Bantam, pp. 103–105.
Coverdale, Miles. 1550. The psalter or psalmes of Dauid after the translacion of the greate Bible ... Available at: http://eebo.chadwyck.com/.
Cowen, M.P. and R.W. Shenton. 1996. *Doctrines of Development.* New York: Routledge.
Cox, Robert. 1987. *Production, Power, and World Order: Social Forces in the Making of History.* New York: Columbia University Press.
Coyle, Diane. 2014. *GDP: A Brief but Affectionate History.* Princeton: Princeton University Press.
Crafts, Nicholas. 2001. "Historical Perspectives on Development." In Gerald M. Meier and Joseph E. Stiglitz, eds. *Frontiers of Development Economic: The Future in Perspective.* New York: The World Bank and Oxford University Press, pp. 301–333.
Crawford, Neta. 2002. *Argument and Change in World Politics.* Cambridge: Cambridge University Press.
Crombie, A.C. and Michael Hoskin. 1970. "The Scientific Movement and the Diffusion of Scientific Ideas." In J.S. Bromley, ed. *New Cambridge Modern History, Vol. 6: The Rise of Great Britain and Russia.* Cambridge: Cambridge University Press, pp. 37–71.
Crook, Paul. 1994. *Darwinism, War, and History: The Debate Over the Biology of War from the "Origin of Species" to the First World War.* Cambridge: Cambridge University Press.
Cross, Mai'a Davis. 2013. "Rethinking Epistemic Communities Twenty Years Later." *Review of International Studies* Vol. 39, No. 1: 137–160.
Crutzen, Paul. 2002. "Geology of Mankind." *Nature* Vol. 415, No. 23: 23.
Culwick, A.T. and G.M. Culwick. 1935. "Culture Contact on the Fringe of Civilization." *Africa* Vol. 8, No. 2: 163–170.
Cunningham, Andrew and Perry Williams. 1993. "De-centring the 'Big Picture': The Origins of Modern Science and the Modern Origins of Science." *British Journal of the History of Science* Vol. 26, No. 4: 407–432.
Czempiel, Ernst-Otto and James N. Rosenau, eds. 1989. *Global Changes and Theoretical Challenges: Approaches to World Politics for the 1990s.* Lexington, MA: Lexington Books.
D'Andrade, Roy. 1995. *The Development of Cognitive Anthropology.* Cambridge: Cambridge University Press.
d'Angeberg, Comte. 1864. *Le Congrès de Vienne et les Traités de 1815, Tome Deuxième.* Paris: Archives Diplomatiques.
Daly, Herman. 1996. *Beyond Growth: The Economics of Sustainable Development.* Boston: Beacon Press.

References

Darwin, Charles. 1861. *On the Origin of Species*, 3rd ed. London: John Murray.
Darwin, John. 2009. *The Empire Project: The Rise and Fall of the British World-System, 1830–1970.* Cambridge: Cambridge University Press.
Daston, Lorraine. 1995. "The Moral Economy of Science." *Osiris* Vol. 10, No. 1: 2–24.
Daston, Lorraine, ed. 2000. *Biographies of Scientific Objects.* Chicago: University of Chicago Press.
Daston, Lorraine and Peter Galison. 2007. *Objectivity.* New York: Zone Books.
de Beaugue, Jean. 1556. *History of the Campagnes.* Edinburgh.
de Castro, Eduardo Viveiros. 1998. "Cosmological Deixis and Amerindian Perspectivism." *Journal of the Royal Anthropology Institute* Vol. 4, No. 3: 469–488.
de la Cadena, Marisol. 2010. "Indigenous Cosmopolitics in the Andes: Conceptual Reflections beyond 'Politics'." *Cultural Anthropology* Vol. 25, No. 2: 334–370.
de Swaan, Abram. 1998. *In Care of the State: Health Care, Education and Welfare in Europe and the USA in the Modern Era.* Cambridge: Polity Press.
Deaton, Angus. 2001. "Counting the World's Poor: Problems and Possible Solutions." *The World Bank Research Observer* Vol. 16, No. 2: 125–147.
   2010. "Price Indexes, Inequality, and the Measurement of World Poverty." In Anand Sughin, Paul Segal, and Joseph Stiglitz, eds. *Debates on the Measurement of Global Poverty.* Oxford: Oxford University Press, pp. 187–222.
DeSombre, Elizabeth R. 2000. *Domestic Sources of International Environmental Policy: Industry, Environmentalists, and U.S. Power.* Cambridge, MA: MIT Press.
Desrosières, Alain. 1998. *The Politics of Large Numbers: A History of Statistical Reasoning.* Trans. Camille Naish. Cambridge, MA: Harvard University Press.
Destro, Anna Maria. 2010. "Cosmology and Mythology." In H. James Blix, ed. *21st Century Anthropology: A Reference Handbook.* London: Sage, pp. 227–234.
Deudney, Daniel. 2007. *Bounding Power: Republican Security Theory From the Polis to the Global Village.* Princeton: Princeton University Press.
Devetak, Richard. 2011. "Law of Nations as Reason of State: Diplomacy and the Balance of Power in Vattel's Law of Nations." *Parergon* Vol. 28, No. 2: 105–128.
Dewey, John. 1922. *Human Nature and Conduct: An Introduction to Social Psychology.* New York: Modern Library.
   1929. *The Quest for Certainty.* New York: G.P. Putnam's Sons.
DiMaggio, Paul J. and Walter W. Powell. 1983. "The Iron Cage Revisited: Institutional Isomorphism and Collective Rationality in Organizational Fields." *American Sociological Review* Vol. 48, No. 2: 147–160.
Dimier, Veronique. 2006. "Three Universities and the British Elite: A Science of Colonial Administration in the UK." *Public Administration* Vol. 84, No. 2: 337–366.
Donnelly, Jack. 2009. "Rethinking Political Structures: From 'Ordering Principles' to 'Vertical Differentiation'—and Beyond." *International Theory* Vol. 1, No. 1: 49–86.
   2012. "The Differentiation of International Societies: An Approach to Structural International Theory." *European Journal of International Relations* Vol. 18, No. 1: 151–176.

Doty, Roxanne. 1996. *Imperial Encounters: The Politics of Representation in North-South Relations*. Minneapolis: University of Minnesota Press.

Douglas, Mary. 1986. *How Institutions Think*. New York: Syracuse University Press.

Drezner, Daniel. 2001. "State Structure, Technological Leadership and the Maintenance of Hegemony." *Review of International Studies* Vol. 27, No. 1: 3–25.

Drezner, Daniel W. and Kathleen R. McNamara. 2013. "International Political Economy, Global Financial Orders and the 2008 Financial Crisis." *Perspectives on Politics* Vol. 11, No. 1: 155–166.

Drori, Gili, John Meyer, Fransisco Ramirez, and Evan Schofer. 2003. *Science in the Modern World Polity: Institutionalization and Globalization*. Stanford: Stanford University Press.

Düppe, T. and R. Weintraub. 2014. "Siting the New Economic Science: The Cowles Commission's Activity Analysis Conference of June 1949." *Science in Context* Vol. 27, No. 3: 453–483.

Earle, Edward Mead. 1986. "Adam Smith, Alexander Hamilton, and Friedrich List: The Economic Foundations of Military Power." In Peter Paret, ed. *The Makers of Modern Strategy*. Princeton: Princeton University Press, pp. 217–261.

Easterly, William. 1999. "The Ghost of Financing Gap: Testing the Growth Model Used in the International Financial Institutions." *Journal of Development Economics* Vol. 60, No. 2: 423–438.

Edwards, Paul N. 1996. *The Closed World: Computers and the Politics of Discourse in Cold War America*. Cambridge, MA: MIT Press.

Ehrich, Cyril. 1973. "Building and Caretaking: Economic Policy in British Tropical Africa, 1890–1960." *The Economic History Review* Vol. 26, No. 4: 649–667.

Eichengreen, Barry. 1996. *Globalizing Capital: A History of the International Monetary System*. Princeton: Princeton University Press.

Eldridge, C.C. 1973. *England's Mission: The Imperial Idea in the Age of Gladstone and Disraeli, 1868–1880*. Chapel Hill: The University of South Carolina Press.

Elliott, J.H. 1968. *Europe Divided, 1559–1598*. New York: Harper & Row.

Elyot, Thomas. 1539. The castel of helth ... Available at: http://eebo.chadwyck.com/.

Engel, Susan. 2012. *The World Bank and the Post-Washington Consensus in Vietnam and Indonesia*. Hoboken: Taylor & Francis.

Engerman, David C. 2004. "The Romance of Economic Development: New Histories of the Cold War." *Diplomatic History* Vol. 28, No. 1: 23–54.

England and Wales. 1554. Anno Mariae primo actes made in the Parliamente begonne and holden at Westminster, the seconde daye of Apryll, in the firste yeare of the reygne of oure moste gratious soueraygne ladye ... Available at: http://eebo.chadwyck.com/.

Estienne, Henri. 1576. Ane meruellous discours vpon the lyfe, deides, and behauiours of Katherine de Medicis ... Available at: http://eebo.chadwyck.com/.

Evangelista, Matthew. 1999. *Unarmed Forces: The Transnational Movement to End the Cold War*. Ithaca, NY: Cornell University Press.

Evrigenis, Ioannis. 2014. *Images of Anarchy: The Rhetoric and Science in Hobbes's State of Nature*. Cambridge: Cambridge University Press.

Fearon, James. 1995. "Rationalist Explanations for War." *International Organization* Vol. 49, No. 3: 379–414.
Fearon, James and Alexander Wendt. 2002. "Rationalism v. Constructivism: A Skeptical View." In Walter Carlsnaes, Thomas Risse-Kappen, and Beth A. Simmons, eds. *Handbook of International Relations*. London: Sage, pp. 52–72.
Ferguson, James. 1990. *The Anti-Politics Machine: "Development," Depoliticization, and Bureaucratic Power in Lesotho*. Minneapolis: University of Minnesota Press.
Ferrier, Auger. 1593 [1549]. A learned astronomical discourse ... Available at: http://eebo.chadwyck.com/.
Final Act of the Vienna Congress. 1815. In T.C. Hansard, superintendent. *The Parliamentary Debates from the Year 1803 to the Present Time*, Vol. 32. London.
Finnemore, Martha. 1993. "International Organizations as Teachers of Norms: The United Nations Educational, Scientific, and Cultural Organization and Science Policy." *International Organization* Vol. 47, No. 4: 565–597.
  1996a. "Norms, Culture, and World Politics: Insights from Sociology's Institutionalism." *International Organization* Vol. 50, No. 2: 325–347.
  1996b. *National Interests in International Society*. Ithaca, NY: Cornell University Press.
  1997. "Redefining Development at the World Bank." In Frederick Cooper and Randall Packard, eds. *International Development and the Social Science*. Berkeley: University of California Press, pp. 203–227.
  2003. *The Purpose of Intervention*. Ithaca, NY: Cornell University Press.
  2009. "Legitimacy, Hypocrisy, and the Social Structure of Unipolarity: Why Being a Unipole Isn't All It's Cracked Up to Be." *World Politics* Vol. 61, No. 1: 58–85.
Finnemore, Martha and Kathryn Sikkink. 1998. "International Norm Dynamics and Political Change." *International Organization* Vol. 52, No. 4: 887–917.
Fioretos, Orfeo. 2011. "Historical Institutionalism in International Relations." *International Organization* Vol. 65, No. 2: 367–399.
Fioretos, Orfeo, Tulia G. Falleti, and Adam Sheingate. 2015. "Historical Institutionalism in Political Science." In Orfeo Fioretos, Tulia G. Falleti, and Adam Sheingate, eds. *The Oxford Book of Historical Institutionalism*. Oxford: Oxford University Press, pp. 3–28.
Floyd, Rita. 2010. *Security and the Environment: Securitisation Theory and US Environmental Security Policy*. Cambridge: Cambridge University Press.
Foley, Stephen F. *et al.* 2013. "The Palaeoanthropocene: The Beginnings of Anthropogenic Environmental Change." *Anthropocene* Vol. 3: 83–88.
Forman, Paul. 1987. "Behind Quantum Electronics: National Security as Basis for Physical Research in the United States, 1940–1960." *Historical Studies in the Physical and Biological Sciences* Vol. 18, No. 1: 149–229.
Fortes, M. 1936. "Culture Contact as a Dynamic Process: An Investigation in the Northern Territories of the Gold Coast." *Africa* Vol. 9, No. 1: 24–55.
Fortun, M. and S.S. Schweber. 1993. "Scientists and the Legacy of World War II: The Case of Operations Research (OR)." *Social Studies of Science* Vol. 23, No. 4: 595–642.
Foster, John Bellamy. 1992. "The Absolute General Law of Environmental Degradation under Capitalism." *Capitalism Nature Socialism* Vol. 3, No. 3: 77–81.

2015. "Marxism and Ecology: Common Fonts of a Great Transition." In *The Great Transition Initiative*. Available at: www.greattransition.org/publication/marxism-and-ecology.
Foucault, Michel. 1970. *The Order of Things*. New York: Vintage.
1972. *The Archaeology of Knowledge and the Discourse on Language*. Trans. A.M. Sheridan Smith. New York: Pantheon.
2007. *Security, Territory, Population: Lectures at the Collège de France, 1977–78*. New York: Palgrave Macmillan.
Fourcade, Marion. 2006. "The Construction of a Global Profession: The Transnationalization of Economics." *American Journal of Sociology* Vol. 112, No. 1: 145–194.
2009. *Economists and Societies: Discipline and Profession in the United States, Britain, and France, 1890s to 1990s*. Princeton: Princeton University Press.
Fourcade-Gourinchas, Marion and Sarah L. Babb. 2002. "The Rebirth of the Liberal Creed: Paths to Neoliberalism in Four Countries." *American Journal of Sociology* Vol. 108, No. 3: 533–579.
Frängsmyr, Tore, J.L. Heilbron, and Robin E. Rider. 1990. *The Quantifying Spirit in the Eighteenth Century*. Cambridge: Cambridge University Press.
Friedberg, Aaron. 2000. *In the Shadow of the Garrison State: America's Antistatism and its Cold War Grand Strategy*. Princeton: Princeton University Press.
Frieden, Jeffry. 1991. "Invested Interests: The Politics of National Economic Policies in a World of Global Finance." *International Organization* Vol. 45, No. 4: 425–452.
Friedland, Nehemia, Giora Keinan, and Yechiela Regev. 1992. "Controlling the Uncontrollable: Effects of Stress on Illusory Control." *Journal of Personality and Social Psychology* Vol. 63, No. 6: 923–931.
Fuchs, Christopher A. 2016. "On Participatory Realism." Available at: arXiv:1601.04360.
Galison, Peter. 1994. "The Ontology of the Enemy: Norbert Wiener and the Cybernetic Vision." *Critical Inquiry* Vol. 21, No. 1: 228–266.
Gallarotti, Giulio M. 2000. "The Advent of the Prosperous Society: The Rise of the Guardian State and Structural Change in the World Economy." *Review of International Political Economy* Vol. 7, No. 1: 1–52.
Galvan, Dennis C. 2004. *The State Must Be Our Master of Fire: How Peasants Craft Culturally Sustainable Development in Senegal*. Berkeley: University of California Press.
Gardner, Richard. 1956. *Sterling-Dollar Diplomacy: Anglo-American Collaboration in the Reconstruction of Multilateral Treaties*. Oxford: Clarendon Press.
Gaukroger, Stephen. 2001. *Francis Bacon and the Transformation of Early-Modern Philosophy*. Cambridge: Cambridge University Press.
2007. *The Emergence of a Scientific Culture: Science and the Shaping of Modernity 1210–1685*. Oxford: Clarendon Press.
2010. *The Collapse of Mechanism and the Rise of Sensibility*. Oxford: Oxford University Press.
Gavin, R.J. 1973. *The Scramble for Africa: Documents on the Berlin West African Conference and Related Subjects, 1884–85*. Ibadan: Ibadan University Press
George, Alexander L. and Andrew Bennett. 2005. *Case Studies and Theory Development in the Social Sciences*. Cambridge, MA: MIT Press.

Gerovitch, Slava. 2002. *From Newspeak to Cyberspeak: A History of Soviet Cybernetics*. Cambridge, MA: MIT Press.
Giddens, Anthony. 1984. *The Constitution of Society: Outline of the Theory of Structuration*. Cambridge: Polity Press.
Gillispie, Charles Goulston. 1951. *Genesis and Geology: A Study of the Relations of Scientific Thought, Natural Theology, and Social Opinion in Great Britain, 1790–1850*. Cambridge, MA: Harvard University Press.
Gilman, Nils. 2003. *Mandarins of the Future: Modernization Theory in Cold War America*. Baltimore: Johns Hopkins University Press.
Gilpin, Robert. 1964. "Natural Scientists in Policy-Making." In Robert Gilpin and Christopher Wright, eds. *Scientists and National Policy Making*. New York: Columbia University Press, pp. 1–18.
  1968. *France in the Age of the Scientific State*. Princeton: Princeton University Press.
  1981. *War and Change in World Politics*. Cambridge: Cambridge University Press.
Goddard, Stacie. 2009. "Brokering Change: Networks and Entrepreneurs in International Politics." *International Theory* Vol. 1, No. 2: 249–281.
  2010. *Indivisible Territory and the Politics of Legitimacy: Jerusalem and Northern Ireland*. Cambridge: Cambridge University Press.
Gorman, Daniel. 2012. *The Emergence of International Society in the 1920s*. Cambridge: Cambridge University Press.
Graham, Loren R. 1993. *Science in Russia and the Soviet Union*. Cambridge: Cambridge University Press.
Grandin, Greg. 2006. *Empire's Workshop*. New York: Henry Holt and Company.
Gray, William Glenn. 2007. "What Did the League Do, Exactly?" *International History Spotlight* Vol. 1. Available at: www.h-net.org/~diplo/IHS.
Grieco, Joseph M. 1988. "Anarchy and the Limits of Cooperation: A Realist Critique of the Newest Liberal Institutionalism." *International Organization* Vol. 42, No. 3: 487–507.
Group of 7 (G7). 2015. "Leaders' Declaration, G7 Summit, 7–8 June 2015." Schloss Elmau, Germany.
Grovogui, Siba N. 1996. *Sovereigns, Quasi Sovereigns, and Africans: Race and Self-Determination in International Law*. Minneapolis: University of Minnesota Press.
  2002. "Regimes of Sovereignty: Rethinking International Morality and the African Condition." *European Journal of International Relations* Vol. 8, No. 3: 315–338.
Grynaviski, Eric. 2014. *Constructive Illusions: Misperceiving the Origins of International Cooperation*. Ithaca, NY: Cornell University Press.
Guerlac, Henry. 1986. "Vauban: The Impact of Science on War." In Peter Paret, ed. *The Makers of Modern Strategy*. Princeton: Princeton University Press, pp. 64–89.
Guha, Ranajit. 1997. *Dominance Without Hegemony: History and Power in Colonial India*. Cambridge, MA: Harvard University Press.
Guillaume, Xavier. 2009. "From Process to Politics." *International Political Sociology* Vol. 3, No. 1: 71–86.
Gulick, Edward V. 1955. *Europe's Classical Balance of Power*. Ithaca, NY: Cornell University Press.
  1970. "The Final Coalition and the Congress of Vienna, 1813–1815." In C. W. Crawley, ed. *The New Cambridge Modern History, Volume 9: War and Peace*

308 References

in an Age of Upheaval, 1793–1830. Cambridge: Cambridge University Press, pp. 639–667.

Guzzini, Stefano. 2013. *The Return of Geopolitics in Europe? Social Mechanisms and Foreign Policy Identity*. Cambridge: Cambridge University Press.

2016. "International Political Sociology, or: The Social Ontology and Power Politics of Process." DIIS Working Paper 2016: 6. Copenhagen: Danish Institute for International Studies.

Haas, Ernst. 1964. *Beyond the Nation-State: Functionalism and International Organization*. Stanford: Stanford University Press.

1990. *When Knowledge is Power: Three Models of Change in International Organizations*. Berkeley: University of California Press.

Haas, Peter M. 1989. "Do Regimes Matter? Epistemic Communities and Mediterranean Pollution Control." *International Organization* Vol. 43, No. 3: 377–403.

1992a. "Introduction: Epistemic Communities and International Policy Coordination." *International Organization* Vol. 46, No. 1: 1–35.

1992b. "Banning Chlorofluorocarbons: Epistemic Community Efforts to Protect Stratospheric Ozone." *International Organization* Vol. 46, No. 1: 187–224.

2014. "Ideas, Experts and Governance." In Monika Ambrus, Karin Arts, Ellen Hey, and Helena Raulus, eds. *The Role of "Experts" in International Decision-Making: Advisors, Decision-Makers or Irrelevant*. Cambridge: Cambridge University Press.

Habermas, Jürgen. 1970. *Toward A Rational Society: Student Protest, Science, and Politics*. Trans. Jeremy J. Shapiro. Boston: Beacon Press.

1984. *The Theory of Communicative Action, Vol. 1: Reason and the Rationalization of Society*. Trans. Thomas McCarthy. Boston: Beacon Press.

1987. *The Theory of Communicative Action, Vol. 2: System and Lifeworld*. Trans. Thomas McCarthy. Boston: Beacon Press.

Hacking, Ian. 1975. *The Emergence of Probability*. Cambridge: Cambridge University Press.

1990. *The Taming of Chance*. Cambridge: Cambridge University Press.

Hailey, Lord. 1938. *An African Survey: A Study of Problems Arising in Africa South of the Sahara*. London: Oxford University Press.

1979 [1940–1942]. *Native Administration and Political Development in British Tropical Africa*. Nendeln/Liechtenstein: Kraus Reprint.

Haldén, Peter. 2013. "Republican Continuities in the Vienna Order and the German Confederation (1815–66)." *European Journal of International Relations* Vol. 19, No. 2: 281–304.

Hall, Peter. 1989. *The Political Power of Economic Ideas: Keynesianism Across Nations*. Princeton: Princeton University Press.

1993. "Policy Paradigms, Social Learning, and the State: The Case of Economic Policymaking in Britain." *Comparative Politics* Vol. 25, No. 3: 275–296.

Hall, Peter and Rosemary Taylor. 1996. "Political Science and the Three New Institutionalisms." *Political Studies* Vol. 44, No. 5: 936–957.

Hall, Rodney Bruce. 1999. *National Collective Identity: Social Constructs and International Systems*. New York: Columbia University Press.

Hamati-Ataya, Inanna. 2012. "Reflectivity, Reflexivity, Reflexivism: IR's 'Reflexive Turn' – and Beyond." *European Journal of International Relations* Vol. 19, No. 4: 669–694.

Hanchard, Michael. 2003. "Acts of Misrecognition: Transnational Black Politics, Anti-imperialism and the Ethnocentrisms of Pierre Bourdieu and Loïc Wacquant." *Theory, Culture & Society* Vol. 20, No. 4: 5–29.

Haslam, Jonathan. 2002. *No Virtue Like Necessity: Realist Though in International Relations since Machiavelli*. New Haven: Yale University Press.

Helleiner, Eric. 1994. *States and the Reemergence of Global Finance: From Bretton Woods to the 1990s*. Ithaca, NY: Cornell University Press.

  2006. "Reinterpreting Bretton Woods: International Development and the Neglected Origins of Embedded Liberalism." *Development and Change* Vol. 36, No. 5: 943–967.

Herbert, Sandra. 2011. *Charles Darwin and the Question of Evolution: A Brief History with Documents*. New York: St. Martin's Press.

Heussler, Robert. 1963. *Yesterday's Rulers: The Making of the British Colonial Service*. Syracuse: Syracuse University Press.

Heyck, Hunter. 2015. *Age of System: Understanding the Development of Modern Social Science*. Baltimore: Johns Hopkins University Press.

Heyck, T.W. 1982. *The Transformation of Intellectual Life in Victorian England*. New York: St. Martin's Press.

Hicks, Alexander. 1999. *Social Democracy and Welfare Capitalism*. Ithaca, NY: Cornell University Press.

Hicks, Norman L. 1979. "Growth vs. Basic Needs: Is There a Trade-Off?" Washington, DC: World Bank Report No. REP139.

Hobbes, Thomas. 1994 [1668]. *Leviathan*. Indianapolis: Hackett.

Hodge, Joseph Morgan. 2007. *Triumph of the Expert: Agrarian Doctrines of Development and the Legacies of British Colonialism*. Athens: Ohio University Press.

Holbraad, Carsten. 1970. *The Concert of Europe: A Study in German and British International Theory, 1815–1914*. London: Longman.

Holsti, K.J. 2004. *Taming the Sovereigns*. Cambridge: Cambridge University Press.

Homer-Dixon, Thomas. 2006. *The Upside of Down*. Toronto: Alfred A. Knopf.

  2009. "The Newest Science." *Alternatives Journal* Vol. 35, No. 4: 8–11.

Homer-Dixon, Thomas, Jonathan Leader Maynard, Matto Mildenberger, Manjana Milkoreit, Steven J. Mocka, Stephen Quilley, Tobias Schröder, and Paul Thagard. 2013. "A Complex Systems Approach to the Study of Ideology: Cognitive-Affective Structures and the Dynamics of Belief Systems." *Journal of Social and Political Psychology* Vol. 1, No. 1: 337–363.

Hopf, Ted. 2002a. *The Social Construction of International Politics*. Ithaca: Cornell University Press.

  2002b. "Making the Future Inevitable: Legitimizing, Naturalizing and Stabilizing. The Transition in Estonia, Ukraine and Uzbekistan." *European Journal of International Relations* Vol. 8, No. 3: 403–436.

  2007. "The Limits of Interpreting Evidence." In Richard Ned Lebow and Mark Lichbach, eds. *Theory and Evidence in Comparative Politics and International Relations*. New York: Palgrave Macmillan, pp. 55–84.

2010. "The Logic of Habit in International Relations." *European Journal of International Relations* Vol. 16, No. 4: 539–561.

2012. *Reconstructing the Cold War*. Oxford: Oxford University Press.

2013. "Commonsense Constructivism and Hegemony in World Politics." *International Organization* Vol. 67, No. 2: 317–354.

Hopf, Ted and Bentley B. Allan. 2016. *Making Identity Count: Building a National Identity Database*. Oxford: Oxford University Press.

Horkheimer, Max and Theodor Adorno. 2002. *Dialectic of Enlightenment: Philosophical Fragments*. Trans. Edmund Jephcott. Stanford: Stanford University Press.

Hornick, Philip Von. 1932 [1684]. "Austria over all if she will." In Arthur Monroe, ed. *Early Economic Thought*. Cambridge, MA: Harvard University Press, pp. 221–244.

Horowitz, Michael C. 2010. *The Diffusion of Military Power: Causes and Consequences for International Relations*. Princeton: Princeton University Press.

Howarth, David. 2000. *Discourse*. Buckingham: Open University Press.

Howe, Anthony. 2007. "Free Trade and Global Order: The Rise and Fall of a Victorian Vision." In Bell, Duncan, ed. *Victorian Visions of Global Order: Empire and International Relations in Nineteenth Century Political Thought*. Cambridge: Cambridge University Press, pp. 26–46.

Howell, Signe. 2002. "Cosmology." In Alan Bernard and Jonathan Spencer, eds. *Routledge Encyclopedia of Social and Cultural Anthropology*. London: Routledge, pp. 196–199.

Huber, Evelyne and John D. Stephens. 2001. *Development and Crisis of the Welfare State*. Chicago: University of Chicago Press.

Hubrigh, Joachim. 1553. [An almanacke and prognostication]. Available at: http://eebo.chadwyck.com/.

Hughes, Helen. 1973. "Trade and Industrialization Policies: The Political Economy of the Second Best." Washington, DC: World Bank Report No. SWP143.

Hunter, Monica. 1934. "Methods of Study of Culture Contact." *Africa* Vol. 7, No. 3: 335–350.

Hurd, Elizabeth. 2015. *Beyond Religious Freedom: The New Global Politics of Religion*. Princeton: Princeton University Press.

Hurd, Ian. 1999. "Legitimacy and Authority in International Politics." *International Organization* Vol. 53, No. 2: 379–408.

Hürni, Bettina. 1980. *The Lending Policy of the World Bank in the 1970s: Analysis and Evaluation*. Boulder: Westview Press.

Hyam, Ronald, ed. 1992. *The Labour Government and the End of Empire 1945–51, 5 Vols. British Documents on the End of Empire*, Series A, Vol. 2. London: HMSO.

Hyam, Ronald. 2010. *Understanding the British Empire*. Cambridge: Cambridge University Press.

Hymans, Jacques E. 2012. *Achieving Nuclear Ambitions: Scientists, Politicians, and Proliferation*. Cambridge: Cambridge University Press.

Ikenberry, G. John. 1992. "A World Economy Restored: Expert Consensus and the Anglo-American Postwar Settlement." *International Organization* Vol. 46, No. 1: 289–321.

References

2001. *After Victory: Institutions, Strategic Restraint, and the Rebuilding of Order After Major Wars*. Princeton: Princeton University Press.

2011. *Liberal Leviathan: The Origins, Crisis, and Transformation of the American World Order*. Princeton: Princeton University Press.

Inayatullah, Naeem and David L. Blaney. 2004. *International Relations and the Problem of Difference*. London: Routledge.

International Monetary Fund (IMF). 2015. *World Economic Outlook: April 2015*. Washington, DC.

Isaacman, Allen F. and Barbara Isaacman. 1976. *The Tradition of Resistance in Mozambique: The Zambesi Valley, 1850–1921*. Berkeley: University of California Press.

Jackson, Patrick Thaddeus. 2006a. *Civilizing the Enemy: German Reconstruction and the Invention of the West*. Ann Arbor: University of Michigan Press.

2006b. "Making Sense of Making Sense: Configurational Analysis and the Double Hermeneutic." In Dvora Yanow and Peregrine Schwartz-Shea, eds. *Interpretation and Method: Empirical Research Methods and the Interpretive Turn*. New York: M.E. Sharpe, pp. 264–280.

Jackson, Patrick Thaddeus and Daniel Nexon 1999. "Relations Before States: Substance, Process and the Study of World Politics." *European Journal of International Relations* Vol. 5, No. 3: 291–332.

Jasanoff, Sheila. 1990. *The Fifth Branch*. Cambridge, MA: Harvard University Press.

Jasanoff, Sheila, ed. 2004. *States of Knowledge: The Co-Production of Science and Social Order*. London: Routledge.

Jepperson, Ronald and John W. Meyer. 2011. "Multiple Levels of Analysis and the Limitations of Methodological Individualism." *Sociological Theory* Vol. 29, No. 1: 54–73.

Jervis, Robert. 1985. "From Balance to Concert: A Study of International Security Cooperation." *World Politics* Vol. 38, No. 1: 58–79.

Joannes. 1554. The discription of the contrey of Aphrique ... Available at: http://eebo.chadwyck.com/.

Johnson, Howard. 1977. "The West Indies and the Conversion of the British Official Classes to the Development Idea." *Journal of Commonwealth and Comparative Politics* Vol. 15, No. 1: 55–83.

Kahler, Miles. 2002. "Bretton Woods and Its Competitors: The Political Economy of Institutional Choice." In David M. Andrews, C. Randall Henning, and Louis W. Pauly, eds. *Governing the World's Money*. Ithaca, NY: Cornell University Press, pp. 38–59.

Kallis, Giorgos. 2017. *In Defense of Degrowth: Opinions and Manifestos*. Ed. Aaron Vansintjan. Open Commons.

Kalpagam, U. 2000. "The Colonial State and Statistical Knowledge." *History of the Human Sciences* Vol. 13, No. 2: 37–55.

Kant, Immanuel. 1991. *Kant: Political Writings*. Ed. H.S. Reiss. Cambridge: Cambridge University Press.

Kapur, Devesh, John P. Lewis, and Richard C. Webb. 1997. *The World Bank: Its First Half Century, Volume 1: History*. Washington, DC: Brookings Institution Press.

Kardam, Nüket. 1993. "Development Approaches and the Role of Policy Advocacy: The Case of the World Bank." *World Development* Vol. 21, No. 11: 1773–1786.
Katzenstein, Peter J. 1996. *Cultural Norms and National Security: Police and Military in Postwar Japan.* Ithaca, NY: Cornell University Press.
Katznelson, Ira. 1996. "Knowledge About What? Policy Intellectuals and the New Liberalism." In Dietrich Rueschemeyer and Theda Skocpol, eds. *States, Social Knowledge, and the Origins of Modern Social Policies.* Princeton: Princeton University Press, pp. 17–47.
  1997. "Structure and Configuration in Comparative Politics." In Mark Irving Lichbach and Alan S. Zuckerman, eds. *Comparative Politics: Rationality, Culture, and Structure.* Cambridge: Cambridge University Press, pp. 81–112.
Kaufman, Stuart, Richard Little, and William Wohlforth, eds. 2007. *The Balance of Power in World History.* New York: Palgrave.
Keck, Margaret and Kathryn Sikkink. 1998. *Activists Beyond Borders.* Ithaca, NY: Cornell University Press.
Keene, Edward. 2002. *Beyond the Anarchical Society.* Cambridge: Cambridge University Press.
  2005. *International Political Thought: A Historical Introduction.* Cambridge: Polity.
  2007. "A Case Study of the Construction of International Hierarchy: British Treaty-Making Against the Slave Trade in the Early Nineteenth Century." *International Organization* Vol. 61, No. 2: 311–339.
Keens-Soper, Maurice. 1978. "The Practice of a States-System." In Michael Donelan, ed. *The Reason of States: A Study in International Political Theory.* London: George Allen & Unwin, pp. 25–44.
Keinan, Giora. 1994. "Effects of Stress and Tolerance of Ambiguity on Magical Thinking." *Journal of Personality and Social Psychology* Vol. 67: 48–55.
Keinan, Giora and Dalia Sivan. 2001. "The Effects of Stress and Desire for Control on the Formation of Causal Attributions." *Journal of Research in Personality* Vol. 35, No. 2: 127–137.
Keohane, Robert. 1984. *After Hegemony.* Princeton: Princeton University Press.
Kertzer, Josh. 2017. "Microfoundations in International Relations." *Conflict Management and Peace Science* Vol. 34, No. 1: 81–97.
Kessler, Oliver and Xavier Guillaume. 2015. "Everyday Practices of International Relations: People in Organizations." *Journal of International Relations and Development* Vol. 15, No. 1: 110–119.
Killingray, David. 1986. "The Maintenance of Law and Order in British Colonial Africa." *African Affairs* Vol. 85, No. 340: 411–437.
Kirk-Greene, Anthony. 1999. *On Crown Service: A History of HM Colonial and Overseas Civil Services.* London: I.B. Tauris.
  2006. *Symbol of Authority: The British District Officer in Africa.* London: I.B. Tauris.
Kirshner, Jonathan. 1999. "The Political Economy of Realism." In Ethan B. Kapstein and Michael Mastanduno, eds. *Unipolar Politics.* New York: Columbia University Press, pp. 69–102.
Kissinger, Henry. 1957. *A World Restored.* Boston: Houghton Mifflin.
  1994. *Diplomacy.* New York: Simon & Schuster.

Kline, Ronald R. 2015. *The Cybernetics Moment, Or Why We Call Our Age the Information Age*. Baltimore: Johns Hopkins University Press.

Koenigsberger, H.G. 1987. *Early Modern Europe, 1500–1789*. New York: Longman.

Konkel, Rob. 2014. "The Monetization of Global Poverty: The Concept of Poverty in World Bank History, 1944–90." *Journal of Global History* Vol. 9, No. 2: 276–300.

Koslowski, Rey and Friedrich V. Kratochwil. 1994. "Understanding Change in International Politics: The Soviet Empire's Demise and the International System." In Richard Ned Lebow and Thomas Risse-Kappen, eds. *International Relations Theory and the End of the Cold War*. New York: Columbia University Press, pp. 127–166.

Koyré, Alexandre. 1957. *From the Closed to the Infinite Universe*. Baltimore: Johns Hopkins University Press.

Kraehe, Enno E. 1992. "A Bipolar Balance of Power." *American Historical Review* Vol. 97, No. 3: 707–715.

Kragh, Helge. 2007. *Conceptions of Cosmos*. Oxford: Oxford University Press.

Krasner, Stephen D. 1991. "Global Communications and National Power: Life on the Pareto Frontier." *World Politics* Vol. 43, No. 3: 336–366.

Kratochwil, Friedrich V. and John Gerard Ruggie. 1986. "International Organization: A State of the Art on an Art of the State." *International Organization* Vol. 40, No. 4: 753–775.

Krueger, Anne O. 1978. *Foreign Trade Regimes and Economic Development: Liberalization Attempts and Consequences*. New York: National Bureau of Economic Research.

Krueger, Anne and Baran Tuncer. 1982. "An Empirical Test of the Infant Industry Argument." *World Bank Reprint Series* No. 284.

Kuhn, Thomas S. 1957. *The Copernican Revolution: Planetary Astronomy in the Development of Western Thought*. Cambridge, MA: Harvard University Press.

Kuklick, Henrika. 1978. "Sins of the Fathers: British Anthropology and African Colonial Administration." *Researches in Sociology of Knowledge, Sciences and Art* Vol. 1: 93–199.

Lal, Deepak. 1974. *Methods of Project Analysis: A Review*. Baltimore: Johns Hopkins University Press.

Lanquet, Thomas. 1548. Coopers chronicle conteininge the whole discourse of the histories … Available at: http://eebo.chadwyck.com/.

Latham, Michael E. 2000. *Modernization as Ideology: American Social Science and "Nation Building" in the Kennedy Era*. Chapel Hill: University of North Carolina Press.

Latour, Bruno. 1987. *Science in Action: How to Follow Scientists and Engineers through Society*. Cambridge, MA: Harvard University Press.

1993. *We Have Never Been Modern*. Trans. Catherine Porter. Cambridge, MA: Harvard University Press.

2005. *Reassembling the Social*. Oxford: Oxford University Press.

2013. *Facing Gaia: A New Enquiry into Natural Religion*. The Gifford Lectures, the University of Edinburgh.

2014. "Agency at the Time of the Anthropocene." *New Literary History* Vol. 45, No. 1: 1–18.

Lavelle, Kathryn C. 2011. *Legislating International Organization: The US Congress, the IMF, and the World Bank*. Oxford: Oxford University Press.

Leach, Edmund. 1966. "On the 'Founding Fathers'." *Current Anthropology* Vol. 7, No. 5: 560–576.

League of Nations. 1920a. "Procès-Verbal of the First Meeting of the Council of the League of Nations." *Official Journal* Vol. 1: 17–26.

1920b. "Procès-Verbal of the Second Session of the Council of the League of Nations." *Official Journal* Vol. 1: 29–59.

1920c. "Procès-Verbal of the Third Session of the Council of the League of Nations." *Official Journal* Vol. 1: 60–64.

1920d. "Institution of an International Bureau for Intellectual Intercourse and Education." *Official Journal* Vol. 1: 445–452.

1967 [1922]. "The Covenant of the League of Nations." In Fred Israel, ed. *Major Peace Treaties of Modern History, 1648–1967, Vol. 4*. New York: Chelsea House Publisher, pp. 1274–1287.

1925. "Minutes of the Thirty-Second Session of the Council." *Official Journal* Vol. 6: 111–260.

1935. "Minutes of the Eighty-Fifth Session of the Council." *Official Journal* Vol. 16: 543–583.

Lebow, Richard Ned. 2008. *A Cultural Theory of International Relations*. Cambridge: Cambridge University Press.

Legro, Jeffrey W. 2005. *Rethinking the World: Great Power Strategies and International Order*. Ithaca, NY: Cornell University Press.

Lepenies, Philip. 2016. *The Power of a Single Number: A Political History of GDP*. New York: Columbia University Press.

Leslie, Stuart W. 1993. *The Cold War and American Science: The Military-Industrial-Academic Complex at MIT and Stanford*. New York: Columbia University Press.

Lévi-Strauss, Claude. 1963. *Structural Anthropology*. Trans. Clair Jacobson and Brooke Grundfest Schoepf. New York: Basic Books.

1966. *The Savage Mind*. Chicago: University of Chicago Press.

Levine, Daniel. 2012. *Recovering International Relations*. Oxford: Oxford University Press.

Levy, Jack. 1994. "Learning and Foreign Policy: Sweeping a Conceptual Minefield." *International Organization* Vol. 48, No. 2: 279–312.

Lewis, David L. 1987. *Race to Fashoda*. New York: Weidenfeld & Nicolson.

Lieber, Keir A. 2005. *War and the Engineers: The Primacy of Politics over Technology*. Ithaca, NY: Cornell University Press.

Lieberman, Robert C. 2002. "Ideas, Institutions, and Political Order: Explaining Political Change." *American Political Science Review* Vol. 96, No. 4: 697–712.

Lilienfeld, Robert. 1978. *The Rise of Systems Theory*. New York: Wiley & Sons.

Little, I.M.D. and J.A. Mirrlees. 1974. *Project Appraisal and Planning for Developing Countries*. New York: Basic Books.

1994. "The Costs and Benefits of Analysis: Project Appraisal and Planning Twenty Years On." In Richard Layard and Stephen Glaister, eds. *Cost-Benefit Analysis, Second Edition*. Cambridge: Cambridge University Press, pp. 199–231.

Little, Richard. 2009. *The Balance of Power in International Relations: Metaphors, Myths and Models*. Cambridge: Cambridge University Press.

Liverani, Mario. 2001. *International Relations in the Ancient Near East, 1600–1100 BC*. New York: Palgrave.

Lorimer, Douglas. 2009. "From Natural Science to Social Science: Race and the Language of Race Relations in Late Victorian and Edwardian Discourse." In Duncan Kelly, ed. *Lineages of Empire*. Oxford: Oxford University Press, pp. 181–212.

Lossky, Andrew. 1970. "International Relations in Europe." In J.S. Bromley, ed. *New Cambridge Modern History, Vol. 6: The Rise of Great Britain and Russia*. Cambridge: Cambridge University Press, pp. 154–192.

Louis XIV. 1924 [1666–1679]. *A King's Lessons in Statecraft*. Ed. Jean Longnon. Trans. H. Wilson. London: T. Fisher Unwin.

Luard, Evan. 1986. *War in International Society*. London: I.B. Tauris & Co.

Lugard, F.D. 1929 [1922]. *The Dual Mandate in British Tropical Africa*, Fourth Edition. London: William Blackwood & Sons.

MacDonald, Julia M. 2015. "Eisenhower's Scientists: Policy Entrepreneurs and the Test-Ban Debate 1954–1958." *Foreign Policy Analysis* Vol. 11, No. 1: 1–21.

MacKenzie, Donald. 1990. *Inventing Accuracy: A Historical Sociology of Nuclear Missile Guidance*. Cambridge, MA: MIT Press.

Madsen, Michael Rask. 2011. "Reflexivity and the Construction of the International Object: The Case of Human Rights." *International Political Sociology* Vol. 5, No. 3: 259–275.

Mahoney, James. 2003. "Strategies of Causal Assessment in Comparative Historical Analysis." In James Mahoney and Dietrich Rueschemeyer, eds. *Comparative Historical Analysis in the Social Sciences*. Cambridge: Cambridge University Press, pp. 337–372.

  2008. "Toward a Unified Theory of Causality." *Comparative Political Studies* Vol. 41, No. 4/5: 412–436.

  2010. "After KKV: The New Methodology of Qualitative Research." *World Politics* Vol. 63, No. 1: 120–147.

Mahoney, James and Kathleen Thelen, eds. 2010. *Explaining Institutional Change: Ambiguity, Agency, and Power*. Cambridge: Cambridge University Press.

Maier, Charles S. 1987. *In Search of Stability*. Cambridge: Cambridge University Press.

  1989. "Alliance and Autonomy: European Identity and U.S. Foreign Policy Objectives in the Truman Years." In Michael J. Lacey, ed. *The Truman Presidency*. New York: Cambridge University Press, pp. 273–298.

Mair, L.P. 1934. "The Place of History in the Study of Culture Contact." *Africa* Vol. 7, No. 4: 415–422.

  1936. "Chieftainship in Modern Africa." *Africa* Vol. 9, No. 3: 305–316.

Malinowski, Bronisław. 1929. "Practical Anthropology." *Africa* Vol. 2, No. 1: 22–38.

  1943. "The Pan-African Problem of Culture Contact." *American Journal of Sociology* Vol. 48, No. 6: 649–665.

  1961 [1922]. *Argonauts of the Western Pacific*. New York: Dutton.

Malinowski, Bronisław, ed. 2014 [1938]. *Methods of Study of Culture Contact in Africa*. London: Read.

Maluf, N.S.R. 1954. "History of Blood Transfusion: The Use of Blood from Antiquity through the Eighteenth Century." *Journal of the History of Medicine and Allied Sciences* Vol. 9, No. 1: 59–107.

Mann, Michael. 1986. *The Sources of Power, Vol. 1, A History of Power from the Beginning to A.D. 1760*. Cambridge: Cambridge University Press.

Mannheim, Karl. 1936. *Collected Works, Vol. 1: Ideology and Utopia*. London: Routledge.

Mantena, Karuna. 2007a. "Mill and the Imperial Predicament." In Nadia Urbinati and Alex Zakaras, eds. *J.S. Mill's Political Thought: A Bicentennial Reassessment*. Cambridge: Cambridge University Press, pp. 298–318.

2007b. "The Crisis of Liberal Imperialism." In Duncan Bell, ed. *Victorian Visions of Global Order: Empire and International Relations in Nineteenth Century Political Thought*. Cambridge: Cambridge University Press, pp. 113–145.

Mantena, Kuruna. 2010. *Alibis of Empire: Henry Maine and the Ends of Liberal Imperialism*. Princeton: Princeton University Press.

Manzione, Joseph. 2000. "Amusing and Amazing and Practical and Military: The Legacy of Scientific Internationalism in American Foreign Policy." *Diplomatic History* Vol. 24, No. 1: 21–55.

March, James G. and Johan P. Olsen. 1998. "The Institutional Dynamics of International Political Orders." *International Organization* Vol. 52, No. 4: 943–969.

Mayer, Maximilian, Mariana Carpes, and Ruth Knoblich. 2014. "The Global Politics of Science and Technology: An Introduction." *Global Science and Technology*, Vol. 1. Heidelberg: Springer, pp. 1–35.

Mayntz, Renate. 2004. "Mechanisms in the Analysis of Social Phenomena." *Philosophy of the Social Sciences* Vol. 34, No. 2: 237–259.

Mazower, Mark. 2012. *Governing the World: The History of an Idea, 1815 to the Present*. New York: Penguin.

McCarthy, Thomas A. 2007. "From Modernism to Messianism: Liberalism Developmentalism and American Exceptionalism." *Constellations* Vol. 14, No. 1: 3–37.

McDonald, Lynn. 1993. *The Early Origins of the Social Sciences*. Montreal: McGill-Queen's University Press.

McMillan, Kevin. 2010. "European Diplomacy and the Origins of Governmentality." In Miguel de Larrinaga and Marc G. Doucet, eds. *Security and Global Governmentality: Globalization, Governance, and the State*. London: Routledge, pp. 23–43.

McNamara, Robert S. 1981. *The McNamara Years at the World Bank: Major Policy Addresses of Robert S. McNamara, 1968–1981*. Baltimore: Johns Hopkins University Press.

McNeill, J.R. 2012. "Global Environmental History: The First 150,000 Years." In J.R. McNeill and Erin Stewart Mauldin, eds. *A Companion to Global Environmental History*. London: Blackwell, pp. 1–17.

McNeill, J.R. and Peter Engelke. 2015. *The Great Acceleration*. Cambridge, MA: Harvard University Press.

McNeill, William H. 1982. *The Pursuit of Power*. Chicago: University of Chicago Press.

Mearsheimer, John J. 2001. *The Tragedy of Great Power Politics*. New York: W.W. Norton & Co.
Merchant, Carol. 1980. *The Death of Nature: Women, Ecology, and the Scientific Revolution*. New York: HarperCollins.
Meseguer, Covadonga. 2005. "Policy Learning, Policy Diffusion, and the Making of a New Order." *Annals of the American Academy of Political and Social Science* Vol. 598, No. 1: 125–144.
Meseguer, Covadonga and Fabrizio Gilardi. 2009. "What is New in the Study of Policy Diffusion?" *Review of International Political Economy* Vol. 16, No. 3: 527–543.
Metternich, Prince Klemens von. 1880a. *Memoirs of Prince Metternich: 1773–1815*, Vol. 1. Ed. Prince Richard Metternich. Trans. Mrs. Alexander Napier. New York: Charles Scribner's Sons.
  1880b. *Memoirs of Prince Metternich: 1773–1815*, Vol. 2. Ed. Prince Richard Metternich. Trans. Mrs. Alexander Napier. New York: Charles Scribner's Sons.
Meyer, John W. 2009. *World Society: The Writings of John W. Meyer*. Ed. Georg Krücken and Gili S. Drori. Oxford: Oxford University Press.
Meyer, John, John Boli, George M. Thomas, and Francisco O. Ramirez. 1997. "World Society and the Nation-State." *American Journal of Sociology* Vol. 103: 144–181.
Milliken, Jennifer. 1999. "The Study of Discourse in International Relations: A Critique of Research and Methods." *European Journal of International Relations* Vol. 5, No. 2: 225–254.
Mills, David. 2002. "British Anthropology at the End of Empire: The Rise and Fall of the Colonial Social Science Research Council, 1944–1962." *Revue d'Histoire des Sciences Humaines* Vol. 1, No. 6: 161–188.
Milner, Helen. 1991. "The Assumption of Anarchy in International Relations Theory: A Critique." *Review of International Studies* Vol. 17, No. 1: 67–85.
Mirowski, Philip. 1989. *More Heat than Light: Economics as Social Physics: Physics as Nature's Economics*. Cambridge: Cambridge University Press.
  1999. "Cyborg Agonistes: Economics Meets Operations Research in Mid-Century." *Social Studies of Science* Vol. 29, No. 5: 685–718.
  2002. *Machine Dreams*. Cambridge: Cambridge University Press.
Mirowski, Philip and Dieter Plehwe. 2009. *The Road from Mont Pelerin: The Making of the Neoliberal Thought Collective*. Cambridge, MA: Harvard University Press.
Mitchell, Timothy. 2002. *Rule of Experts: Egypt, Techno-Politics, Modernity*. Berkeley: University of California Press.
  2005. "Economists and the Economy in the Twentieth Century." In George Steinmetz, ed. *The Politics of Method in the Human Sciences: Positivism and its Epistemological Others*. Durham, NC: Duke University Press, pp. 126–141.
  2008. "Culture and Economy." In Tony Bennett and John Frow, eds. *The Sage Handbook of Cultural Analysis*. London: Sage, pp. 447–466.
  2014. "Econometality: How the Future Entered Government." *Critical Inquiry* Vol. 40, No. 4: 479–507.
Mitzen, Jennifer. 2013. *Power in Concert: The Nineteenth-Century Origins of Global Governance*. Chicago: University of Chicago Press.

Monteiro, Nuno P. 2014. *Theory of Unipolar Politics*. Cambridge: Cambridge University Press.
Moravcsik, Andrew. 1997. "Taking Preferences Seriously: A Liberal Theory of International Politics." *International Organization* Vol. 51, No. 4: 513–553.
Morgan, Mary. 2003. "Economics." In Theodore M. Porter and Dorothy Ross, eds. *The Cambridge History of Science, Volume 7: The Modern Social Sciences*. Cambridge: Cambridge University Press, pp. 275–305.
Morgenthau, Hans J. 1946. *Scientific Man versus Power Politics*. Chicago: University of Chicago Press.
  2006 [1948]. *Politics Among Nations, Seventh Edition*. New York: McGraw-Hill.
Mornay, Philip du Plessis. 1969 [1572] "Vindiciae contra tyrannos." *In Constitutionalism and Resistance in the Sixteenth Century: Three Treatises by Hotman, Beza, and Mornay*. Trans. Julian H. Franklin. New York: Pegasus.
Morris, James P., Nancy K. Squires, Charles S. Taber, and Milton Lodge. 2003. "Activation of Political Attitudes: A Psychophysiological Examination of the Hot Cognition Hypothesis." *Political Psychology* Vol. 24, No. 4: 727–745.
Motter, Adilson E., Alessandro P.S. de Moura, Ying-Cheng Lai, and Partha Dasgupta. 2002. "Topology of the Conceptual Network of Language." *Physical Review E* Vol. 65, No. 065102(R): 1–4.
Mukerji, Chandra. 1997. *Territorial Ambitions and the Garden of Versailles*. Cambridge: Cambridge University Press.
Muthu, Sankar. 2003. *Enlightenment Against Empire*. Princeton: Princeton University Press.
Neill, Deborah J. 2014. "Science and Civilizing Missions: Germans and the Transnational Community of Tropical Medicine." In Bradley Naranch and Geoff Eley, eds. *German Colonialism in a Global Age*. Durham, NC: Duke University Press, pp. 74–92.
Nelson, Stephen. 2014. "Playing Favorites: How Shared Beliefs Shape the IMF's Lending Decisions." *International Organization* Vol. 68, No. 2: 297–328.
  2017. *The Currency of Confidence: How Economic Beliefs Shape the IMF's Relationship with Its Borrowers*. Ithaca, NY: Cornell University Press.
Newburg, Colin. 1999. "Great Britain and the Partition of Africa, 1870–1914." In Andrew Porter, Alaine Low, and William Roger Louis, eds. *The Oxford History of the British Empire, Vol. III: The Nineteenth Century*. Oxford University Press, pp. 624–650.
Newton, Isaac. 1934 [1729]. *Mathematical Principles of Natural Philosophy*. Trans. Andrew Motte. Revised by Florian Cajori. Berkeley: University of California Press.
Nexon, Daniel H. 2009. *The Struggle for Power in Early Modern Europe: Religious Conflict, Dynastic Empires, and International Change*. Princeton: Princeton University Press.
Nielson, Daniel L. and Michael J. Tierney. 2005. "Delegation to International Organizations: Agency Theory and World Bank Reform." *International Organizations* Vol. 57, No. 2: 241–276.
Nielson, Daniel L., Michael J. Tierney, and Catherine E. Weaver. 2006. "Bridging the Rationalist-Constructivist Divide: Re-Engineering the Culture of the World Bank." *Journal of International Relations and Development* Vol. 9, No. 2: 107–139.

North Atlantic Treaty Organization (NATO). 2014. *The Secretary-General's Annual Report*. Brussels.
Nostradamus. 1559. An excellent treatise, shevving suche perillious, and contagious infirmities ... Available at: http://eebo.chadwyck.com/.
Organization for Economic Cooperation and Development (OECD). 2015. *The Secretary-General's Report to Ministers*. Paris.
Organski, A.F.K. and Jacek Kugler. 1980. *The War Ledger*. Chicago: University of Chicago Press.
Osiander, Andreas. 1994. *The States System of Europe, 1640–1990: Peacemaking and the Conditions of International Stability*. Oxford: Clarendon Press.
Oxford English Dictionary (OED). 2011 [2001]. "Calculate." Third Edition, March 2001; online version June 2011.
Oxford English Dictionary (OED). 2011 [2001]. "Counterpoise." Third Edition, March 2001; online version June 2011.
Oxford English Dictionary (OED). 2011 [2001]. "Maxim." Third Edition, March 2001; online version June 2011.
Pagden, Anthony. 1995. *Lords of All the World: Ideologies of Empire in Britain, France and Spain, 1400–1800*. New Haven: Yale University Press.
Park, Katherine and Lorraine Daston. 2006. "Introduction: Age of the New." In Katherine Park and Lorraine Daston, eds. *The Cambridge History of Science*, Vol. 3: *Early Modern Science*. Cambridge: Cambridge University Press, pp. 1–20.
Parker, Geoffrey. 1988. *The Military Revolution*. Cambridge: Cambridge University Press.
1998. *The Grand Strategy of Philip II*. New Haven: Yale University Press.
Parkinson, Cosmo. 1945. *The Colonial Office From Within, 1909–1945*. London: Faber & Faber.
Partridge, Michael and David Gillard, eds. 1996. *West Africa: Diplomacy of Imperialism, 1868–1895*. British Documents on Foreign Affairs: Reports and Papers from the Confidential Print, Part I, Series G, Vol. 19. University Publications of America.
Pedersen, Susan. 2007. "Back to the League of Nations." *The American Historical Review* Vol. 112, No. 4: 1091–1117.
Perham, Margery. 1937. *Native Administration in Nigeria*. London: Oxford University Press.
Petty, William. 1690. Political Arithmetick, or A discourse concerning the extent and value of lands, buildings ... Available at: http://eebo.chadwyck.com/.
1691. The Political Anatomy of Ireland. Available at: http://eebo.chadwyck.com/.
Phillips, Andrew. 2011. *War, Religion and Empire: The Transformation of International Orders*. Cambridge: Cambridge University Press.
Philpott, Daniel. 2001. *Revolutions in Sovereignty: How Ideas Shaped Modern International Relations: How Ideas Shaped Modern International Relations*. Princeton: Princeton University Press.
Pierson, Paul. 2003. "Big, Slow-Moving, and ... Invisible: Macrosocial Processes in the Study of Comparative Politics." In James Mahoney and Dietrich

Rueschemeyer, eds. *Comparative Historical Analysis in the Social Sciences*. Cambridge: Cambridge University Press, pp. 177–207.
  2004. *Politics in Time*. Princeton: Princeton University Press.
Pitt, H.G. 1970. "The Pacification of Utrecht." In J.S. Bromley, ed. *New Cambridge Modern History, Vol. 6: The Rise of Great Britain and Russia*. Cambridge: Cambridge University Press, pp. 446–479.
Pitts, Jennifer. 2005. *A Turn to Empire: The Rise of Imperial Liberalism in Britain and France*. Princeton: Princeton University Press.
Poggi, Gianfranco. 2006. *Weber*. Malden, MA: Polity.
Poiger, Uta G. 2000. *Jazz, Rock, and Rebels: Cold War Politics and American Culture in a Divided Germany*. Berkeley: University of California Press.
Ponet, John. 1556. A shorte treatise of politike pouuer ... Available at: http://eebo.chadwyck.com/.
Popper, Karl. 1959. *The Logic of Discovery*. London: Routledge.
Porter, Roy. 1979. "Creation and Credence: The Career of Theories of the Earth in Britain, 1660–1820." In Barry Barnes and Steven Shapin, eds. *Natural Order: Historical Studies of Scientific Culture*. London: Sage, pp. 73–92.
Porter, Theodore M. 1986. *The Rise of Statistical Thinking, 1820–1900*. Princeton: Princeton University Press.
Porter, Theodore. 2003. "Genres and Objects of Social Inquiry, From the Enlightenment to 1890." In Theodore M. Porter and Dorothy Ross, eds. *The Cambridge History of Science, Volume 7: The Modern Social Sciences*. Cambridge: Cambridge University Press, pp. 11–39.
Pouliot, Vincent. 2009. *International Security in Practice*. Cambridge: Cambridge University Press.
Povinelli, Elizabeth. 1995. "Do Rocks Listen? The Cultural Politics of Apprehending Australian Aboriginal Labor." *American Anthropologist* Vol. 97, No. 3: 505–518.
  2002. *The Cunning of Recognition*. Durham, NC: Duke University Press.
Price, Richard P. 1997. *The Chemical Weapons Taboo*. Ithaca, NY: Cornell University Press.
Purdey, Stephen J. 2010. *Economic Growth, the Environment, and International Relations*. London: Routledge.
Queen Anne. 1904 [1712]. "Queen Anne's Account of the Terms of the Treaty of Utrecht." In James Robinson, ed. *Readings in Modern European History, Vol. 1*. New York: Ginn and Company, pp. 51–53.
Rabb, Theodore K. 1975. *The Struggle for Stability in Early Modern Europe*. Oxford: Oxford University Press.
Rabinow, Paul and William M. Sullivan, eds. 1987. *Interpretive Social Sciences: A Second Look*. Los Angeles: University of California Press.
Ragep, F.J. 2007. "Copernicus and His Islamic Predecessors." *History of Science* Vol. 45: 65–81.
Ranger, Terence. 1977. "The People in African Resistance." *Journal of Southern African Studies* Vol. 1, No. 1: 125–146.
Reddy, Sanjay G. and Thomas Pogge. 2010. "How Not to Count the Poor." In Anand Sughin, Paul Segal, and Joseph Stiglitz, eds. *Debates on the Measurement of Global Poverty*. Oxford: Oxford University Press, pp. 42–81.

Redman, Deborah. 1997. *The Rise of Political Economy as a Science: Methodology and the Classical Economists*. Cambridge, MA: MIT Press.
Reiss, Timothy. 1982. *The Discourse of Modernism*. Ithaca, NY: Cornell University Press.
Reus-Smit, Christian. 1999. *The Moral Purpose of the State*. Princeton: Princeton University Press.
Rhodes, Richard. 1986. *The Making of the Atomic Bomb*. New York: Simon & Schuster.
Richards, Audrey I. 1935. "The Village Census in the Study of Culture Contact." *Africa* Vol. 8, No. 1: 20–33.
Richelieu, Cardinal. 1961 [1635]. *Political Testament*. Ed. and trans. Henry Bertram Hill. Madison: University of Wisconsin Press.
  1988 [1625]. "Memorandum for the King." In Richard Bonney, ed. *Society and Government Under Richelieu and Mazarin*. London: Macmillan, pp. 4–6.
Ridley, Lancelot. 1540. A commentary in Englyshe vpon Saynte Paules epystle to the Ephesyans ... Available at: http://eebo.chadwyck.com/.
Rindzevičiūtė, Eglė. 2016. *The Power of Systems: How the Policy Sciences Opened Up the Cold War World*. Ithaca, NY: Cornell University Press.
Risse, Thomas. 2000. "Let's Argue: Communicative Action in World Politics." *International Organization* Vol. 54, No. 1: 1–39.
  2010. *A Community of Europeans?* Ithaca, NY: Cornell University Press.
Rockström, Johan, et al. 2009. "A Safe Operating Space for Humanity." *Nature* Vol. 461: 472–475.
Rogers, G.A.J. 2007. "Hobbes and His Contemporaries." In Patricia Springbord, ed. *The Cambridge Companion to Hobbes's Leviathan*. Cambridge: Cambridge University Press, pp. 413–440.
Rohan, Henri. 1640. A treatise of the interest of the princes and states of Christendome. Available at: http://eebo.chadwyck.com/.
Romer, Paul M. 1989. "What Determines the Rate of Growth and Technological Change?" World Bank Policy, Planning, and Research Working Paper No. WPS 279.
Rosenberg, Emily, ed. 2012. *A World Connecting, 1870–1945*. Cambridge, MA: The Belknap Press of Harvard University Press.
Rosenstein-Rodan, P.N. 1943. "Problems of Industrialisation of Eastern and Southern Europe." *The Economic Journal* Vol. 53, No. 210/11: 202–211.
Ross, Dorothy. 1991. *The Origins of American Social Science*. Cambridge: Cambridge University Press.
  2003. "Changing Contours of the Social Science Disciplines." In Theodore M. Porter and Dorothy Ross, eds. *The Cambridge History of Science, Volume 7: The Modern Social Sciences*. Cambridge: Cambridge University Press, pp. 203–237.
Rostow, W.W. 1990 [1960]. *The Stages of Economic Growth: A Non-Communist Manifesto, Third Edition*. Cambridge: Cambridge University Press.
Rowntree, B. Seebohm. 1901. *Poverty: A Study of Town Life*. London: Macmillan and Co.
Ruddiman, William F. 2013. "The Anthropocene." *Annual Review of Earth and Planetary Sciences* Vol. 41: 45–68.
Rudwick, Martin J.S. 2005. *Bursting the Limits of Time: The Reconstruction of Geohistory in the Age of Revolution*. Chicago: University of Chicago Press.

Rueschemeyer, Dietrich and Theda Skocpol, eds. 1996. *States, Social Knowledge, and the Origins of Modern Social Policies*. Princeton: Princeton University Press.
Ruggie, John G. 1975. "International Responses to Technology: Concepts and Trends." *International Organization* Vol. 29, No. 3: 557–583.
   1982. "International Regimes, Transactions, and Change: Embedded Liberalism in the Postwar Economic Order." *International Organization* Vol. 36, No. 2: 379–415.
   1983. "Continuity and Transformation in the World Polity: Toward a Neorealist Synthesis." *World Politics* Vol. 35, No. 2: 261–285.
   1989. "International Structure and International Transformation: Space, Time, and Method." In Ernst-Otto Czempiel and James Rosenau, eds. *Global Changes and Theoretical Challenges: Approaches to World Politics for the 1990s*. Lexington, MA: Lexington Books, pp. 21–35.
   1993. "Territoriality and Beyond: Problematizing Modernity in International Relations." *International Organization* Vol. 47, No. 1: 139–174.
   2004. "Reconstituting the Global Public Domain: Issues, Actors, and Practices." *European Journal of International Relations* Vol. 10, No. 4: 499–531.
Said, Edward. 1978. *Orientalism*. New York: Vintage Books.
Saliba, George. 2007. *Islamic Science and the Making of the European Renaissance*. Cambridge, MA: MIT Press.
Sanger, Clyde. 1995. *Malcolm MacDonald: Bringing an End to Empire*. Montreal: McGill-Queen's University Press.
Saull, Richard. 2012. "Rethinking Hegemony: Uneven Development, Historical Blocs, and the World Economic Crisis." *International Studies Quarterly* Vol. 56: 323–338.
Schmidt, Vivien A. 2008. "Discursive Institutionalism: The Explanatory Power of Ideas and Discourse." *Annual Review of Political Science* Vol. 11: 303–326.
Schroeder, Paul. 1989. "The Nineteenth Century System: Balance of Power or Political Equilibrium?" *Review of International Studies* Vol. 15: 135–153.
   1992. "Did the Vienna Settlement Rest on a Balance of Power?" *The American Historical Review* Vol. 97, No. 3: 683–706.
   1994a. *The Transformation of European Politics, 1763–1848*. Oxford: Clarendon Press.
   1994b. "Historical Reality Vs. Neo-Realist Theory." *International Security* Vol. 19, No. 1: 108–148.
Schweller, Randall L. 2001. "The Problem of International Order Revisited: A Review Essay." *International Security* Vol. 26, No. 1: 161–186.
   2004. "Unanswered Threats: A Neoclassical Realist Theory of Underbalancing." *International Security* Vol. 29, No. 2: 159–201.
Scott, H.M. 2006. *The Birth of a Great Power System, 1740–1815*. London: Routledge.
Scott, James C. 1998. *Seeing Like a State: How Certain Schemes to Improve the Human Condition Have Failed*. New Haven: Yale University Press.
Seabrooke, Leonard. 2014. "Epistemic Arbitrage: Transnational Professional Knowledge in Action." *Journal of Professions and Organization* Vol. 1, No. 1: 49–64.
Searle, John R. 1995. *The Construction of Social Reality*. New York: The Free Press.
Selznick, Philip. 1984 [1957]. *Leadership in Administration*. Berkeley: University of California Press.

References

Sen, Amartya. 1999. *Development as Freedom*. New York: Anchor Books.
Sending, Ole Jacob. 2002. "Constitution, Choice and Change: Problems with the 'Logic of Appropriateness' and its Use in Constructivist Theory." *European Journal of International Relations* Vol. 8, No. 4: 443–470.
  2015. *The Politics of Expertise: Competing for Authority in Global Governance*. Ann Arbor: University of Michigan Press.
Sewell, William H., Jr. 2005. *Logics of History*. Chicago: University of Chicago Press.
Shanghai Cooperation Organisation (SCO). 2014. "Joint Communique of the Results of the 13th Meeting of the Council of Heads of Government." Astana, Kazakhstan.
Shapin, Steven. 1996. *The Scientific Revolution*. Chicago: University of Chicago Press.
Shapin, Steven and Simon Schaffer. 1985. *Leviathan and the Air-Pump: Hobbes, Boyle, and the Experimental Life*. Princeton: Princeton University Press.
Shapley, Deborah. 1993. *Promise and Power: The Life and Times of Robert McNamara*. Boston: Little, Brown, and Co.
Sharma, Patrick. 2013. "Bureaucratic Imperatives and Policy Outcomes: The Origins of World Bank Structural Adjustment Lending." *Review of International Political Economy* Vol. 20, No. 4: 667–686.
  2017. *McNamara's Other War: The World Bank and International Development*. Philadelphia: University of Pennsylvania Press.
Sharma, Vivek. 2005. The Impact of Institutions on Conflict and Cooperation in Early Modern Europe. PhD Dissertation, New York University.
Sheehan, Michael. 1996. *The Balance of Power: History and Theory*. London: Routledge.
Silberman, Bernard. 1983. *Cages of Reason: The Rise of the Rational State in France, Japan, the United States, and Great Britain*. Chicago: University of Chicago Press.
Simmons, Beth. 1994. *Who Adjusts? Domestic Sources of Foreign Economic Policy During the Interwar Years*. Princeton: Princeton University Press.
Simmons, Beth and Zachary Elkins. 2003. "Globalization and Policy Diffusion: Explaining Three Decades of Liberalization." In Miles Kahler and David A. Lake, eds. *Governance in a Global Economy*. Princeton: Princeton University Press, pp. 275–304.
Simmons, Beth A., Frank Dobbin, and Geoffrey Garrett, eds. 2008. *The Global Diffusion of Markets and Democracy*. Cambridge: Cambridge University Press.
Singer, David. 1963. "International Conflict: Three Levels of Analysis." *World Politics* Vol. 12, No. 3: 453–461.
Skinner, Quentin. 1978. *The Foundations of Modern Political Thought*, 2 Vols. Cambridge: Cambridge University Press.
  1996. *Reason and Rhetoric in the Philosophy of Hobbes*. Cambridge: Cambridge University Press.
Skolnikoff, Eugene B. 1993. *The Elusive Transformation: Science, Technology, and the Evolution of International Politics*. Princeton: Princeton University Press.
Smith, Adam. 1976 [1759]. *The Theory of Moral Sentiments*. Eds. D.D. Raphael and A.L. Macfie. Oxford: Oxford University Press.

1976 [1776]. *An Inquiry into the Nature and Causes of the Wealth of Nations.* Ed. R.H. Campbell, A.S. Skinner, and W.B. Todd. Oxford: Oxford University Press.

Smith, Roger. 1997. *The Norton History of the Human Sciences.* New York: W.W. Norton. & Co.

Snidal, Duncan. 1985. "The Limits of Hegemonic Stability Theory." *International Organization* Vol. 39: 579–614.

Sofka, James R. 1998. "Metternich's Theory of European Order: A Political Agenda for 'Perpetual Peace.'" *The Review of Politics* Vol. 60: 115–150.

Solé, Ricard V., Bernat Corominas-Murtra, Sergi Valverde, and Luc Steels. 2010. "Language Networks: Their Structure, Function, and Evolution." *Complexity* Vol. 15, No. 6: 20–26.

Solingen, Etel. 2012. "Of Dominoes and Firewalls: The Domestic, Regional, and Global Politics of International Diffusion." *International Studies Quarterly* Vol. 56, No. 4: 631–644.

Soll, Jacob. 2009. *The Information Master: Jean-Baptiste Colbert's Secret State Intelligence System.* Ann Arbor: University of Michigan Press.

Solow, Robert M. 1956. "A Contribution to the Theory of Economic Growth." *The Quarterly Journal of Economics* Vol. 70, No. 1: 65–94.

  1957. "Technical Change and the Aggregate Production Function." *The Review of Economics and Statistics* Vol. 39, No. 3: 312–320.

Srinivasan, T.N. 1977a. "Development, Poverty, and Basic Human Needs: Some Issues." Washington, DC: World Bank Report No. REP76.

  1977b. "Poverty: Some Measurement Problems." Washington, DC: World Bank Report No. REP77.

St. Clair, Asunción Lera. 2006a. "The World Bank as a Transnational Expertised Institution." *Global Governance* Vol. 12, No. 1: 77–95.

  2006b. "Global Poverty: The Co-Production of Knowledge and Politics." *Global Social Policy* Vol. 6, No. 1: 57–77.

Staples, Amy. 2006. *The Birth of Development: How the World Bank, Food and Agriculture Organization, and World Health Organization Changed the World, 1945–1965.* Kent, OH: The Kent State University Press.

Star, S.L. and J.R. Griesemer. 1989. "Institutional Ecology, 'Translations' and Boundary Objects: Amateurs and Professionals in Berkeley's Museum of Vertebrate Zoology, 1907–39." *Social Studies of Science* Vol. 19, No. 3: 387–420.

Stedman Jones, Daniel. 2012. *Masters of the Universe: Hayek, Friedman, and the Birth of Neoliberal Politics.* Princeton: Princeton University Press.

Steffen, Will, Jacques Grinevald, Paul Crutzen, and John McNeill. 2011. "The Anthropocene: Conceptual and Historical Perspectives." *Philosophical Transactions of the Royal Society* Vol. 369, No. 1938: 842–867.

Steinmetz, George. 2007. *The Devil's Handwriting: Precoloniality and the German Colonial State in Qingdao, Samoa, and Southwest Africa.* Chicago: University of Chicago Press.

Stern, Nicholas and Francisco Ferreira. 1997. "The World Bank as 'Intellectual Actor.'" In Devesh Kapur, John P. Lewis, and Richard C. Webb, eds. *The World Bank: Its First Half Century,* Vol. 2: Perspectives. Washington, DC: Brookings Institution Press, pp. 523–609.

References

Stocking, Jr., George W. 1987. *Victorian Anthropology*. New York: Simon & Schuster.
  1995. *After Tylor: British Social Anthropology 1888–1951*. Madison: University of Wisconsin Press.
Strang, David. 1991. "Adding Social Structure to Diffusion Models: An Event History Framework." *Sociological Methods & Research* Vol. 19, No. 3: 324–353.
Strauss, Leo. 1936. *The Political Philosophy of Hobbes*. Trans. Elsa M. Sinclair. Oxford: Clarendon Press.
Streeten, Paul, Frances Stewart, Shahid Javid Burki, Alan Berg, Norman L. Hicks, Hollis B. Chenery, and Mahbub ul Haq. 1980. *Poverty and Basic Needs*. Washington, DC: World Bank.
Stutchey, Benedikt. 2005. *Science Across the European Empires, 1800–1950*. Oxford: Oxford University Press.
Suddaby, Roy and Thierry Viale. 2011. "Professional and Field-Level Change: Institutional Work and the Professional Project." *Current Sociology* Vol. 59, No. 4: 423–442.
Suzuki, Shogo. 2009. *Civilization and Empire: China and Japan's Encounter with European International Society*. London: Routledge.
Swidler, Ann. 1986. "Culture in Action: Symbols and Strategies." *American Sociological Review* Vol. 51, No. 2: 273–286.
  2001. "What Anchors Cultural Practices." In Theodore R. Schatzki, Karin Knorr Cetina, and Eike von Savigny, eds. *The Practice Turn in Contemporary Theory*. London: Routledge, pp. 83–101.
Taber, Charles. 1998. "The Interpretation of Foreign Policy Events: A Cognitive Process Theory." In Donald A. Sylvan and James F. Voss, eds. *Problem Representation in Foreign Policy Decision-Making*. Cambridge: Cambridge University Press, pp. 29–52.
Talleyrand, Prince Charles-Maurice. 1881. *The Correspondence of Prince Talleyrand and King Louis XVI During the Congress of Vienna*. New York: Harper & Brothers.
Taylor, Charles. 1985. *Philosophy and the Human Sciences: Philosophical Papers*, Vol. 2. Cambridge: Cambridge University Press.
  1987. "Interpretation and the Sciences of Man." In Paul Rabinow and William M. Sullivan, eds. *Interpretive Social Sciences: A Second Look*. Los Angeles: University of California Press, pp. 33–81.
Taylor, Ian. 2010. *The International Relations of Sub-Saharan Africa*. New York: Continuum.
Taylor, Mark Z. 2012. "Toward an International Relations Theory of National Innovation Rates." *Security Studies* Vol. 21, No. 1: 113–152.
Temple, William. 1814 [1690]. "Observations upon the United Provinces of the Netherlands." In *The Works of Sir William Temple, Bart. Complete. In Four Volumes*. London: S. Hamilton.
Thelen, Kathleen. 2004. *How Institutions Evolve: The Political Economy of Skills in Germany, Britain, the United States, and Japan*. Cambridge: Cambridge University Press.
Thompson, William R. 1990. "Long Waves, Technological Innovation, and Relative Decline." *International Organization* Vol. 44, No. 2: 201–233.

Tilley, Helen. 2011. *Africa as a Living Laboratory: Empire, Development, and the Problem of Scientific Knowledge, 1870–1950*. Chicago: University of Chicago Press.

Tilley, Helen and Robert Gordon. 2007. *Ordering Africa: Anthropology, European Imperialism, and the Politics of Knowledge*. Manchester: Manchester University Press.

Tilly, Charles. 1984. *Big Structures, Large Processes, Huge Comparisons*. New York: Russell Sage Foundation.

  1992. *Coercion, Capital, and European States, AD 990–1992*. Cambridge, MA: Blackwell.

Toulmin, Stephen. 1990. *Cosmopolis: The Hidden Agenda of Modernity*. New York: The Free Press.

Toye, John. 2009. "Solow in the Tropics." *History of Political Economy* Vol. 41, S1: 221–240.

Treaty of Utrecht 1973 [1713]. In W.N. Hargreaves-Mawdsley, ed. *Spain Under the Bourbons, 1700–1833*. Columbia: University of South Carolina Press.

True, Jacqui and Michael Mintrom. 2001. "Transnational Networks and Policy Diffusion: The Case of Gender Mainstreaming." *International Studies Quarterly* Vol. 45, No. 1: 27–57.

Tuck, Richard. 1993. *Philosophy and Government, 1572–1651*. Cambridge: Cambridge University Press.

Tylor, Edward B. 1871. *Primitive Cultures: Researches into the Development of Mythology, Philosophy, Religion, Art, and Custom*. London: John Murray.

The Universal Magazine of Knowledge and Pleasure. 1749. Vol. IV. London: John Hinton.

Vagts, Alfred. 1948. "The Balance of Power: Growth of an Idea." *World Politics* Vol. 1, No. 1: 82–101.

Van Creveld, Martin. 1989. *Technology and War*. New York: St. Martin's Press.

Verdier, Daniel. 2002. *Moving Money: Banking and Finance in the Industrialized World*. Cambridge: Cambridge University Press.

Vernon, James. 2007. *Hunger: A Modern History*. Cambridge, MA: Belknap Press.

Viret, Pierre. 1548. A verie familiar [and] fruiteful exposition of the.xii. articles of the Christian faieth ... Available at: http://eebo.chadwyck.com/.

von Laue, Theodore H. 1976. "Anthropology and Power: R.S. Rattray Among the Ashanti." *African Affairs* Vol. 75, No. 298: 33–54.

Wade, Robert H. 1996. "Japan, the World Bank, and the Art of Paradigm Maintenance: The East Asian Miracle in Political Perspective." *New Left Review* Vol. 217, May–June: 3–36.

  2002. "US Hegemony and the World Bank: The Fight over People and Ideas." *Review of International Political Economy* Vol. 9, No. 2: 201–229.

Wagner, Günter. 1936. "The Study of Culture Contact and the Determination of Policy." *Africa* Vol. 9, No. 3: 317–331.

Wagner, Peter. 2000. "'An Entirely New Object of Consciousness, of Volition, of Thought': The Coming into Being and (almost) Passing Away of 'Society' as a Scientific Object." In Lorraine Daston, ed. *Biographies of Scientific Objects*. Chicago: University of Chicago Press, pp. 132–157.

2003a. "The Uses of the Social Sciences." In Theodore M. Porter and Dorothy Ross, eds. *The Cambridge History of Science, Volume 7: The Modern Social Sciences*. Cambridge: Cambridge University Press, pp. 535–552.

2003b. "Social Science and Social Planning During the Twentieth Century." In Theodore M. Porter and Dorothy Ross, eds. *The Cambridge History of Science, Volume 7: The Modern Social Sciences*. Cambridge: Cambridge University Press, pp. 591–607.

Walker, R.B.J. 1993. *Inside/Outside: International Relations as Political Theory*. Cambridge: Cambridge University Press.

Waltz, Kenneth N. 1979. *Theory of International Politics*. New York: McGraw-Hill.

Walzer, Michael. 1971. *Revolution of the Saints: A Study in the Origins of Radical Politics*. New York: Atheneum.

Wapner, Paul. 1996. *Environmental Activism and World Civic Politics*. Albany: State University of New York Press.

Waters, Colin N. *et al.* 2016. "The Anthropocene is Functionally and Stratigraphically Distinct from the Holocene." *Nature* Vol. 351, No. 6269: 137–147.

Weaver, Catherine. 2008. *Hypocrisy Trap*. Princeton: Princeton University Press.

Weber, Max. 1946. *From Max Weber: Essays in Sociology*. Trans. H.H. Gerth and C. Wright Mills. Oxford: Oxford University Press.

  1949. *The Methodology of the Social Sciences*. Trans. Edward A. Shils and Henry A. Finch. Glencoe: Free Press.

  1958. *The Protestant Work Ethic*. Trans. Talcott Parsons. New York: Charles & Scribner's Sons.

  1978. *Economy and Society: An Outline of Interpretive Sociology*. Ed. Guenther Roth and Claus Wittich. Berkeley: University of California Press.

  2004. *The Vocation Lectures*. Eds. David S. Owen and Tracy B. Strong. Trans. Rodney Livingstone. Indianapolis: Hackett.

Webster, Charles. 1931. *The Foreign Policy of Castlereagh, 1812–1815*. London: G. Bell & Sons.

  1947. *The Foreign Policy of Castlereagh, 1815–1822*. London: G. Bell & Sons.

Welch, David. 2005. *Painful Choices: A Theory of Foreign Policy Change*. Princeton: Princeton University Press.

Weldes, Jutta. 1999. *Constructing National Interests: The United States and the Cuban Missile Crisis*. Minneapolis: University of Minnesota Press.

Wendt, Alexander. 1987. "The Agent-Structure Problem in International Relations Theory." *International Organization* Vol. 41, No. 3: 335–370.

  1999. *Social Theory of International Politics*. Cambridge: Cambridge University of Press.

  2015. *Quantum Mind and Social Science: Unifying Physical and Social Ontology*. Cambridge: Cambridge University Press.

Wendt, Alexander and Raymond Duvall. 1989. "Institutions and International Order." In Ernst-Otto Czempiel and James N. Rosenau, eds. *Global Changes and Theoretical Challenges: Approaches to World Politics for the 1990s*. Lexington: Lexington Books, pp. 51–73.

Westad, Odd Arne. 2000. "The New International History of the Cold War: Three (Possible) Paradigms." *Diplomatic History* Vol. 24, No. 4: 551–565.

2007. *The Global Cold War: Third World Interventions and the Making of Our Times*. Cambridge: Cambridge University Press.

Widmaier, Wesley W., Mark Blyth, and Leonard Seabrooke. 2007. "Exogenous Shocks or Endogenous Constructions? The Meanings of Wars and Crises." *International Studies Quarterly* Vol. 51, No. 4: 747–759.

Wight, Martin. 1966. "The Balance of Power." In Herbert Butterfield and Martin Wight, eds. *Diplomatic Investigations: Essays in the Theory of International Politics*. Cambridge, MA: Harvard University Press, pp. 132–148.

Wilson, Peter. 2012. "The English School Meets the Chicago School: The Case for a Grounded Theory of International Institutions." *International Studies Review* Vol. 14, No. 4: 567–590.

Wittrock, Björn and Peter Wagner. 1996. "Social Science and the Building of the Early Welfare State: Toward a Comparison of Statist and Non-Statist Western Societies." In Dietrich Rueschemeyer and Theda Skocpol, eds. *States, Social Knowledge, and the Origins of Modern Social Policies*. Princeton: Princeton University Press, pp. 90–114.

Woods, Ngaire. 2006. *The Globalizers: The IMF, the World Bank, and Their Borrowers*. Ithaca, NY: Cornell University Press.

Worboys, Michael. 1988. "The Discovery of Colonial Malnutrition Between the Wars." In David Arnold, ed. *Imperial Medicine and Indigenous Societies*. Manchester: Manchester University Press, pp. 208–224.

World Bank. 1947. *International Bank for Reconstruction and Development Second Annual Report 1946–1947*. Washington, DC.

1948a. *International Bank for Reconstruction and Development Third Annual Report 1947–1948*. Washington, DC.

1948b. "Brazil: Country Study." Washington, DC: World Bank Report No. L-27.

1951. "Proposed Loans to the Belgian Congo and the Kingdom of Belgium." Washington, DC: World Bank Report No. P-20.

1953. "Report on the Desirability of Establishing a Bank for the Development of Lebanon's Industry and Agriculture." Washington, DC: World Bank Report No. Z-3a.

1955. "World Economic Growth: Retrospect and Prospects." Washington, DC: World Bank Report No. EC-47a.

1957. *The World Bank: Policies and Operations*. Washington, DC: World Bank.

1960. "Colombia: Appraisal of the Expansion Program of Empresa De Energia De Bogota." Washington, DC: World Bank Report No. TO-219a.

1963. "The Economy of the Ivory Coast." Washington, DC: World Bank Report No. AF-6a.

1965a. "Proposed Loan to Iran for a Road Project." Washington, DC: World Bank Report No. P-427.

1965b. "Proposed Loan to India for the Second Kothagudem Power Project." Washington, DC: World Bank Report No. P-435.

1965c. "Proposed Credit to the Islamic Republic of Pakistan for and Agricultural Development Bank Project." Washington, DC: World Bank Report No. P-747.

1965d. "Economic Position and Prospects of Argentina. Annex III: Agriculture." Washington, DC: World Bank Report No. WH-144.

1968a. *World Bank: Policies and Operations*. Washington, DC: World Bank.

1968b. "Current Economic Position and Prospects of Chile." Washington, DC: World Bank Report No. WH-186a.
1969. "The Fertilizer Program in India." Washington, DC: World Bank Report No. 5A-5.
1974. *World Bank: Policies and Operations.* Washington, DC: World Bank.
1976. "Measurement of the Health Benefits of Investments in Water Supply: Report of an Expert Panel." World Bank P.U. Report No. PUN 20.
1977. "Annual Meeting of the Board of Governors: Summary Proceedings." Washington, DC.
1978. *World Development Report 1978.* Washington, DC: World Bank.
1979. "Annual Meeting of the Board of Governors: Summary Proceedings." Belgrade, Yugoslavia.
1980a. *World Development Report 1980.* Washington, DC: World Bank.
1980b. "Economic Memorandum on Argentina." Washington, DC: World Bank Report No. 2988-AR.
1981a. *Accelerated Development in Sub-Saharan Africa: An Agenda for Action [The Berg Report].* Washington, DC: World Bank.
1981b. "Sri Lanka: Proposed Credit… for a Mahaweli Ganga Development Project III." Washington, DC: World Bank Report No. P-3082-CE.
1983. "Zimbabwe: Proposed Credit … for a Rural Afforestation Project." Report No. P-3426a-ZIM.
1985a. *World Development Report 1985.* Washington, DC: World Bank.
1985b. "Pakistan: Recent Economic Developments and Structural Adjustment." Washington, DC: World Bank Report No. 5347-PAK.
1985c. "Chile: Proposed Structural Adjustment Loan." Washington, DC: World Bank Report No. P-4131-CH.
1986. *World Development Report 1986.* Washington, DC: World Bank.
1987. *World Development Report 1987.* Washington, DC: World Bank.
1990. *World Development Report 1990.* Washington, DC: World Bank.
1993a. *World Development Report 1993.* Washington, DC: World Bank.
1993b. *The East Asian Miracle: Economic Growth and Public Policy.* Washington, DC: World Bank Report No. 11293-IND.
1994. "Poland: Growth with Equity, Policies for the 1990s." Washington, DC: World Bank Report No. 13039-POL.
1996. "Botswana: Tuli Block Roads Project." Washington, DC: World Bank Report No. 16193.
1998. "Brazil: Minas Gerais State Privatization." Washington, DC: World Bank Report No. 16466-BR.
2000. "Swaziland: Reducing Poverty Through Shared Growth." Washington, DC: World Bank Report No. 19658-SW.
2003. "Implementation Completion Report … for the Small Towns Water and Sanitation Project." Washington, DC: World Bank Report No. 27529.
2015. *World Development Report: Mind, Society, and Behavior.* Washington, DC.
World Trade Organization (WTO). 2014. *World Trade Report: Trade and Development.* Washington, DC.
Wright, Moorhead, ed. 1975. *Theory and Practice of the Balance of Power, 1486–1914: Selected European Writings.* London: J.M. Dent & Sons.

Wright, Quincy. 1930. *Mandates Under the League of Nations*. Chicago: University of Chicago Press.
Wuthnow, Robert. 1979. "The Emergence of Modern Science and World System Theory." *Theory and Society* Vol. 8, No. 2: 215–243.
Yarrow, Andrew L. 2010. *Measuring America: How Economic Growth Came to Define American Greatness in the Late Twentieth Century*. Amherst: University of Massachusetts Press.
Yates, Francis A. 1975. *Astraea: The Imperial Theme in the Sixteenth Century*. Boston: Routledge and Kegan Paul.
Young, Robert J.C. 2001. *Postcolonialism*. Malden, MA: Blackwell.
Zalasiewicz, J., Mark Williams, Alan Haywood, and Michael Ellis. 2009. "The Anthropocene: A New Epoch of Geological Time?" *Philosophical Transactions of the Royal Society A: Mathematical, Physical and Engineering Sciences* 369/1938: 830–841.
Zimmerman, Andrew. 2010. *Alabama in Africa: Booker T. Washington, the German Empire, and the Globalization of the New South*. Princeton: Princeton University Press.
Zucker, Lynne G. 1977. "The Role of Institutionalization in Cultural Persistence." *American Sociological Review* Vol. 42, No. 5: 726–743.

Index

absolute poverty, 234, 238, 240, 249
absolute time, 93, 95, 139, 210
 absence of, 108
Académie des Sciences, 97
African Institute of Languages and
 Cultures, 160, 196
agency, 41, 42
 creativity, 50
Alexander, Emperor of Russia, 1, 118,
 121, 126
alternative explanations, 71–72, 78, 142,
 203, 209, 246
An African Survey, 191, 197
anthropocene, 278–282
anthropology, 24, 142, 150, 157–163, 170,
 see also functionalism, diffusionism,
 evolutionism
 in the British Colonial Office, 160, 174
 influence of science on, 157
aristocratic dynasticism, 22, 76, 80–81,
 88–89, 111
Aristotelian cosmology, 22, 75, 86,
 90–91, 101
Ashley, Richard K., 13, 30, 58, 69,
 276n. 27
associational change, 12, 21, 41, 45, 46–53,
 54, 67, 95–97, 114–117, 141, 154,
 171–181, 191–202, 211, 226, 234–250,
 253, 262, 264, 269–272, 278, 280
 two-stage model, 46, 67, 176
associations, *see* associational change
 and cosmological shifts, 69
 and international structure, 44
 definition, 43
 historical view of, 44
 and international structure, 44
astrology, 82, 92
astronomy, 92, 93, 94, 157
 medieval astronomy, 75, 90

Bacon, Francis, 97, 137
balance of power, 22, 75, 100n.134, 103,
 105–109, 117–131, 155

calculation of, 117, 126–131
construction of, 123, 126, 129
as counterpoise, 101, 104
as discursive configuration, 135
as equal power, 104, 105
as equilibrium, 121
erosion of, 204
explanations for, 119, 131
institutionalization of, 103
and interests, 101
as land and people, 23, 113, 116,
 128, 136
mechanistic view, 106
as secondary institution, 78, 102,
 122, 126
systemic view, 105, 106n.175, 114,
 122, 123
Balassa, Bela, 233, 242
Bartelson, Jens, 10, 98, 101
basic human needs, 211, 235, 238–242
Berg report, 245
Berger, Peter and Thomas
 Luckmann, 36, 55
Berlin Conference (1885), 24, 140,
 154–156
Bertalanffy, Ludwig von, 213
Bhagwati, Jagdish, 233, 243
biological determinism, 149, 150n.59, 158
biology, 137, 148, 216, 248, 260, 278n.36
blood, 81, 109, 111, 124
Bohr, Niels, 213
Bolingbroke, Henry Saint John,
 Viscount, 104
Boyle, Robert, 1, 22, 112
Bretton Woods, 223–224, 227
Briand, Aristide, 188
Britain, 117, 118, 133, 135, 142, 146,
 151–152, 223
British Colonial Office, 24, 51, 141, 160,
 174–181, 190–201, 232
brokers, 22, 49, 50, 97, 115, 156, 161,
 224, 263, 264, 268, 269, 282
Brougham, Henry, 107

331

332 Index

Bruno, Giordano, 92
Buell, Raymond, 179, 193
Buffon, Georges Louis Leclerc, Comte de, 23, 146
Bukovansky, Mlada, 16
Bull, Hedley, 274
bureaucratization, 26, 44, 115, 156
Bush, Vannevar, 212–213
Buzan, Barry, 6, 30, 31, 31n.10

Caine, Sydney, 198
calculation, 23, 113, 116, 125, 129, 135, 216
  absence of, 86
Calvin, John, 83
Cambridge University, 151, 224
Cameron, D.C., 179, 180
capitalism, 26, 205, 262
case selection, 70
Castlereagh, Robert Stewart, Viscount, 23, 118, 119, 121, 123, 125, 126, 132
casuistry, 84, 91
Cateau-Cambrésis, Treaty of, 87
Chamberlain, Joseph, 172–173
Chambers, Robert, 23, 139, 147, 147n.44
change, *see* international change
Charles V, Holy Roman Emperor, 88, 89
Chaumont, Treaty of, 118, 119
Chenery, Hollis, 230, 239, 240
China, 276–278
Chinese cosmology, 37, 132
Christian cosmology, 23, 34, 37, 76, 80, 81, 143, 146, *see also* science and religion
  and natural philosophy, 64, 83
civilization, 23, 136, 140, 144, 145, 154, 155, 158, 159, 170, 171, 172, 180, 181, 183, 185, 204
classification, 52, 117, 146, 205, 258
climate change, 259, 277, 278, 280, 282
clockwork image, 86, 94, 95
coding scheme, 71, 286–289
coercion, 47, 211, 275
Colbert, Jean-Baptiste, 96, 100n.134
Cold War, 204, 209, 222, 224, 225, 250, 266, 268, 280, 281
Colonial Development Act, 173
Colonial Development and Welfare Act, 203
Colonial Development and Welfare Bill, 199
colonial order, 153–156
colonial science, 163
colonial self-sufficiency, 142, 172, 173, 199, 203

Colonial Social Science Research Council, 199
compensations, 122, 127, 134
Comte, Auguste, 145
Condorcet, Nicolas, Marquis de, 144, 158, 209
configurations, 11, 41, 43, 51, 53, 71, *see* discursive configurations
  explanatory logic, 26, 71, 71n.164
conservatism, 118, 145n.30
constitutive mechanisms, 70, 176, 205, 211, 237
constructivism, 6, 12, 40, 58, 265
containment, 122
contestation, 42, 49, 50, 67, 281, 282, *see* cosmological contestation
contextual view of social action, 40–42, 57, 73
control, 22, 23, 24, 67, 77, 109, 113, 114, 117, 120, 131, 138, 165, 186, 208, 216, 220, 222, 250n.275, 282
convenances, *see* compensations
Copernicus, Nicolas, 1, 22, 76, 90
cosmological contestation, 14, 38, 39, 68, 205, 225, 276
cosmological shifts, 4, 5, 10, 20, 21, 54, 63–65, 77, 117, 267
  in the anthropocene, 278
  Copernican revolution, 65, 69, 90–95
  cybernetic-systems thinking, 65, 212–220
  definition, 63
  historical sciences, 65, 143–151
  in the anthropocene, 278
  materialism and mechanism, 109–114
cosmology, 10–12, 39, 51, 58, 120, 132, 135, 155, 165, 208, 211, 219, 224, 232, 242, 247, 248, 260, 271, 279, *see also* scientific cosmology
  and associations, 14
  and change, 12
  definition, 11
  distinct from ontology and episteme, 11
  in European discourses (1550), 83–87
  and hegemony, 59
  in international discourses (2015), 257–260
  and international order, 10, 33
  in the League of Nations (1920–1935), 186–188
  and political order, 36
  and purposes, 35, 53, 68
  at Utrecht (1713), 102–105
  at the Vienna Congress (1815), 124

Index

Cowles Commission, 218, 222
crisis, 37, 46, 52, 140, 159, 171, 174, 251, 270–272, 280
Curzon, George, Marquess, 185
Cuvier, Georges, 23, 147
cybernetics, 24, 213, 218, 229
cybernetic-systems thinking, 25, 208, 216, 237, 243, 248, 257
cyclical time, 2, 84, 108, 143

Darwin, Charles, 23, 147, 148, 181
Darwin, Erasmus, 147
Darwinian ideas, 2, 24, 64, 279
  erosion of, 200
  influence on anthropology, 157
Dawes Commission, 217
democratic politics, 222
Descartes, René, 1, 92, 94, 109
destiny, 11, 260
determinism, 69, 95, 108, 110, 117, 139, 151, 157, 164, 167, 170, *see also* biological determinism
  erosion of, 156, 165
detribalization, 195–202
development, 24, 140, 151, 154, 202, 225, 255, 259, 275, *see also* evolutionary development, state-led development
  civilizational development, 24, 153, 183, 258
  colonial development, 24, 141, 204
  as contingent process, 160
  economic development, 169, 200, 202, 228, 230, 233
  multidimensional development, 230, 261
  as natural law, 148, 149, 151
  as scientific and technological progress, 194
diffusion, 47, 49
  as distinct from horizontal change, 60
diffusionism (anthropological tradition), 159, 160n.107, 177n.191
discourse, 30, 38
  definition, 40
  and international structure, 45
discourse analysis, 71, 285–289
discourse of state purpose, 32, 40, 62, 78
  at the Berlin Conference (1885), 154
  definition, 45
  in European discourses (1550), 80–82
  in international discourse (2015), 254–257
  in the League of Nations (1920–1935), 183–186
  micro-level basis, 40
  at the Vienna Congress (1815), 120–124, 131, 133
  at Utrecht (1713), 102–105

discursive configurations, 41, 42, 50, 73, 80, 150, 165, 181, 183, 237, 254, 288
discursive reconfiguration, 21, 41, 68, 173, 211, 289
disenchantment, 18, 19, 38, 64
distribution of ideas, 61
divine law, 69, 76, 80, 81, 146
divine providentialism, 22, 76, 80, 87, 89, 93, 108, 111, 145, 150
  erosion of, 109, 148
  and grand strategy, 90
  and natural philosophy, 83
  reconfiguration, 90
dynastic law, 81, 82, 121, 123

early modern European order, 79–90
East Asian Miracle report, 247, 249
ecological sciences, 279, 280
economic stability, 185, 210, 221, 222, 223
economics, 66, 148n.46, 205, 208, 210, 218n.52, 224, 226, 230, 236, 237, 239, 242, 246, 250, 252, 261, 269, *see also* neoclassical economics, neoliberal economics
  influence of natural sciences on, 216–218
economists, 61, 66, 211, 214, 236
economy, the, 169, 186n.244, 210, 217–218
  control of, 222
education policy, 191, 193, 194
  scientific education, 194
Elizabeth I, Queen of England, 87, 97
emulation, 47, 115
engineering, 24, 212, 218
English School, 6, 30
Enlightenment, the, 16, 23, 62, 84, 106, 116, 137, 142, 144
episteme, 11, 83, 92, 109, 117, 120, 124, 143, 164, 186, 228, 237, 243, 251, 258, 260, *see also* epistemic modernism, rationalist episteme, representational episteme, patterning episteme
  definition, 34
  and purpose, 69
epistemic communities, 17, 49, 60, 270
epistemic modernism, 165, 170, 186, 190, 202, 210, 226, 232, 262
  challenges to, 279, 280
eugenics, 150
evolutionary development, 150, 170, 176, 179, 181, 184

evolutionism (anthropological tradition), 157, 177, *see also* sociocultural evolution
experiments, 164
expert authority, 135, 187, 212, 218, 250n.275
expert knowledge, 141, 163–170, 174–176, 185, 187, 191–195

falsification, 71
Fermi, Enrico, 207
field of forces, 23, 105, 108, 109, 114, 164, 208, 213
Finnemore, Martha, 9, 32n.14, 132
Food and Agricultural Organization, 142
food policy, 167, 192, 200
Foucault, Michel, 137
France, 120, 134, 152
Frederick the Great, 107, 116
free trade, 23, 152, 154, 154n.84, 223, 243, 255
functionalism (anthropological tradition), 159, 160n.107, 161, 177n.191, 191, 195–197
future time, 186, 220, 229, 248

Galileo Galilei, 1, 92, 109
Galton, Francis, 150
game theory, 214
general equilibrium analysis, 218, 243
generative structure, 33, 39
geologic time, 146, 278, 279
geology, 23, 64, 137, 142, 146–147, 157, 158, 248, 278
Gilpin, Robert, 5, 6, 30, 55, 56, 275
global south, 152, 200, 204, 225, 252
global value chains, 256, 258
glory, 77, 88, 89, 100, 104, 113, 116
gold standard, 223
governmentality, 2n.4, 137
gradual change, 3, 30, 60, 224, 251, 264
great power, concept of, 116
gross domestic product, 212, 240, 256, 257, 261
gross national product, 212, 231, 235, 244, 261
growth, 2, 24, 25, 41, 170, 210, 219, 221, 229, 234, 242, 245, 246, 255, 261
  cosmological basis, 2, 260
  as a discursive configuration, 210, 227
growth-based order, 221–224
Guizot, Francois, 145
Gulick, Edward V., 75, 132, 134

Habsburg-Valois rivalry, 87–90
Hailey, Lord Malcolm, 191, 197, 199

Hardenberg, Karl August von, 119, 127, 129
Harrod-Domar model, 230
Harvey, William, 22, 111
hegemonic imposition, 13, 58–60, 221–224, 226, 272–273, 276
hegemony, 13, 59, 133
  American, 25, 26, 203, 209, 211, 221–224, 254, 262, 281
  British, 23, 26, 140, 202
  Chinese, 277
  historical view of, 60, 60n.134
  as leadership, 60
heliocentric universe, 22, 75, 91
Henry II, King of France, 87
Hicks, Norman, 238
high modernism, 24n.104, 166n.138, 201
historical institutionalism, 21n.97, 46, 270
historical sciences, 23, 72, 137, 139, 143, 146, 279, *see also* biology, geology
history and philosophy of science, 72n.168
Hobbes, Thomas, 22, 110
honour, 82, 116
Hooke, Robert, 92
horizontal change, 13, 58, 60–61, 95, 272–273, 281
  definition, 60
Hughes, Helen, 243
human agency, 109, 117, 125, 137, 138, 165, 208
human capital, 248–249
human nature, 87, 95, 110, 150, 155, 164, 216, 219, 242, 260, 279
Hutton, James, 23, 146, 159
hybrid change, 61–62, 153, 181, 272–273

improvement, 2, 23, 24, 118, 135, 139, 153, 172, 185, 202, 205
indemnity, 122, 122n.274
indigenous knowledge, 205
indigenous anti-colonial resistance, 140, 140n.7, 169, 171, 172, 174, 191
indirect rule, 178, 179, 195
industrial revolution, 23, 26, 141, 204
infant industry protection, 243
institutionalization, *see* recursive institutionalization
  definition, 55
interests, 98–102, 113, 121
  and the balance of power, 103, 104
international change, 39
  mechanisms and processes, 6–8
  theories of, 3, 9, 264–266
international discourses, 12, 43
  definition, 45
International Labour Organization, 142

Index

International Monetary Fund, 204, 227, 250
international order, 5, 5n.16, 31, 31n.7, *see also* growth-based order, balance of power order, colonial order, laissez-faire liberal order, early modern European order
   definition, 5, 31
   hierarchical structure of, 7
   primary and secondary institutions, 7, 31
   structure of, 53
Islamic science, 90n.75

Kaunitz, Prince, 107, 116
Kepler, Johannes, 64
Keynes, John Maynard, 217, 221, 223
Kolmogorov, Andrey, 219
Krueger, Anne, 233, 243, 246
Kuznets, Simon, 217

labour, 154n.84, 167, 171, 179, 191, 193
   forced labour, 161, 167
laissez-faire liberal order, 152, 153–156, 158
   erosion of, 171
Lamarck, Jean-Baptiste, 147
League of Nations, 167, 181–190, 202, 224, 258
   design of, 182
   expert committees, 185
learning, 47, 48, 97, 211, 227, 245
legal ideas, 105, 108, 121, 183, 188, 254, *see also* divine law
legibility, 10, 62, 63, 128, 164, 242
legitimation, 13, 14, 19, 21, 35, 36, 39, 49, 50, 51, 82, 96, 132, 199, 201, 246–250
liberal ideas and norms, 62, 153, 155, 165, 176, 182, 227, 254, 261, *see also* laissez-faire liberal order
   erosion of laissez-faire beliefs, 201
linear time, 143, 144, 148, 228, 232
   challenges to, 279
Linnaeus, Carl, 146
London School of Economics, 160, 224
Louis XIV, 96, 100, 102
Lugard, Frederick D., Baron, 178, 180, 199

MacDonald, Malcolm, 199
macro-level, 53–62, 70
   definition, 40
Maine, Henry, 158
Mair, Lucy P., 196

Malinowski, Bronisław, 159–163, 191, 195–197
Mandate System, 183, 184, 188, 203, 204
MANIAC, 215
Mantena, Karuna, 158, 158n.102, 159n.103, 178n.198, 179n. 211
marriage, 82, 88, 109, 113
materialism, 22, 23, 86, 86n.48, 95, 99, 100n.134, 110, 116, 117, 126, 217
McNamara, Robert S., 25, 211, 234–237, 238, 243, 244
measurement, 22, 113, 116, 126, 129, *see also* quantification, statistics
mechanism, 22, 72, 86n.48, 92, 94, 95, 101, 109, 126, 164, 210, 216
   absence of, 86
meso-level, 21, 29, 46–53, 70, 77, 132, 141, 289
   definition, 39
Mesopotamian cosmology, 37
methodological individualism, 16n.69
Metternich, Prince Klemens von, 23, 115, 119, 122, 125, 126–131, 135
Meyer, John, *see* World Polity School
microfoundations, 40
micro-level, 39–46, 40n.46
   definition, 39
Mill, J.S., 23, 145, 151
mind-body dualism, 86n.48
Mitchell, Timothy, 169, 210, 220, 229
Mitzen, Jennifer, 132, 133n.319
modeling, 24, 208, 212, 216, 220
modernism, 24, 24n.104, 69, 187, 220, 224, 248, 258, 282, *see also* high modernism
   in the anthropocene, 280
   versus rationalism, 186
modernity, 200, 205, 225, 228, 258, 259, 261, 275
modernization theory, 232
monarchical legitimacy, 118, 123
Morgenthau, Hans, 182
multilevel theory, 39, 264
   linking micro, meso, and macro, 8–10, 42–46

Napoleon Bonaparte, 118
national accounting, 217, 218, 240
National Bureau of Economic Research, 217, 233, 243
National Defense Research Committee, 212, 214, 215
natural history, *see* historical sciences
natural law, 22, 69, 93, 105, 145, 154, 190

natural philosophy, 64, 79, 92
natural providentialism, 23, 82, 95, 103, 108, 109, 145, 150
naturalization, 25, 35, 36, 51, 52, 132, 148, 190, 231, 246–250, 261, 275
definition, 53
neoclassical economics, 25, 209, 219, 222, 231–233, 236, 238, 246, 251, 261, 262
neoliberal economics, 233
neoliberal policy, 233, 242–246, 252
New Deal planning, 224
Newton, Isaac, 1, 64, 93–95
Newtonian ideas, 23, 69, 93, 104, 105–109, 143, 151
Nexon, Daniel H., 9, 43
Nostradamus, 82, 86
nuclear weapons, 208, 212, 268, 276, 280

object-centred thinking, 25, 208, 260
objects, 24, 157, 163, 164, 191, 195, 197, 208
Office of Scientific Research and Development, 212
ontology, 11, 52, 84, 86, 92, 109, 117, 120, 124, 164, 190, 229, 237, 243, 251, 257, 260
and purpose, 69
definition, 34
order-building moments, 3, 6, 29, 56, 60, 210, 224
orientalism, 19
Orr, John Boyd, 204
orrery, 94
Oxford University, 151, 159, 224

Paris, Treaty of (1814), 118, 119
partition of Polish and Saxon lands, 127, 129
Passfield, Sydney Webb, Baron, 180
patterning episteme, 35, 84, 90, 91, 98
erosion, 92
peaceful change, 276
Pearson, Karl, 150
Perham, Margery, 199
Petty, William, 22, 112, 113, 116
Philip II, King of Spain, 87, 89
Philip, Duke of Anjou, 102, 103
Phillips, Andrew, 7, 11, 37
physics, 24, 207, 212, 214, 218
Polish-Saxon question, 119, 126–131
political economy, 26n.107, 109–114, 137, 163
political-economic factors, 26n.107, 221, 222, 252

poverty, 167, 234
poverty alleviation, 25, 212, 221, 235, 238–242, 244, 255, 256
power, 30, 42, 57, 58, 89
cultural-epistemic, 59, 226
definition, 57n.125
military-economic, 226
organizational power, 51, 211, 237
state power, 26, 30, 59, 62, 218
practice theory, 8, 40
prediction, 215, 216, 219, 220
problem-solving, 49, 51, 97, 191, 197, 269, 273, 280
process-tracing, 71, 289
productivism, 200, 222, 225, 227, 252
progress, 2, 24, 93, 139, 148, 150, 154, 155, 159, 165, 170, 181, 187, 190, 195, 209, 232, 259, 260
absence of, 84
projections, 220, 240, 244, 257, 260
providentialism, 69, 176, *see also* divine providentialism, natural providentialism
Prussia, 118, 119, 121, 133
public health, 166, 185, 191, 204

quantification, 167, 201, 212, 226, 235, 239, 240, 241, 244, 261
quantum mechanics, 267–269, 278, 282
Quetelet, Adolphe, 165

race, 158, 166, 178, *see also* biological determinism, scientific racism
Radcliffe-Brown, A.R., 160, 174
raison d'état, *see* reason of state
RAND, 218
rationalism, 22, 23, 69, 117, 248
in the anthropocene, 280
versus modernism, 186
rationalist episteme, 99, 103, 109, 113, 117, 126, 141
rationalization, 18, 19, 31, 283
Rattray, R.S., 177
realism, 57, 60n.135, 209
reason, 99, 103, 105, 144
reason of state, 99
recursive institutionalization, 21, 22, 24, 53–62, 77, 95, 117, 135, 143, 153, 172, 181, 209, 226, 250, 262, 264, 266–274
generality, 73, 267, 274
hegemonic pathway, *see* hegemonic imposition
horizontal pathway, *see* horizontal change

hybrid pathway, *see* hybrid change
recursivity, 45, 55, 58, 62
relationalism, 8, 9, 12n.51, 40, 40n.50, 43
religious ideas, 22, 105, *see also* science and religion, Christian cosmology
representational constraints, 25, 220, 237, 241, 245, 284
representational episteme, 93, 95, 98, 113
republican thought, 105, 108
research design, 70–73
resemblance episteme, *see* patterning episteme
resettlement, 201
resources (troops, money, arms, and land), 82, 120
Reus-Smit, Christian, 6, 7, 10, 16, 30, 32
Revised Minimum Standard Model, 230
Richelieu, Cardinal de, 99
Rockefeller Foundation, 161, 218
Rohan, Henri Duke de, 100
Romer, Paul, 249
Roosevelt, Franklin Delano, 221
Rosenstein-Rodan, Paul, 230, 230n.135
Rostow, Walt, 232
Rowntree, Seebohm, 167, 242
Royal Society of London for Improving Natural Knowledge, 97
Ruggie, John G., 10, 13, 30, 32, 32n.14, 210, 277
Russia, 118, 119, 121, 133

Samuelson, Paul, 219, 222, 233
science, 14–20
  definition, 14, 20
  instrumental view of, 17
  and meaning, 18
  plural and historical view of, 20, 64
  as productive power, 13, 16, 18
science and religion, 64, 93, 95, 143, 146
scientific and technological progress, 24, 25, 145, 158, 180, 190, 193, 204, 209, 212, 237, 247, 248, 249, 256, 258, 260, 277, 280
scientific cosmology, 12, 15, 18, 38, 65–70
  challenges to, 277
scientific management, 234, 235
scientific racism, 157, 160
Scott, James C., 10, 52, 165, 281
Scottish Enlightenment, 144
security concerns, 122, 155, 184, 202, 222, 254, 282
Sewell, William H., 15n.60, 42n.61, 66
Smith, Adam, 23, 144, 151, 158, 169, 181
social cost–benefit analysis, 241
social forces, 164, 166, 170

social knowledge, 157, 163–170
social sciences, 190, 282–284, *see also* anthropology, economics, social knowledge
  influence of natural sciences on, 164, 218
  need for reflexivity in, 282–284
  as productive power, 283
society, 158, 160, 165, 166
  social bonds, 160, 179
sociocultural evolution, 24, 147, 150, 157, *see also* evolutionism
sociology of knowledge, 265
Solow, Robert, 231, 233, 247
sovereignty, 7, 109, 189, 203, 213
Soviet Union, 204, 209, 219, 225
spiritual-temporal dualism, 84, 91
Srinivasan, T.N., 238
stage theory of human history, 84, 145, 150, 151, 158
stagnationism, 210, 221, 222
state purpose, 2, 3n.7, 9n.36, 32, 32n.14, 103, 126, 139, 172, 190, 220, 248, *see also* balance of power, civilization, evolutionary development, growth, improvement, state-led development
  definition, 32
  as distinct from goals, 35
  institutionalization of, 54
state-centric assumptions of, 43
state-led development, 24, 142, 157, 162, 190–201, 202, 262
statistical commission, 23, 119, 126–131, 134
statistical offices, 116
statistics, 113, 116, 127, 136, 164, 220, 230, 242
status, 121
strategic deployment, 21, 46, 95–97, 114–117, 171–176, 211, 234–237, 289
  definition, 48
  of scientific ideas, 67
Streeten, Paul, 238
structural adjustment, 243
supervenience, 44n.69

Talleyrand, Prince Charles Maurice, 121, 122, 124, 126, 127, 128, 129, 136
technocracy, 182
technology transfer, 233, 247, 255, 258
Temple, William, 22, 97, 112
temporality, *see* time
territory, 23, 112, 114, 122, 123, 126, 128, 134, 136
text selection, 79n.13, 80n.16, 141n.10, 210n.17, 289–293

Thomas, N.W., 177
time, 11, 23, 170, 208, 260, 279, *see also* future time, geologic time, linear time, absolute time, cyclical time
  and purposes, 68
trade-led growth, 212, 243, 245, 246, 248
transnational actors, 13, 43, 45, 67, 115
  aristocratic class, 44, 96, 123, 224
  knowledge networks, 13, 54, 59, 60, 61, 62, 66, 96, 190, 209, 219, 226, 262, 265, 266
  social movements, 49
trusteeship, 24, 178, 179, 184, 194, 200, *see also* Mandate System
Turgot, Anne Robert Jacques, 144
Tylor, Edward Burnett, 157, 159, 177

ul Haq, Mahbub, 238
uncertainty, 52, 67, 270, 273
United States, 211, 217, 221–224, 225, 246, 276–278, 281
universalism, 59, 69, 225, 247, 258, 262

universities, 115
Utrecht (1713), Treaty of, 102–105

Versailles Conference (1919), 24, 140
Vienna Congress (1815), 1, 23, 69, 117–131
vitalism, 83
von Neumann, John, 215–216

War of Spanish Succession, 102
Weber, Max, 18, 19, 31, 36, 36n.28, 63, 283
welfare, 114, 154, 155, 164, 183, 184, 186, 190, 202, 221, 222, 223, 227, 255
welfare state, 24, 172, 199
White, Harry, 223
Wiener, Norbert, 213, 215, 219
William III, King of England, 105
World Bank, 25, 26, 51, 142, 204, 210, 226, 227–231, 233, 234–250, 250n.275, 257
World Polity School, 19, 20, 67

Cambridge Studies in International Relations

132 Nuno P. Monteiro
   Theory of unipolar politics
131 Jonathan D. Caverley
   Democratic militarism
   Voting, wealth, and war
130 David Jason Karp
   Responsibility for human rights
   Transnational corporations in imperfect states
129 Friedrich Kratochwil
   The status of law in world society
   Meditations on the role and rule of law
128 Michael G. Findley, Daniel L. Nielson and J. C. Sharman
   Global shell games
   Experiments in transnational relations, crime, and terrorism
127 Jordan Branch
   The cartographic state
   Maps, territory, and the origins of sovereignty
126 Thomas Risse, Stephen C. Ropp and Kathryn Sikkink (eds.)
   The persistent power of human rights
   From commitment to compliance
125 K. M. Fierke
   Political self-sacrifice
   Agency, body and emotion in international relations
124 Stefano Guzzini
   The return of geopolitics in Europe?
   Social mechanisms and foreign policy identity crises
123 Bear F. Braumoeller
   The great powers and the international system
   Systemic theory in empirical perspective
122 Jonathan Joseph
   The social in the global
   Social theory, governmentality and global politics
121 Brian C. Rathbun
   Trust in international cooperation
   International security institutions, domestic politics and American multilateralism
120 A. Maurits van der Veen
   Ideas, interests and foreign aid

119 Emanuel Adler and Vincent Pouliot (eds.)
     International practices
118 Ayṣe Zarakol
     After defeat
     How the East learned to live with the West
117 Andrew Phillips
     War, religion and empire
     The transformation of international orders
116 Joshua Busby
     Moral movements and foreign policy
115 Séverine Autesserre
     The trouble with the Congo
     Local violence and the failure of international peacebuilding
114 Deborah D. Avant, Martha Finnemore and Susan K. Sell (eds.)
     Who governs the globe?
113 Vincent Pouliot
     International security in practice
     The politics of NATO-Russia diplomacy
112 Columba Peoples
     Justifying ballistic missile defence
     Technology, security and culture
111 Paul Sharp
     Diplomatic theory of international relations
110 John A. Vasquez
     The war puzzle revisited
109 Rodney Bruce Hall
     Central banking as global governance
     Constructing financial credibility
108 Milja Kurki
     Causation in international relations
     Reclaiming causal analysis
107 Richard M. Price
     Moral limit and possibility in world politics
106 Emma Haddad
     The refugee in international society
     Between sovereigns
105 Ken Booth
     Theory of world security
104 Benjamin Miller
     States, nations and the great powers
     The sources of regional war and peace
103 Beate Jahn (ed.)
     Classical theory in international relations
102 Andrew Linklater and Hidemi Suganami
     The English School of international relations
     A contemporary reassessment
101 Colin Wight
     Agents, structures and international relations
     Politics as ontology

100 Michael C. Williams
The realist tradition and the limits of international relations
99 Ivan Arreguín-Toft
How the weak win wars
A theory of asymmetric conflict
98 Michael Barnett and Raymond Duvall (eds.)
Power in global governance
97 Yale H. Ferguson and Richard W. Mansbach
Remapping global politics
History's revenge and future shock
96 Christian Reus-Smit (ed.)
The politics of international law
95 Barry Buzan
From international to world society?
English School theory and the social structure of globalisation
94 K. J. Holsti
Taming the sovereigns
Institutional change in international politics
93 Bruce Cronin
Institutions for the common good
International protection regimes in international security
92 Paul Keal
European conquest and the rights of indigenous peoples
The moral backwardness of international society
91 Barry Buzan and Ole Wæver
Regions and powers
The structure of international security
90 A. Claire Cutler
Private power and global authority
Transnational merchant law in the global political economy
89 Patrick M. Morgan
Deterrence now
88 Susan Sell
Private power, public law
The globalization of intellectual property rights
87 Nina Tannenwald
The nuclear taboo
The United States and the non-use of nuclear weapons since 1945
86 Linda Weiss
States in the global economy
Bringing domestic institutions back in
85 Rodney Bruce Hall and Thomas J. Biersteker (eds.)
The emergence of private authority in global governance
84 Heather Rae
State identities and the homogenisation of peoples
83 Maja Zehfuss
Constructivism in international relations
The politics of reality

82 Paul K. Ruth and Todd Allee
The democratic peace and territorial conflict in the twentieth century
81 Neta C. Crawford
Argument and change in world politics
Ethics, decolonization and humanitarian intervention
80 Douglas Lemke
Regions of war and peace
79 Richard Shapcott
Justice, community and dialogue in international relations
78 Phil Steinberg
The social construction of the ocean
77 Christine Sylvester
Feminist international relations
An unfinished journey
76 Kenneth A. Schultz
Democracy and coercive diplomacy
75 David Houghton
US foreign policy and the Iran hostage crisis
74 Cecilia Albin
Justice and fairness in international negotiation
73 Martin Shaw
Theory of the global state
Globality as an unfinished revolution
72 Frank C. Zagare and D. Marc Kilgour
Perfect deterrence
71 Robert O'Brien, Anne Marie Goetz, Jan Aart Scholte and Marc Williams
Contesting global governance
Multilateral economic institutions and global social movements
70 Roland Bleiker
Popular dissent, human agency and global politics
69 Bill McSweeney
Security, identity and interests
A sociology of international relations
68 Molly Cochran
Normative theory in international relations
A pragmatic approach
67 Alexander Wendt
Social theory of international politics
66 Thomas Risse, Stephen C. Ropp and Kathryn Sikkink (eds.)
The power of human rights
International norms and domestic change
65 Daniel W. Drezner
The sanctions paradox
Economic statecraft and international relations
64 Viva Ona Bartkus
The dynamic of secession
63 John A. Vasquez
The power of power politics
From classical realism to neotraditionalism

62 Emanuel Adler and Michael Barnett (eds.)
    Security communities
61 Charles Jones
    E. H. Carr and international relations
    A duty to lie
60 Jeffrey W. Knopf
    Domestic society and international cooperation
    The impact of protest on US arms control policy
59 Nicholas Greenwood Onuf
    The republican legacy in international thought
58 Daniel S. Geller and J. David Singer
    Nations at war
    A scientific study of international conflict
57 Randall D. Germain
    The international organization of credit
    States and global finance in the world economy
56 N. Piers Ludlow
    Dealing with Britain
    The Six and the first UK application to the EEC
55 Andreas Hasenclever, Peter Mayer and Volker Rittberger
    Theories of international regimes
54 Miranda A. Schreurs and Elizabeth C. Economy (eds.)
    The internationalization of environmental protection
53 James N. Rosenau
    Along the domestic-foreign frontier
    Exploring governance in a turbulent world
52 John M. Hobson
    The wealth of states
    A comparative sociology of international economic and political change
51 Kalevi J. Holsti
    The state, war, and the state of war
50 Christopher Clapham
    Africa and the international system
    The politics of state survival
49 Susan Strange
    The retreat of the state
    The diffusion of power in the world economy
48 William I. Robinson
    Promoting polyarchy
    Globalization, US intervention, and hegemony
47 Roger Spegele
    Political realism in international theory
46 Thomas J. Biersteker and Cynthia Weber (eds.)
    State sovereignty as social construct
45 Mervyn Frost
    Ethics in international relations
    A constitutive theory

44 Mark W. Zacher with Brent A. Sutton
   Governing global networks
   International regimes for transportation and communications

43 Mark Neufeld
   The restructuring of international relations theory

42 Thomas Risse-Kappen (ed.)
   Bringing transnational relations back in
   Non-state actors, domestic structures and international institutions

41 Hayward R. Alker
   Rediscoveries and reformulations
   Humanistic methodologies for international studies

40 Robert W. Cox with Timothy J. Sinclair
   Approaches to world order

39 Jens Bartelson
   A genealogy of sovereignty

38 Mark Rupert
   Producing hegemony
   The politics of mass production and American global power

37 Cynthia Weber
   Simulating sovereignty
   Intervention, the state and symbolic exchange

36 Gary Goertz
   Contexts of international politics

35 James L. Richardson
   Crisis diplomacy
   The Great Powers since the mid-nineteenth century

34 Bradley S. Klein
   Strategic studies and world order
   The global politics of deterrence

33 T. V. Paul
   Asymmetric conflicts
   War initiation by weaker powers

32 Christine Sylvester
   Feminist theory and international relations in a postmodern era

31 Peter J. Schraeder
   US foreign policy toward Africa
   Incrementalism, crisis and change

30 Graham Spinardi
   From Polaris to Trident
   The development of US Fleet Ballistic Missile technology

29 David A. Welch
   Justice and the genesis of war

28 Russell J. Leng
   Interstate crisis behavior, 1816–1980
   Realism versus reciprocity

27 John A. Vasquez
   The war puzzle

26 Stephen Gill (ed.)
Gramsci, historical materialism and international relations

25 Mike Bowker and Robin Brown (eds.)
From cold war to collapse
Theory and world politics in the 1980s

24 R. B. J. Walker
Inside/outside
International relations as political theory

23 Edward Reiss
The strategic defense initiative

22 Keith Krause
Arms and the state
Patterns of military production and trade

21 Roger Buckley
US-Japan alliance diplomacy 1945–1990

20 James N. Rosenau and Ernst-Otto Czempiel (eds.)
Governance without government
Order and change in world politics

19 Michael Nicholson
Rationality and the analysis of international conflict

18 John Stopford and Susan Strange
Rival states, rival firms
Competition for world market shares

17 Terry Nardin and David R. Mapel (eds.)
Traditions of international ethics

16 Charles F. Doran
Systems in crisis
New imperatives of high politics at century's end

15 Deon Geldenhuys
Isolated states
A comparative analysis

14 Kalevi J. Holsti
Peace and war
Armed conflicts and international order 1648–1989

13 Saki Dockrill
Britain's policy for West German rearmament 1950–1955

12 Robert H. Jackson
Quasi-states
Sovereignty, international relations and the third world

11 James Barber and John Barratt
South Africa's foreign policy
The search for status and security 1945–1988

10 James Mayall
Nationalism and international society

9 William Bloom
Personal identity, national identity and international relations

8 Zeev Maoz
National choices and international processes

7   Ian Clark
    The hierarchy of states
    Reform and resistance in the international order
6   Hidemi Suganami
    The domestic analogy and world order proposals
5   Stephen Gill
    American hegemony and the Trilateral Commission
4   Michael C. Pugh
    The ANZUS crisis, nuclear visiting and deterrence
3   Michael Nicholson
    Formal theories in international relations
2   Friedrich V. Kratochwil
    Rules, norms, and decisions
    On the conditions of practical and legal reasoning in international relations and domestic affairs
1   Myles L. C. Robertson
    Soviet policy towards Japan
    An analysis of trends in the 1970s and 1980s

CPSIA information can be obtained
at www.ICGtesting.com
Printed in the USA
LVHW051817170519
618254LV00010BA/215/P